**High-Throughput Screening
in Drug Discovery**

Edited by
Jörg Hüser

Methods and Principles in Medicinal Chemistry

Edited by R. Mannhold, H. Kubinyi, G. Folkers
Editorial Board
H.-D. Höltje, H. Timmerman, J. Vacca, H. van de Waterbeemd, T. Wieland

Related Titles

G. Cruciani (ed.)
Molecular Interaction Fields
Vol. 27

2005
ISBN 3-527-31087-8

M. Hamacher, K. Marcus, K. Stühler, A. van Hall, B. Warscheid, H. E. Meyer (eds.)
Proteomics in Drug Design
Vol. 28

2005
ISBN 3-527-31226-9

D. Triggle, M. Gopalakrishnan, D. Rampe, W. Zheng (eds.)
Voltage-Gated Ion Channels as Drug Targets
Vol. 29

2006
ISBN 3-527-31258-7

D. Rognan (ed.)
GPCR Modelling and Ligand Design
Vol. 30

2006
ISBN 3-527-31284-6

D. A. Smith, H. van de Waterbeemd, D. K. Walker
Pharmacokinetics and Metabolism in Drug Research, 2nd Ed.
Vol. 31

2006
ISBN 3-527-31368-0

T. Langer, R. D. Hoffmann (eds.)
Pharmacophore and Pharmacophore Searches
Vol. 32

2006
ISBN 3-527-31250-1

E. Francotte, W. Lindner (eds.)
Chirality in Drug Research
Vol. 33

2006
ISBN 3-527-31076-2

W. Jahnke, D. A. Erlanson (eds.)
Fragment-based Approaches in Drug Discovery
Vol. 34

2006
ISBN 3-527-31291-9

High-Throughput Screening in Drug Discovery

Edited by
Jörg Hüser

WILEY-VCH Verlag GmbH & Co. KGaA

Series Editors

Prof. Dr. Raimund Mannhold
Biomedical Research Center
Molecular Drug Research Group
Heinrich-Heine-Universität
Universitätsstrasse 1
40225 Düsseldorf
mannhold@uni-duesseldorf.de

Prof. Dr. Hugo Kubinyi
Donnersbergstrasse 9
67256 Weinheim am Sand
Germany
kubinyi@t-online.de

Prof. Dr. Gerd Folkers
Collegium Helveticum
STW/ETH Zürich
8092 Zürich
Switzerland
folkers@collegium.ethz.ch

Volume Editor

Dr. Jörg Hüser
Molecular Screening
Technology
Bayer HealthCare AG
Pharma Discovery Research Europe
42096 Wuppertal
Germany
Joerg.hueser@bayerhealthcare.com

■ All books published by Wiley-VCH are carefully produced. Nevertheless, authors, editors, and publisher do not warrant the information contained in these books, including this book, to be free of errors. Readers are advised to keep in mind that statements, data, illustrations, procedural details or other items may inadvertently be inaccurate.

Library of Congress Card No.:
applied for

British Library Cataloguing-in-Publication Data
A catalogue record for this book is available from the British Library.

Bibliographic information published by Die Deutsche Bibliothek
Die Deutsche Bibliothek lists this publication in the Deutsche Nationalbibliografie; detailed biliographic data is available in the Internet at <http://dnb.ddb.de>.

© 2006 WILEY-VCH Verlag GmbH & Co. KGaA, Weinheim

All rights reserved (including those of translation into other languages). No part of this book may be reproduced in any form – by photoprinting, microfilm, or any other means – nor transmitted or translated into a machine language without written permission from the publishers.
Registered names, trademarks, etc. used in this book, even when not specifically marked as such, are not to be considered unprotected by law.

Typesetting   Dörr + Schiller GmbH, Stuttgart
Printing   betz-druck GmbH, Darmstadt
Binding   Littges & Dopf
Cover Design   Grafik-Design Schulz, Fußgönheim

Printed in the Federal Republic of Germany
Printed on acid-free paper

ISBN-13   978-3-527-31283-2
ISBN 10   3-527-31283-8

Foreword

Random screening of comprehensive compound collections constitutes a major source of novel lead structures reflected by industry's ongoing commitment to invest in extensive compound libraries and screening technologies. During the last decade, High-Throughput Screening (HTS) has evolved to become an innovative multidisciplinary branch in biological chemistry combining aspects of medicinal chemistry, biology, and laboratory automation. While basic medicinal chemistry techniques and views are largely conserved throughout industry, HTS strategies differ to a great extend. Each strategy can be justified by scientific rationale. However, it is also the result of different scientific backgrounds, different therapeutic areas, different technical expertise within a group, and different ways HTS is integrated within the overall discovery process in a given organization. For most strategies, the close interrelation between HTS and the molecular target approach to drug discovery renders the validity of a disease link for a selected biomolecular target an essential prerequisite for success. As a consequence, a critical assessment of HTS has to incorporate also reflections on the discovery process from target selection to appropriate screening cascades. A different approach employing phenotypic readouts, e.g. cell proliferation, has a long tradition in screening, particularly for chemotherapeutic principles in cancer and antiinfectives research. Similarly, chemical genetics makes use of small molecule perturbation of specific cellular responses to unravel the underlying gene and pathways function. Within the later paradigm, High-Throughput Screening techniques have gained increasing relevance also in academic research.

The current book presents a collection of review-style papers written by experts in the field intended to provide insights into selected aspects of the experimental lead discovery process in High-Throughput Screening. It is by no means claimed to comprehensively cover the entire field. A number of aspects have been discussed in previous volumes within this series on "Methods and Principles in Medicinal Chemistry". It complements this book series by illustrating HTS as one of the technologies of great relevance to the medicinal chemist and molecular pharmacologist working in pharmaceutical or academic research.

I am personally thankful to the Series Editors not only for providing the opportunity to present High-Throughput Screening within a single dedicated volume, but also for their patience during the preparation of this volume. In addition, the continuous support of my colleagues, Stefan Mundt, Nils Griebenow and Peter Nell, is gratefully acknowledged.

Wuppertal, July 2006
Jörg Hüser

List of Contents

Preface *XIV*

List of Contributors *XVI*

Part I   Concept of Screening

1   Chemical Genetics: Use of High-throughput Screening to Identify Small-molecule Modulators of Proteins Involved in Cellular Pathways with the Aim of Uncovering Protein Function
Sa L. Chiang
1.1   Introduction   *1*
1.2   Classical and Chemical Genetics   *1*
1.2.1   Forward and Reverse Screens   *3*
1.3   Identifying Bioactive Molecules   *4*
1.4   Target Identification   *5*
1.4.1   Hypothesis-driven Target Identification   *5*
1.4.2   Affinity-based Target Identification   *6*
1.4.3   Genomic Methods of Target Identification   *7*
1.4.4   Proteomic Methods   *9*
1.5   Discovery for Basic Research Versus Pharmacotherapy Goals   *10*
1.6   Chemical Genetic Screens in the Academic Setting   *11*
1.7   Conclusions   *12*

2   High-throughput Screening for Targeted Lead Discovery
Jörg Hüser, Emanuel Lohrmann, Bernd Kalthof, Nils Burkhardt, Ulf Brüggemeier, and Martin Bechem
2.1   Chemical Libraries for High-throughput Screening   *15*
2.2   Properties of Lead Structures   *17*
2.3   Challenges to High-throughput Screening   *19*
2.4   Assay Technologies for High-throughput Screening   *21*
2.5   Laboratory Automation   *24*
2.6   From Target Selection to Confirmed Hits – the HTS Workflow and its Vocabulary   *25*

2.7	Separating Specific Modulators from Off-Target Effects	29
2.8	Data Analysis and Screening Results	32
2.9	Conclusions	34

Part II Automation Technologies

3   Tools and Technologies that Facilitate Automated Screening
John Comley
3.1   Introduction – the Necessity to Automate   37
3.1.1   Compound Libraries   37
3.1.2   Targets and Data Points   38
3.1.3   Main Issues Facing HTS Groups Today   38
3.1.4   Benefits of Miniaturization   39
3.1.5   Benefits of Automated HTS   39
3.1.6   Screening Strategies   40
3.1.7   Ultra HTS (UHTS)   40
3.2   Sample Carriers   41
3.2.1   A Brief History of the Microplate   41
3.2.2   Microplate Usage Today   41
3.2.3   Microplate Arrays   42
3.2.4   Non-microplate Alternatives   43
3.2.4.1 Labchips   43
3.2.4.2 LabCDs   43
3.2.4.3 LabBrick   44
3.2.4.4 Arrayed Compound Screening   44
3.3   Liquid Handling Tools   45
3.3.1   Main Microplate Dispense Mechanisms   45
3.3.1.1 Pin Tools   45
3.3.1.2 Air and Positive Displacement   45
3.3.1.3 Peristaltic   46
3.3.1.4 Solenoid-syringe   47
3.3.1.5 Solenoid-pressure bottle   47
3.3.1.6 Capillary Sipper   48
3.3.1.7 Piezoelectric   48
3.3.1.8 Acoustic Transducer   48
3.3.2   HTS Liquid Handling Applications and Dispensing Technologies Used   49
3.3.2.1 Bulk Reagent and Cell Addition   49
3.3.2.2 Compound Reformatting and Nanoliter Dispensing   50
3.3.2.3 Cherry Picking and Serial Dilution   51
3.3.2.4 Microplate Washing   52
3.4   Detection Technologies   53
3.4.1   Main Detection Modalities Used in HTS   53
3.4.2   Plate Readers   54
3.4.3   Plate Imagers   55

3.4.3.1 Macro-imaging   56
3.4.3.2 Micro-imaging   57
3.4.4   Dispense and Read Devices   60
3.4.5   Other Detection Technologies   60
3.4.6   Automation of Detection Technologies   61
3.4.7   Potential Sources of Reading Error   61
3.5   Laboratory Robotics   62
3.5.1   Traditional Workstations   64
3.5.2   Robotic Sample Processors   64
3.5.3   Plate Storage Devices   64
3.5.4   Plate Moving Devices   65
3.5.5   Fully Integrated Robotic Systems   65
3.5.6   Turnkey Workstations   66
3.5.7   Automated Cell Culture Systems   66
3.5.8   Compound Management Systems   67
3.5.8.1 Current Practice in Compound Management   67
3.5.8.2 Plate-based versus Tube-based Liquid Compound Storage   68
3.5.8.3 Associated Automated Instrumentation   70
3.5.8.4 Sample Integrity and QC Testing   70

Part III Assay Technologies

**4   Functional Cell-based Assays for Targeted Lead Discovery
          in High-throughput Screening**
          Jörg Hüser, Bernd Kalthof, and Jochen Strayle
4.1   Introduction   75
4.2   Reporter Gene Technologies   78
4.3   Membrane Potential Indicators   82
4.4 Ca^{2+} Indicators   88
4.5   Conclusions   90

5   Biochemical Assays for High-throughput Screening
          *William D. Mallender, Michael Bembenek, Lawrence R. Dick, Michael Kuranda,
          Ping Li, Saurabh Menon, Eneida Pardo and Tom Parsons*
5.1   General Considerations for Biochemical High-throughput Screening   93
5.2   Expression and Purification of Recombinant Enzymes   95
5.2.1   Design of Expression Constructs   98
5.2.2   Expression Assessment and Optimization   99
5.2.3   Purification   99
5.3   Peptidases   102
5.3.1   Application of Fluorogenic Substrates to Configure Peptidase
         Screens   102
5.3.2   The Value of Continuous Assays   107
5.4   Oxidoreductases   107
5.4.1   NAD(P)-dependent Oxidoreductases   108

5.4.2	Non-NAD(P) Cofactor-dependent Oxidoreductases	*110*
5.4.3	Oxidases or Oxygen-utilizing Oxidoreductases	*111*
5.4.4	General Considerations	*113*
5.5	Transferases, Synthetases and Lipid-modifying Enzymes	*114*
5.5.1	Streptavidin–Biotin Capture	*114*
5.5.2	Ionic Capture	*117*
5.5.3	Hydrophobic Capture	*117*
5.6	Kinases	*120*
5.6.1	Streptavidin–Biotin Capture	*120*
5.6.2	Homogeneous Time-resolved Fluorescence (HTRF)	*122*
5.6.3	Pyruvate Kinase–Lactate Dehydrogenase Assay System	*122*
5.7	Pitfalls and Reasons for Assay Development Failures	*125*

6 **Image-based High-content Screening – A View from Basic Sciences**
Peter Lipp and Lars Kaestner

6.1	Introduction	*129*
6.2	HCS Systems Employing Confocal Optical Technologies	*132*
6.3	Single-point Scanning Technology	*134*
6.4	Line Scanning Technology	*136*
6.5	Multi-beam Technology	*138*
6.6	Structured Illumination	*145*
6.7	Summary and Perspectives	*147*

Part IV Data Analysis

7 **Methods for Statistical Analysis, Quality Assurance and Management of Primary High-throughput Screening Data**
Hanspeter Gubler

7.1	Introduction	*151*
7.1.1	Overview	*151*
7.1.2	Problems during the Analysis of Primary HTS Data	*151*
7.2	Statistical Considerations in Assay Development	*155*
7.3	Data Acquisition, Data Preprocessing, and HTS Data Analysis Environment	*158*
7.4	Data Normalization	*160*
7.5	Robust Statistics in HTS Data Analysis	*163*
7.5.1	The General Problem	*163*
7.5.2	Threshold Setting – a Simple Model	*164*
7.5.3	Threshold Setting – a Complex Model	*165*
7.5.4	The Most Important Robust Estimation Methods for Data Summaries	*165*
7.5.5	An Illustrative Example: Performance of Location and Scale Estimators on Typical HTS Data Activity Distributions	*167*
7.5.6	A Robust Outlier Detection Method	*168*
7.5.7	Outlier-resistant Versions of Simple HTS Data Quality Indicators	*170*

7.6	Measures of HTS Data Quality, Signaling of Possible QC Problems, Visualizations *170*	
7.6.1	Trends and Change Points *171*	
7.6.2	Positional Effects – Summary Statistics and Views *175*	
7.6.3	Positional Effects – Heat Maps, Trellis Plots, and Assay Maps *177*	
7.6.4	Distribution Densities – Histograms, Smoothed Distributions *180*	
7.6.5	Numerical Diagnostics and Fully Automated QC Assessment *181*	
7.6.6	Possible Sources of Systematic Errors and Trends *182*	
7.7	Correction of Position-dependent Response Effects *182*	
7.7.1	The General Problem *182*	
7.7.2	Plate Averaging (Multiple Plates) *184*	
7.7.3	Median Polish Smoothing (Single Plates, Multiple Plates) *185*	
7.7.4	Change Point Detection (Multiple Plates, Plate Sequences) *186*	
7.7.5	Parametric (Polynomial) Surface Fitting (Single Plates, Multiple Plates) *187*	
7.7.6	Nonparametric Surface Fitting, Local Regression, and Smoothing (Single Plates, Multiple Plates) *189*	
7.7.7	Expansion into Orthogonal Basis Functions (Single Plates) *190*	
7.7.8	Empirical Orthogonal Function (EOF) Analysis, Singular Value Decomposition (SVD) (Multiple Plates) *191*	
7.7.9	Some Remarks on the Correction of Individual Plates versus Complete Plate Sets *194*	
7.7.10	Position-dependent Correction of Background Response Surface *195*	
7.8	Hit Identification and Hit Scoring *199*	
7.9	Conclusion *201*	

8 Chemoinformatic Tools for High-throughput Screening Data Analysis
Peter G. Nell and Stefan M. Mundt

- 8.1 Introduction *207*
- 8.1.1 Definition of Chemoinformatics *208*
- 8.1.2 High-throughput Screening *208*
- 8.1.2.1 Random Screening *209*
- 8.1.2.2 Sequential Screening *209*
- 8.2 Workflow of High-throughput Screening and Use of Chemoinformatics *211*
- 8.3 Chemoinformatic Methods Used in HTS Workflow *214*
- 8.3.1 Substructure Search/Similarity Search *214*
- 8.3.1.1 Structural Descriptors/Fingerprints *215*
- 8.3.1.2 Measures of Similarity *217*
- 8.3.2 Clustering *218*
- 8.3.2.1 Hierarchical Clustering *219*
- 8.3.2.2 Nonhierarchical Clustering *219*
- 8.3.2.3 Partitioning Methods *220*
- 8.3.2.4 Principal Components Analysis *221*
- 8.3.3 Maximum Common Substructure – Distill *222*

8.3.4   Mode of Action – Profiling   *223*
8.3.5   Artificial Neural Networks (ANNs)   *224*
8.3.6   Decision Trees/Recursive Partitioning   *226*
8.3.7   Reduced Graph-based Methods   *226*
8.3.7.1   Reduced Graph Theory Analysis: Baytree   *227*
8.3.8   Fragment-based Methods – Structural Units Analysis   *230*
8.4   Chemoinformatic Methods in the Design of a Screening Library   *231*
8.4.1   Drug and Lead Likeness   *231*
8.4.2   ADME Parameters   *233*
8.4.2.1   Absorption   *233*
8.4.2.2   Distribution   *234*
8.4.2.3   Metabolism   *234*
8.4.2.4   Excretion   *235*
8.4.2.5   Toxicity   *235*
8.4.3   Diversity   *236*
8.5   Integrated Software Packages   *238*
8.5.1   Commercially Available Packages   *239*
8.5.1.1   Accelrys: DIVA®   *239*
8.5.1.2   BioSolveIT: HTSview   *239*
8.5.1.3   SciTegic: Pipeline Pilot™   *240*
8.5.1.4   Bioreason: ClassPharmer™ Suite   *241*
8.5.1.5   Spotfire Lead Discovery   *242*
8.5.1.6   LeadScope   *243*
8.5.1.7   OmniViz   *244*
8.5.1.8   SARNavigator   *245*
8.5.2   In-house Packages   *246*
8.6   Conclusions   *247*

9   Combinatorial Chemistry and High-throughput Screening
Roger A. Smith and Nils Griebenow
9.1   Introduction   *259*
9.2   Categories of Compound Libraries for High-throughput Screening   *260*
9.3   Synthesis Techniques and Library Formats   *261*
9.3.1   Solid-phase Synthesis   *262*
9.3.1.1   Parallel Solid-phase Synthesis Techniques and Tools   *265*
9.3.1.2   Pool/Split Techniques with Encoding   *265*
9.3.2   Solution-phase Synthesis   *269*
9.3.2.1   Polymer-supported Reagents and Scavengers   *269*
9.3.2.2   Extraction Techniques for Purification   *271*
9.3.2.3   Purification by Chromatography   *273*
9.3.3   Library Formats   *273*
9.3.3.1   One-bead One-compound Libraries   *274*
9.3.3.2   Pre-encoded Libraries   *275*
9.3.3.3   Spatially Addressable Libraries   *276*
9.4   Library Design and Profiling Approaches   *277*

9.5	Impact of Combinatorial Libraries on Drug Discovery	277
9.5.1	Lead Identification 278
9.5.2	Lead Optimization 283
9.5.3	Clinical Drug Candidates 286
9.6	Conclusion 290

10 High-throughput Screening and Data Analysis
Jeremy S. Caldwell and Jeff Janes
10.1 Introduction 297
10.2 Analysis of Cellular Screening Data 298
10.2.1 Quality Control and Analysis 298
10.2.2 Enrichment for Hits 301
10.2.3 Meta-data Analysis 302
10.3 Massively Parallel Cellular Screens 305
10.3.1 Data Analysis of Multidimensional Datasets 307
10.3.2 Multidimensional Cellular Profiling for MOA prediction 311
10.3.3 Cellular Profiling in Lead Exploration 314
10.4 Systematic Serendipity 317
10.5 Conclusion 320

Appendix 323

Index 333

Preface

Biological "trial and error" testing of large collections of small molecules for a specific pharmacological effect is the classical route to discover novel lead compounds which subsequently serve as templates for further optimization in medicinal chemistry programs. During the past two decades, the advent of recombinant DNA technologies together with improved assay techniques and high-performance laboratory automation dramatically changed the pharmacological screening process. Nowadays, high-throughput screening (HTS) has gained attention also in the academic environment. Using HTS of compound collections in combination with cell-based tests, small molecule modulators of relevant biochemical and signal transduction pathways should be identified. Following identification of the biological target, the small molecule modulator (inhibitor or activator) substitutes mutational analysis to unravel the target protein function. The vision of this approach, referred to as "Chemical Genetics", is to identify a small molecule partner for every gene product. In the future, pharmaceutical drug research might be stimulated by Chemical Genetics by revealing novel drug targets and initial lead structures.

The present volume provides fascinating insights into this important part of the early drug discovery process. Four most important issues relevant to HTS are covered: a) concepts of pathway/phenotypic versus target-based screening, b) automation technologies, c) assay technologies, and d) data analysis.

Part I contrasts the two approaches of "Chemical Genetics" using pathway assays dependent on a initially not defined set of possible drug receptor sites and the target-based lead finding process commonly used in pharmaceutical research. Caroline Shamu uses case studies to introduce the basic concept, the assay technologies and selected results. Jörg Hüser and colleagues summarize the concepts of target-directed screening for lead discovery and discuss strengths and weaknesses of random/diversity screening when compared to knowledge-based *in silico* methodologies.

In **Part II**, John Comley provides a general overview of laboratory automation technologies covering assay carriers, liquid handling automats, signal detection instrumentation, and robotic integration.

Part III focuses on HTS assay technologies. For the discovery of novel lead candidates the choice of the appropiate assay technology and its technical realization will determine the overall quality of the screening experiment. There are two general appoaches for pharmacological assays: Assays measuring the binding of a candidate molecule to the target receptor ("binding tests") and assays monitoring the function of a target (or pathway) to visualize a possible modulation by small molecules. In chapter 4, Jörg Hüser and colleagues focus on principles of functional cell-based test systems. Improved readout techniques together with the rich molecular biology toolbox have rendered cell-based assays an important methodology for the targeted discovery of pharmacological lead compounds in HTS. Designed cell-based HTS assays ideally combine high specificity for and superior sensitivity towards the targeted receptor. In addition, measuring receptor function rather than binding allows one to monitor all possible drug-receptor interactions including allosteric modulation and reveals additional information on ligand efficacy, i.e. agonism or antagonism. Chapter 5 by William Mallender and colleagues gives an overview on functional biochemical tests, providing examples for the most important drug target classes approached by enzyme assays (proteases, kinases and others). In chapter 6, Peter Lipp and Lars Kästner introduce and critically discuss "Image-based High-content Screening", a recently emerging technology using subcellular imaging for target-based and pathway assays.

The last set of papers (**Part IV**) covers various aspects of HTS data analysis. Chapter 7, by Hanspeter Gubler, introduces concepts of HTS data management, assay quality assurance, and analysis. It touches on some fundamental statistical consideration relevant to handling large sets of compound activity data. Chapter 8 (Peter Nell and Stefan Mundt) illustrates the use of chemoinformatic tools, e.g. structural clustering of active compound sets, aiming to discriminate between specific hits and compounds acting through unspecific mechanisms and providing clues for preliminary SARs or pharmacophoric elements (i.e. molecular fragments contributing to activity). In chapter 9, Roger Smith and Nils Griebenow discuss pro's and con's of focused library screening, including methodologies and strategies to design such subsets of the available compound file and its use for lead discovery. The data analysis section is concluded by a paper by Jeremy Caldwell describing the consequent exploitation of large activity databases derived from functional cell-based screening by data mining technologies to reveal molecules acting through unexpected mechanisms.

The series editors are grateful to Jörg Hüser for his enthusiasm to organize this volume and to work with such a fine selection of authors. We believe that this book adds a fascinating new facet to our series on "Methods and Principles in Medicinal Chemistry". Last, but not least we thank the publisher Wiley-VCH, in particular Renate Dötzer and Dr. Frank Weinreich, for their valuable contributions to this project and to the entire series.

July 2006

Raimund Mannhold, Düsseldorf
Hugo Kubinyi, Weisenheim am Sand
Gerd Folkers, Zürich

List of Contributors

Martin Bechem
Molecular Screening Technology
Bayer HealthCare AG
Pharma Discovery Research Europe
42096 Wuppertal
Germany

Michael Bembenek
Department of High Throughput
Biochemistry
Millennium Pharmaceuticals, Inc.
Cambridge, MA 02139
USA

Ulf Brüggemeier
PH-R&D Molecular Screening
Technology
Bayer HealthCare AG
Research Center Wuppertal
42096 Wuppertal
Germany

Nils Burkhardt
Molecular Screening Technology
Bayer HealthCare AG
Pharma Discovery Research Europe
42096 Wuppertal
Germany

Jeremy S. Caldwell
10675 John Jay Hopkins Drive
San Diego, CA 92121
USA

Sa L. Chiang
Seeley Madd 504, Harvard Medical
School
250 Longwood Avenue
Boston, Ma 02115
USA

John Comley
HTStec Limited
12 Cockle Close, Newton
Cambridge, CB2 5TW
UK

Larry Dick
Department of Oncology Biochemistry
Millennium Pharmaceuticals, Inc.
Cambridge, MA 02139
USA

Nils Griebenow
Medicinal Chemistry
Bayer HealthCare AG
Pharma Discovery Research Europe
42096 Wuppertal
Germany

List of Contributors

Hanspeter Gubler
Discovery Technologies
Novartis Institutes for BioMedical Research
WSJ-350.P.15
4002 Basel
Switzerland

Jörg Hüser
Molecular Screening Technology
Bayer HealthCare AG
Pharma Discovery Research Europe
42096 Wuppertal
Germany

Jeff Janes
10675 John Jay Hopkins Drive
San Diego, CA 92121
USA

Lars Kaestner
Molecular Cell Biology
University Hospital
Medical Faculty
Saarland University
66421 Homburg/Saar
Germany

Bernd Kalthof
Molecular Screening Technology
Bayer HealthCare AG
Pharma Discovery Research Europe
42096 Wuppertal
Germany

Michael Kuranda
Department of High Throughput Biochemistry
Millennium Pharmaceuticals, Inc.
Cambridge, MA 02139
USA

Ping Li
Department of Protein Sciences
Millennium Pharmaceuticals, Inc.
Cambridge, MA 02139
USA

Peter Lipp
Molecular Cell Biology
University Hospital
Medical Faculty
Saarland University
66421 Homburg/Saar
Germany

Emanuel Lohrmann
Molecular Screening Technology
Bayer HealthCare AG
Pharma Discovery Research Europe
42096 Wuppertal
Germany

William D. Mallender
Department of Oncology Biochemistry
Millennium Pharmaceuticals, Inc.
Cambridge, MA 02139
USA

Saurabh Menon
Department of High Throughput Biochemistry
Millennium Pharmaceuticals, Inc.
Cambridge, MA 02139
USA

Stefan M. Mundt
Molecular Screening Technology
Bayer HealthCare AG
Pharma Discovery Research Europe
42096 Wuppertal
Germany

Peter G. Nell
Medicinal Chemistry
Bayer HealthCare AG
Pharma Discovery Research Europe
42096 Wuppertal
Germany

Eneida Pardo
Department of High Throughput Biochemistry
Millennium Pharmaceuticals, Inc.
Cambridge, MA 02139
USA

Tom Parsons
Department of Protein Sciences
Millennium Pharmaceuticals, Inc.
Cambridge, MA 02139
USA

Roger A. Smith
Department of Chemistry Research
Pharmaceuticals Division
Bayer HealthCare
400 Morgan Lane
West Haven, CT 06516
USA

Jochen Strayle
PH-R&D Molecular Screening Technology
Bayer HealthCare
400 Morgan Lane
West Haven, CT 06516
USA

Part I
Concept of Screening

1
Chemical Genetics: Use of High-throughput Screening to Identify Small-molecule Modulators of Proteins Involved in Cellular Pathways with the Aim of Uncovering Protein Function

Sa L. Chiang

1.1
Introduction

Understanding cellular pathways and their molecular mechanisms is one of the longstanding goals of scientific research. However, the tools that researchers utilize to study biological processes become progressively more sophisticated as technology and our knowledge of biology advance. This chapter discusses one of the most recent and exciting developments in this area: chemical genetics and the use of small, bioactive molecules to characterize the components of cellular pathways and their functions.

1.2
Classical and Chemical Genetics

All genetic approaches depend on the ability to perturb gene function and to correlate phenotypic changes with specific changes in gene function. Classical genetics relies on physical perturbation of a gene, through methods such as irradiation, chemical mutagenesis or insertional mutagenesis (e.g. via the use of transposons). There are also numerous molecular biology techniques for creating directed mutations that allow highly specific modifications at the level of the gene or even the nucleotide in a variety of experimental systems. These methods are well established and readers should consult a standard genetics text for the relevant descriptions.

    A common feature of all classical genetic methods is that they cause a permanent change in the structure of a gene. Therefore, except for the few situations noted below, the phenotypes arising from classical genetic mutations are irreversible. A notable disadvantage of this irreversibility is that it hinders the study of genes that are essential for viability. Irreversibility also makes it difficult to study the effects of temporal variations in gene expression or protein function. Placing genes under the control of inducible promoters has gone some way towards solving these problems, but inducible promoters, of course, act at the level of transcription and the researcher may not obtain sufficient control over the activity of the encoded protein. Temperature-sensitive mutations may provide some control over protein activity, but they are not easy to construct and generally cannot be used in animal models. Pleiotropic effects caused by temperature shifts may also complicate analysis.

In the last decade, researchers have increasingly explored the use of low molecular weight chemical entities to modulate and characterize protein function. These methods are generally analogous to classical genetic approaches and have accordingly been termed *chemical genetic* strategies [1, 2]. Whereas classical genetics uses physical modification of a gene to perturb protein function, chemical genetics employs specific, biologically active small-molecule modulators to perturb protein function.

It is relatively simple to imagine mechanisms by which small molecules may modulate protein function. Enzymatic activity can be affected by the binding of small molecules to active or allosteric sites and protein–ligand interactions may be disrupted by small molecules that interfere with binding between interaction partners. Alternatively, interaction partners could be brought together more effectively by ligands that bind to both. Interactions between small molecules and their targets may be reversible or irreversible and protein function can be either diminished or enhanced, depending on the situation.

One advantage of small molecules is that they can be used in biological systems where there is little or no ability for classical genetic manipulation. Moreover, since many small molecules will not interact irreversibly with their targets, chemical genetics is expected to provide enhanced opportunities to create conditional phenotypes. Theoretically, a high degree of control over protein function could be afforded by simply adding or washing away a small-molecule modulator. Indeed, several widely used inducible promoter systems (lac, arabinose, tetracycline) employ small-molecule inducers. In addition, conditional phenotypes induced by small molecules can potentially be studied in animal models, where temperature-dependent phenotypes generally cannot be used.

In actuality, the ability of small molecules to induce and reverse phenotypes will depend on factors such as binding kinetics and the physical accessibility of the target. More important, most protein targets have no known small-molecule modulators of their activity. Obtaining a specific and potent small-molecule modulator for a chosen target often requires structure-based drug design or full-scale high-throughput screening (HTS) and medicinal chemistry optimization may also be necessary. Owing to this "front-end" effort, directed perturbations in many systems are currently more difficult to achieve with chemical genetics than with classical genetics.

1.2.1
Forward and Reverse Screens

Many classical genetic screens begin with mutagenesis of organisms or populations of cells, followed by attempts to associate the resulting phenotypic changes with specific genes. These approaches are termed *forward genetic* strategies and can be generally characterized as starting with phenotypes and progressing towards the identification of the genes that are responsible for those phenotypes. With the advent of molecular biology, directed mutagenesis techniques became available and these advances allowed *reverse genetic* studies, which begin with the introduction of programmed or directed mutations into a known target gene, followed by analysis of the resulting phenotypes to obtain information about the function of that gene.

Both forward and reverse genetic strategies have also been employed for chemical screens. In a forward chemical genetic screen, chemical libraries are screened for compounds that produce a phenotype of interest, typically in a cell-based assay. An example of forward chemical genetics is the strategy that was employed to identify inhibitors of mitotic spindle bipolarity [3]. Monastrol was identified through a chemical genetic screen employing a whole-cell immunodetection assay to screen a library of 16 320 compounds for those that increase nucleolin phosphorylation, a phenotype predicted for cells experiencing mitotic arrest. A total of 139 compounds found to increase nucleolin phosphorylation were subjected to further analysis, resulting in the identification of five compounds that affect the structure of the mitotic spindle. One of these induced the formation of a monoastral microtubule array and was accordingly named monastrol. Monastrol has since been employed in multiple studies as a tool for investigating the process of cell division.

Conversely, in a reverse chemical genetic screen, small-molecule libraries might be screened for compounds that bind to a purified protein target, modulate the activity of the target or affect the target's ability to interact with other proteins. Such compounds could then be used in cell-based assays to characterize the function of the target protein in cellular pathways. For instance, a luminescence-based reverse chemical genetic screen was employed to identify inhibitors of rabbit muscle myosin II subfragment (S1) actin-stimulated ATPase activity [4]. The most potent compound identified (*N*-benzyl-*p*-toluenesulfonamide; BTS) inhibited S1 ATPase activity with an IC_{50} of ~5 µM and BTS also inhibited the activity of skeletal muscle myosin in a gliding-filament assay. Subsequent studies demonstrated that BTS inhibits the binding of myosin–ADP to actin and also affects various properties of rabbit and frog muscle preparations.

A similar reverse chemical genetic screen identified an inhibitor (blebbistatin) of nonmuscle myosin II [5]. Blebbistatin was found to inhibit a variety of activities in whole vertebrate cells, including directed cell migration and cytokinesis. The use of blebbistatin and additional drugs (including monastrol) to manipulate the mitotic process in whole cells led to the discovery that ubiquitin-dependent proteolysis is required for exit from the cytokinetic phase of the cell cycle. Blebbistatin is thus another example of how small molecules identified via HTS can be used to study the function of both individual proteins and major cellular processes.

The above points are summarized in Table 1.1.

Table 1.1 Attributes of classical genetic, chemical genetic, and RNAi approaches.

	Classical genetics	Chemical genetics	RNAi
Nature of perturbation	Permanent genetic change; true null is usually possible; heritable	Transient or irreversible, depending on situation; not heritable	Transient; generally not heritable
Directed perturbation	Possible in many systems, can be accomplished at the single-nucleotide level	Requires identification of a specific effector; may require synthetic chemistry effort to optimize	Yes; some occasional off-target effects
Conditional perturbation	Possible in some situations (temperature-sensitive alleles, inducible expression)	Possible, depending on specific situation	Yes, although difficult to control temporally
Target identification	Often simple, depending on situation	Often difficult	Target is known

1.3
Identifying Bioactive Molecules

Biologically active small molecules have often been discovered by testing a single compound at a time, but such an approach is obviously highly inefficient and cost-intensive in terms of both reagents and personnel time. As a result, HTS technologies have been developed to screen large numbers of compounds simultaneously, typically through the miniaturization and automation of assay protocols. What constitutes high throughput will vary depending on technical considerations and the screener's economic situation, but generally, the number of compounds involved may range from tens of thousands to several million. With current technologies, throughput is high enough that screening this number of compounds can be accomplished within several weeks, if not within several days.

A large number of screening technologies are available today for identifying bioactive small molecules; in fact, there are too many methods to be discussed adequately in this chapter. A very general discussion is provided here, but the reader is referred to other chapters in this volume for greater detail.

High-throughput screening assays can be divided into two main classes: "pure protein" and "cell-based". Pure protein screens generally have optical assay readouts that monitor enzymatic or binding activity. For instance, fluorescence polarization or FRET techniques are commonly used to screen for compounds that affect binding between protein partners. In pure protein assays, every compound screened should have equal access to the target. However, the membrane permeability characteristics of any active molecules that are identified may subsequently pose a major concern if the target is intracellular.

Readouts for cell-based screens may rely on reporter gene systems (e.g. luciferase, β-lactamase), cell density, cell viability or cell morphology. Screen readouts can also be divided into two broad classes: uniform well readout acquired via plate readers and images acquired via automated microscopy. Screens involving image-based readouts are usually technically more difficult and their computational analysis more challenging than plate-reader screens. However, image-based readouts can provide a vast amount of information and microscopy screens are therefore often referred to as *high-content screens* [6–8]. The development of equipment and data analysis techniques for automated imaging screens is an area of active research, but application of early technologies has already yielded promising results [9–15].

1.4
Target Identification

When biologically active small molecules are identified through chemical screens, particularly forward genetic screens, a substantial amount of work is often required to identify the molecular target. Ignorance of the target does not preclude clinical or research use of the molecule; indeed, many clinical agents have been used effectively even when their targets were not known (e.g. aspirin, nitroglycerin, fumagillin, epoxomycin). Nevertheless, defining the molecular mechanism of action is vital to understanding the biological principles involved and also for potentially creating more potent or specific molecules through synthetic chemistry approaches.

Target identification is currently a major area of research in chemical genetics and Tochtrop and King have recently provided an excellent discussion of this topic [16]. This section will therefore emphasize HTS-related examples and also discuss some recent work not included in that review.

1.4.1
Hypothesis-driven Target Identification

For well-characterized cellular pathways, it is sometimes possible to deduce the target of a small molecule by comparing data from across the field. For instance, characteristic phenotypes induced by the small molecule may permit the assignment of the target to a previously identified complementation group. Subsequent hypothesis-driven testing of potential targets can then be undertaken.

This deductive approach was used successfully to define the target of monastrol, a small-molecule inhibitor of mitotic spindle bipolarity [3]. As noted above, monastrol was identified in a forward chemical genetic screen for compounds that affect the structure of the mitotic spindle. Monastrol induces the formation of a monoastral microtubule array and, since previous studies had noted that antibody inhibition of the Eg5 kinesin also causes monoaster formation, it was postulated that Eg5 might be a target of monastrol. Subsequent work demonstrated that monastrol reversibly inhibits Eg5-driven microtubule motility *in vitro* but is not a general inhibitor of motor proteins.

Hypothesis-driven target identification was also used in a chemical screen for inhibitors of SARS coronavirus (SARS-CoV) replication [17]. In this study, a library of 50 240 compounds was screened by imaging for compounds that inhibit the cytopathic effect (CPE) of SARS-CoV towards Vero cells. A total of 104 compounds demonstrating effective inhibition of CPE and viral plaque formation were then tested *in vitro* against two protein targets known to affect SARS-CoV replication (M^{pro} protease and the NTPase/helicase), and also in a pseudotype virus assay for inhibition of S protein–ACE2-mediated entry of SARS-CoV into 293T cells. Two inhibitors of M^{pro}, seven inhibitors of Hel and 18 inhibitors of viral entry were identified and each of these three classes contained at least one inhibitor active in the low micromolar range. The authors subsequently assayed the 104 compounds against other common RNA viruses and found that most were specifically active against SARS-CoV, and approximately 3% were active against all viruses tested.

1.4.2
Affinity-based Target Identification

Affinity-based methods such as affinity labeling, affinity chromatography and crosslinking are also commonly used target identification strategies. Since these approaches often require synthetic modification of the molecule (e.g. addition of linkers or immunoreactive epitopes), care must be taken that the modifications do not interfere with the molecule–target interaction. Nonspecific interactions may also complicate the analysis and controls must be designed to address this issue. Despite such limitations, however, these strategies are of tremendous utility in target identification.

With affinity labeling, molecules may be radioactively or chemically labeled; they may be synthetically modified with reactive groups to promote covalent attachment to the target or tagged with specific moieties to facilitate detection. Standard protein fractionation and detection techniques can then be used to identify proteins from crude extracts that are specifically labeled by the molecule. Examples of drugs whose cellular targets were determined by such methods include acetylcholine, the anti-angiogenic agent fumagillin [18] and the antifungal lipopetide echinocandin [19].

Affinity purification of putative targets from cellular extracts is accomplished with molecules immobilized on a solid support used either as a column matrix or as a bead slurry. Small molecules for which cellular targets were identified through affinity purification include the immunosuppressant FK506 [20], the kinase inhibitor purvalanol [21] and the anti-inflammatory compound SB 203580 [22].

Additional studies involving affinity chromatography were instrumental in further characterizing the biochemistry of FK506 and purifying the target of the structurally related compound rapamycin [23].

A natural technological development of affinity-based target identification involves probing proteome chips with small molecules [24]. A recent application of this approach [25] is discussed in Section 1.4.4. Conversely, it is also possible to probe small-molecule microarrays with a known protein to identify specific interactions [26].

1.4.3
Genomic Methods of Target Identification

Yeast three-hybrid system   The yeast two-hybrid system has been widely used to study protein–protein interactions and the approach was adapted to create a yeast three-hybrid system for identifying protein–ligand interactions [27]. The method relies on the use of a hybrid "bait" ligand consisting of the query molecule linked to a known ligand. A protein fusion between the known ligand's receptor and a DNA-binding domain serves as the "hook", while the "fish" is a protein fusion between a transactivation domain and the target protein. If the target protein interacts with the query molecule, the "fish" and "hook" will be brought together by the "bait" and transcription from a reporter gene is activated. Hence it should be possible to identify the target of the query molecule by cloning a library into the "fish" domain and screening or selecting for activity of the reporter.

The three-hybrid proof-of-principle study was conducted with a dexamethasone–FK506 hybrid ligand and a Jurkat cDNA library. This experiment successfully identified two variants of human FKBP12 as targets of FK506. Another proof-of-principle study employed a dexamethasone–methotrexate hybrid ligand to screen a mouse cDNA library and identified dihydrofolate reductase as a target of methotrexate [28].

A three-hybrid approach was recently used to identify targets of various kinase inhibitors [29]. For each inhibitor, a hybrid ligand was synthesized by attachment to methotrexate and the hook protein was a LexA–DHFR fusion. A three-hybrid screen of a human cDNA library with a purvalanol B–methotrexate ligand identified several (but not all) known purvalanol targets and also several new candidate targets, and all identified targets were kinases. Affinity chromatography and enzymatic assays confirmed 12 of 16 novel candidate targets identified in the cDNA screens.

Induced haploinsufficiency   One genomic approach to target identification relies on the premise that altering the gene dosage of the target will affect sensitivity to the small molecule. For example, inactivating one copy of the target gene in a diploid organism would in many cases be expected to increase sensitivity. This method of identifying drug targets through *induced haploinsufficiency* was established by Giaever et al. [30], who constructed and screened a set of 233 heterozygous yeast deletion mutants for those demonstrating increased sensitivity to known drugs. Each mutant was chromosomally tagged with a unique oligonucleotide and the mutants were pooled and grown in the presence of tunicamycin, at a level of drug that is sublethal for wild type. The relative number of each mutant in the pool was monitored at various time points by polymerase chain reaction (PCR) amplification and fluorescence labeling of all tags in the pool, followed by hybridization of the PCR-generated probes to an oligonucleotide microarray. The fluorescence intensity generated at each spot on the array permitted quantitation of the relative abundance of each corresponding heterozygote in the pool. Mutants unaffected by tunicamycin showed no reduction in signal over time, whereas tunicamycin-sensitive heterozygotes showed varying decreases in signal. In this study, one known and two new loci were identified and confirmed as involved in tunicamycin resistance.

A subsequent study extended this approach to test 78 compounds against a pool of 3503 yeast heterozygotes, representing over half the yeast genome [31]. Most of the compounds tested had known activities and many also had known targets. Of 20 compounds with known targets, in most cases this method correctly identified the target or members of the target complex. Targets were also identified for a number of compounds with previously unknown targets in yeast.

Recently, a complementary approach of using gene overexpression to identify small-molecule targets has been pioneered [32]. A plasmid-based yeast genomic library was introduced into yeast and 7296 individual transformants were arrayed in 384-well plates. The arrayed library was then replicated on to solid agar containing an inhibitor. The plasmid inserts from resistant strains were then recovered and sequenced. Pkc1 was identified via this method as a target for the inhibitor and this was subsequently confirmed by affinity chromatography and genetic and biochemical assays.

Expression profiling Expression profiling was recently employed in identifying the targets of a class of small-molecule antagonists of FK506 in yeast cells subjected to salt stress [33]. These molecules (termed SFKs for suppressors of FK506) were identified in a chemical genetic screen for molecules that rescue yeast growth in the presence of high salt and FK506 [34]. Expression profiling results obtained with SFK2-treated yeast suggested that the Ald6 p pathway was a target of SFK2 and haploinsufficency screening supported this hypothesis. Overexpression of ALD6 was found to suppress the effects of SFK2–SFK4 on growth and the ability of SFKs to inhibit Ald6 p *in vitro* was subsequently demonstrated.

In addition to using DNA microarray technologies in conjunction with haploinsufficiency studies, it is likely that gene expression profiling will be used increasingly as a primary means of identifying the targets of bioactive small molecules. Many research groups have used expression profiling to identify characteristic patterns of gene expression ("fingerprints") that are associated with certain disease states or biological pathways. If the patterns are sufficiently unique, then it is sometimes possible to assign previously uncharacterized mutants to specific cellular pathways or complementation groups based on their expression profiles. Analogously, by profiling cells grown with and without a small-molecule modulator, it may be possible to identify genes or pathways that are affected by the small molecule.

The utility of expression profiling as a target identification method has been explored in a study involving 300 full-genome expression profiles in yeast generated with 276 deletion mutants, 11 regulatable alleles of essential genes and treatment with 13 compounds with known targets [35]. Of the 276 deletion mutants, 69 contained deletions of open reading frames of unknown function. To test the predictive value of the expression profiles, the profile associated with the well-defined ergosterol pathway was used to assign function to an uncharacterized ORF (YER044 c) and to identify the target of dyclonine. Subsequent biochemical and genetic experiments confirmed that both YER044 c and dyclonine affect sterol biosynthesis. The expression profiles associated with disruptions in cell wall function, mitochondrial function or protein synthesis were also used to assign function to several uncharacterized ORFs.

In combination with RNAi RNA interference (RNAi) refers to sequence-specific gene silencing triggered by the presence of double-stranded RNA within cells [36, 37]. This phenomenon has been recognized in multiple organisms, including plants, *Drosophila*, mice and humans, and researchers have rapidly employed RNAi as a tool for genome-wide studies in a variety of organisms. RNAi-based screens are logistically very similar to small-molecule screens and provide a powerful complement to both classical and chemical genetic methods.

Recently, parallel chemical genetic and RNAi screens were performed to identify cytokinesis inhibitors and their targets [38]. Over 51 000 small molecules and 19 470 dsRNAs (representing >90% of the annotated *Drosophila* genome) were screened by imaging in *Drosophila* Kc$_{167}$ cells and assessed for the ability to increase the frequency of binucleate cells. This study resulted in the identification of 50 small-molecule cytokinesis inhibitors and 214 genes that are important for cytokinesis. Analysis of compounds and dsRNAs that induced similar phenotypes led to the finding that one of the small-molecule inhibitors (binucleine 2) affects the Aurora B pathway.

1.4.4
Proteomic Methods

A proteomic approach to target identification was recently reported by Huang et al. [25], who used a chemical genetic screen to isolate small-molecule enhancers and inhibitors of rapamycin's effect on yeast growth. Biotinylated inhibitor molecules were used to probe yeast proteome chips, followed by detection with fluorescently labeled streptavidin. This resulted in the identification of several putative target proteins, the role of which in rapamycin sensitivity was assessed using mutants bearing deletions in those genes. The deletion of one of these genes (YBR077C) resulted in rapamycin hypersensitivity. Deletion of another gene (Tep1 p) had no effect on rapamycin sensitivity, but deleting the Apl2 p gene, which encodes a previously identified interaction partner of Tep1 p, did result in rapamycin hypersensitivity.

A proteomic method based on two-dimensional gel electrophoresis was recently used to identify the target of the synthetic bengamide analog LAF389 [39]. Cells were grown in the presence and absence of a natural bengamide, bengamide E, and proteins isolated from both populations were separated on 2D gels. One protein, a 14-3-3 isoform, was noted to display a bengamide-dependent change in charge. A similar effect was observed with LAF389 and the affected isoform was identified as 14-3-3 γ. Analysis of the altered 14-3-3 γ and additional studies established that LAF389 directly inhibits methionine aminopeptidases, resulting in retention of the initiator methionine in 14-3-3 γ.

1.5
Discovery for Basic Research Versus Pharmacotherapy Goals

The required characteristics of a small-molecule research tool are much less stringent than those for a lead molecule destined for clinical development. First, since a research tool will be used *in vitro* or in animal models, safety and regulatory issues are limited. ADMET constraints are not an issue for *in vitro* research and they are obviously more easily met for experimental animals than human patients, in whom adverse effects are tolerated only under extraordinary circumstances. This "experiment versus patient" issue also affects specificity and efficacy requirements, as well as the motivation to undertake subsequent structure–activity relationship (SAR) studies. For a research tool, the parameters that define adequate specificity and efficacy considerations will vary depending on whether the molecule is to be used in enzymatic, cell-based or animal studies and on the experiment's time course and sensitivity of detection. Minor or short-lived effects can be extremely useful in the basic research setting, but they will generally not suffice for a therapeutic agent and substantial SAR studies may be needed to optimize molecules intended as therapeutic leads. Molecules used as research tools often do not undergo an intensive SAR effort, as increased specificity or efficacy may not be needed.

Molecules that have been discovered through primary screens and used for basic research typically have IC_{50} values ranging from the low micromolar to approximately 50 µM and the upper end of this range likely reflects the compound concentrations employed during primary screens. Two such molecules that have been used in multiple basic research studies without further structural modification are monastrol and blebbistatin, which have IC_{50} values of 14 and 2 µM, respectively [3, 5]. In contrast, nanomolar IC_{50}s are considered desirable for therapeutic agents.

1.6
Chemical Genetic Screens in the Academic Setting

During the last decade, HTS methods have been actively pursued by the biotechnology and pharmaceutical industries to accelerate the identification of lead compounds for drug discovery. Owing to the financial resources needed to support even a modest HTS effort, these methods were largely unavailable to academic investigators. Therefore, to promote chemical genetics as an academic discipline, the Institute of Chemistry and Cell Biology/Initiative for Chemical Genetics (ICCB/ICG) at Harvard Medical School established one of the first academic HTS facilities, using an organizational model that relies on a single screening facility used by multiple investigators. This model (termed Investigator-initiated Screening) has been highly successful, facilitating screening projects for more than 80 different research groups from throughout the USA and abroad. In contrast to industry efforts towards drug discovery, the interest of the ICCB/ICG was primarily the identification of bioactive molecules for use as research tools. It should noted, however, that bioactive molecules discovered in academic screens may also show promise as pharmaceutical leads and that "high-risk" lead discovery (e.g. for targets that are not yet validated or that are considered economically nonviable for the private sector) may be better suited to academia than industry.

Under the Investigator-initiated Screening program, individual researchers propose and carry out the majority of the work for their own screening projects, including assay development, reagent dispensing, assay readout and subsequent data analysis. The screening facility staff assist by maintaining and providing access to compound collections and screening robots, training screeners for independent operation of some machines and providing some informatics and chemistry support. Data from all screens conducted at the ICCB/ICG were deposited in a nonpublic database and comparison of information across screens was used to help screeners eliminate uninteresting screening positives from further analysis. For example, fluorescent compounds that score as positive in fluorescence-based screens are most likely irrelevant to the actual screen target and compounds that score as positive in multiple cell-based screens may have nonspecific effects.

Most academic screens have been performed in 384-well format. Through spring 2005, the ICCB/ICG screening facility performed an average of 12–15 screening sessions per week, screening 14 080 compound wells (20 plates in duplicate) in a typical session. A typical investigator-initiated screening project screened 50 000–100 000 compound wells in duplicate. At this capacity, the facility could comfortably initiate 40 new screens each year.

Although academic investigators share some goals with their industry counterparts, there are certain key differences. For example, most industrial screening programs target a relatively small number of disease-relevant pathways and proteins, whereas academic investigators may wish to use chemical genetics to study a wide variety of biological pathways in diverse organisms. This scientific heterogeneity among academic groups is somewhat unpredictable and tends to result in a greater variety of assay protocols than typically encountered in an industry setting. For this reason, the ICCB/ICG screening facility was specifically designed to have the flexibility to accommodate many different types of assays, generally by employing screening instruments that can be used for as wide a variety of assay types as possible and that can work either in stand-alone mode or in custom configurations with other pieces of equipment for automation of sequential steps in a screening protocol.

Compound selection is also affected by the diverse nature of academic screens. Computational methods are employed in both academia and industry during the compound selection process to identify compounds that are "drug-like" or "lead-like" and to eliminate molecules that are unstable, toxic or otherwise unsuitable for screening. However, since academic screens may address any area of biology, the ICCB/ICG also chose to acquire compound libraries of maximal diversity and to avoid "targeted libraries" that are predicted to contain a relatively high proportion of compounds likely to act on specific pathways (e.g. kinase-targeted libraries). The ICCB/ICG libraries also contained compounds donated by chemists in exchange for data acquired by screening those compounds, thus extending the collaborative model to the wider community of synthetic chemists.

Finally, the many different types of primary screen data (including both numerical and imaging data) acquired under this organizational model present a particular challenge for data analysis and archiving purposes. Each dataset will have a unique definition of screening positives and may need highly individualized analysis techniques. Any cross-screen data comparison in this situation also requires a fairly detailed understanding of each screen being considered.

1.7
Conclusions

The use of HTS and small molecules to study cellular processes has recently begun to gain momentum. The substantial costs and logistic issues associated with establishing even a modest screening effort has deterred most academic researchers (and potentially also small biotech firms) from pursuing HTS, but it is hoped that more and more investigators will soon be able to avail themselves of this exciting new technology. Collaborative efforts such as the ICCB/ICG and the multiple screening centers currently being established by the National Institutes of Health will be vital in this respect and commercial efforts could play also a large role in putting HTS within the reach of smaller groups. A similar "popularization" of DNA microarray technologies was achieved over the last decade using both private and public sector resources and microarray experiments are now common in both academic and industry laboratories. Chemical genetics and HTS admittedly require a higher level of investment, but the potential gains in both basic science and clinical medicine are immense.

References

1 T.J. Mitchison, *Chem. Biol.* **1994**, *1*, 3–6.
2 S.L. Schreiber, *Bioorg. Med. Chem.* **1998**, 6, 1127–1152.
3 T.U. Mayer, et al., *Science* **1999**, *286*, 971–974.
4 A. Cheung, et al., *Nat. Cell. Biol.* **2002**, *4*, 83–88.
5 A.F. Straight, et al., *Science* **2003**, *299*, 1743–1747.
6 V.C. Abraham, D. L. Taylor, J. R. Haskins, *Trends. Biotechnol.* **2004**, *22*, 15–22.
7 R.A. Blake, *Curr. Opin. Pharmacol.* **2001**, *1*, 533–539.
8 J.C. Yarrow, et al., *Comb. Chem. High. Throughput Screen.* **2003**, *6*, 279–286.
9 Y. Feng, et al., *Proc. Natl. Acad. Sci. USA* **2003**, *100*, 6469–6474.
10 T.J. Mitchison, *ChemBiochem* **2005**, *6*, 33–39.
11 T.J. Nieland, et al., *Proc. Natl. Acad. Sci. USA* **2002**, *99*, 15422–15427.
12 Z.E. Perlman, T. J. Mitchison, T. U. Mayer, *ChemBiochem*. **2005**, *6*, 218.
13 Z.E. Perlman, et al., *Science* **2004**, *306*, 1194–1198.
14 N. Venkatesh, et al., *Proc. Natl. Acad. Sci. USA* **2004**, *101*, 8969–8974.
15 J.C. Yarrow, et al., *Chem. Biol.* **2005**, *12*, 385–395.
16 G.P. Tochtrop, R. W. King, *Comb. Chem. High. Throughput Screen.* **2004**, *7*, 677–688.
17 R.Y. Kao, et al., *Chem. Biol.* **2004**, *11*, 1293–1299.
18 N. Sin, et al., *Proc. Natl. Acad. Sci. USA* **1997**, *94*, 6099–6103.
19 J.A. Radding, S.A. Heidler, W. W. Turner, *Antimicrob. Agents Chemother.* **1998**, *42*, 1187–1194.
20 M.W. Harding, et al., *Nature* **1989**, *341*, 758–760.
21 M. Knockaert, et al., *Chem. Biol.* **2000**, *7*, 411–22.
22 K. Godl, et al., *Proc. Natl. Acad. Sci. USA* **2003**, *100*, 15434–15439.
23 E.J. Brown, et al., *Nature* **1994**, *369*, 756–758.
24 G. MacBeath, S. L. Schreiber, *Science* **2000**, *289*, 1760–1763.
25 J. Huang, et al., *Proc. Natl. Acad. Sci. USA* **2004**, *101*, 16594–16599.
26 A.N. Koehler, A. F. Shamji, S. L. Schreiber, *J. Am. Chem. Soc.* **2003**, *125*, 8420–8421.
27 E.J. Licitra, J. O. Liu, *Proc. Natl. Acad. Sci. USA* **1996**, *93*, 12817–12821.
28 D.C. Henthorn, A. A. Jaxa-Chamiec, E. Meldrum, *Biochem. Pharmacol.* **2002**, *63*, 1619–1628.
29 F. Becker, et al., *Chem. Biol.* **2004**, *11*, 211–223.
30 G. Giaever, et al., *Nat. Genet.* **1999**, *21*, 278–283.
31 P.Y. Lum, et al., *Cell* **2004**, *116*, 121–137.
32 H. Luesch, et al., *Chem. Biol.* **2005**, *12*, 55–63.
33 R.A. Butcher and S. L. Schreiber, *Proc. Natl. Acad. Sci. USA* **2004**, *101*, 7868–7873.
34 R.A. Butcher, S. L. Schreiber, *Chem. Biol.* **2003**, *10*, 521–531.
35 T.R. Hughes, et al., *Cell* **2000**, *102*, 109–126.
36 G.J. Hannon, J. J. Rossi, *Nature* **2004**, *431*, 371–378.
37 C.C. Mello, D. Conte Jr., *Nature* **2004**, *431*, 338–342.
38 U.S. Eggert, et al., *Public Library of Science Biology* **2004**, *2*, e379.
39 H. Towbin, et al., *J. Biol. Chem.* **2003**, *278*, 52964–52971.

2
High-throughput Screening for Targeted Lead Discovery

Jörg Hüser, Emanuel Lohrmann, Bernd Kalthof, Nils Burkhardt, Ulf Brüggemeier, and Martin Bechem

2.1
Chemical Libraries for High-throughput Screening

Experimental screening through chemical collections has been a major route for pharmaceutical research to discover novel compounds with interesting biological activity. Historically, the major source of chemical diversity for screening has been derived from natural products of plant or microbial origin. Because natural products are thought of as being the result of a molecular evolution directing molecules towards selected protein folds, these compounds are often considered as "privileged motifs" with intrinsically optimized scaffolds. However, testing of natural product extracts suffers from a number of difficulties which are less compatible with high-throughput testing of samples [1]. Extract screening necessitates the tedious and time- and resource-intensive identification of the active principle. In addition, the resulting compounds frequently are not readily accessible by chemical synthesis, thus slowing or preventing chemical derivatization in many projects. Finally, the importance of antibacterial drug discovery, which has been a traditional area for natural product research in industry, has declined over recent years because of the chronic lack of success in lead discovery. In contrast, recent advances in organic synthesis, and combinatorial chemistry in particular, made available large libraries of small synthetic compounds. The established chemical access to these small synthetic molecules offers the possibility of fast expansion of the chemical space around active compounds by structural derivatization. Moreover, *a priori* knowledge of the chemical structures of the compounds in the screening library allows one to apply sophisticated chemoinformatic and data mining tools to enhance the performance of high-throughput screening (HTS) further. As a consequence, today most large pharmaceutical companies rely predominantly on libraries made up of synthetic compounds for lead discovery (Fig. 2.1). Bridging the two approaches, a number of experimental strategies have been proposed employing structure templates derived from natural products in the generation of combinatorial chemistry libraries [2, 3]. Acknowledging that syn-

High-Throughput Screening in Drug Discovery. Edited by Jörg Hüser
Copyright © 2006 WILEY-VCH Verlag GmbH & Co. KGaA, Weinheim
ISBN: 3-527-31283-8

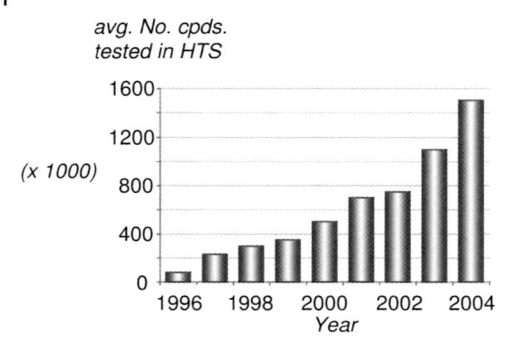

Figure 2.1 The development of the Bayer HTS library. In 1996, Bayer decided to strengthen its small-molecule lead discovery by significantly expanding its compound library. Through several collaborations with chemistry providers and a total investment of more than €100 million, the number of compounds tested in HTS steadily increased. Today >1.5 million single compounds were routinely tested in all screening programs. The logistic efforts of compound plating on to microtiter plates and the storage and retrieval of larger compound samples for pharmacological follow-up studies in large compound repositories limit the growth rates of corporate libraries.

thetic chemical libraries contain compounds selected randomly rather than shaped through a molecular evolution process also means that screening in most cases can provide only compounds with weak activity and numerous other liabilities. In this concept, hit and lead optimization serves as an artificial evolutionary process to yield drug candidates.

The notion that for target-based discovery strategies size and diversity of the compound collection employed for HTS will directly impact success rates is undebated. Screening libraries represent only a minute fraction of the chemical space satisfying the structural [4] and physicochemical requirements [5] expected from drug candidates. They are typically comprised of compounds derived from past medicinal chemistry programs, "targeted libraries" derived from target structure or pharmacophore information and randomly assembled "discovery libraries". A large number of strategies for library design have been published trying to weigh aspects of combinatorial synthesis, statistical considerations (e.g. [6]) and molecular recognition models (e.g. [7]). Apart from enriching chemical collections with biologically relevant pharmacophoric motifs, chemical diversity has been in the focus of discussions. To this end, the target employed in screening and its discriminatory power in ligand recognition will add a final measure of diversity (cf. [8]). Finally, costs and the technical feasibility regarding storage and logistics for HTS delimit the scope of library expansion. Recent surveys (e.g. HTS 2003 published by HighTech Business Decisions and Compound Management Trends 2004 published by HTStec) reveal that collections of large pharmaceutical companies today average around 1 million compounds with an ongoing trend towards further expansion. As indicated earlier, library design particularly through means of combinatorial chemistry has witnessed a significant evolution from its early days where synthetic feasibility was the main criterion to today's focus on synthesizing "drug-like" compounds. Miniaturization of assays in high-density microtiter plates, most prominently the 1536-well plate with assay volumes between 5 and 10 µl, has been a prerequisite for testing of large-scale compound collections at affordable cost.

Screening of only a preselected ("focused") subset of compounds has been proposed as an alternative strategy to random screening (e.g. [9]). In this paradigm, information derived from the structure of the targeted receptor site or from pharmacophore models is used to filter either virtual or physically available libraries through a virtual screening process (cf. [10]). The procedure is intended to reduce the number of compounds for biological testing and to increase the probability of a "focused library" to contain active compounds. Alternatively, structural motifs deduced from "privileged structures", i.e. recurring motifs with activity in a defined target class or even devoid of any target family correlation (see [11] for a review), have been employed as templates for library design with the aim of increasing hit probability. There are numerous literature reports on strategies to design focused libraries with major emphasis on kinases and G protein-coupled receptors. In addition, most compound suppliers have included focused libraries in their offerings. Some of these studies claim significantly increased hit rates in biological assays. In the end, however, most critical issues, e.g. specificity and possibly selectivity of the compound effect and the degree of novelty for patentability, remained open.

A rational selection or filtering process for compounds which still have to be synthesized or purchased is undisputed. However, if HTS can provide tests with appropriate sensitivity and sufficient throughput at acceptable cost, any preselection is far outweighed by its inherent risk of missing meaningful actives in the file. After all, our predictive capabilities are still very limited and screening success stories frequently center around the unexpected finding.

2.2
Properties of Lead Structures

Lead structures provide entry points into medicinal chemistry programs. A lead structure displays a desired biological activity but does not yet combine all pharmacodynamic and pharmacokinetic properties required for therapeutic use. In the past, traditional medicines, toxins and endogenous ligands have served as templates for the medicinal chemist in optimization programs. Iterative cycles of synthesis of structural analogs and their pharmacological testing serve to construct structure–activity relationships (SARs), which provide the means for guiding the identification of the most suitable molecule for therapeutic application. Lead structures have to display many properties to excite chemists or research managements sufficiently to invest in significant synthesis work. First, the biological effect should be significant. Therefore, drug affinity and, for agonists, their efficacy must allow reproducible quantitative detection in an appropriate test system. Moreover, efficacy should not be restricted to an artificial *in vitro* test system, but must be clearly detectable in a disease-relevant isolated organ or *in vivo* animal model.

Second, the drug should act through a specific interaction with a corresponding receptor to facilitate the understanding of the underlying SAR and to reduce the risk of unwanted side-effects in the course of the project. In addition, binding to the receptor should be preferentially reversible because low-complexity but reactive compounds with only moderate affinity will tend to interact promiscuously with a great variety of proteins. Third, the early lead structure should ideally be a member of a compound series containing structurally related molecules with graded activity indicative of an initial SAR. Fourth, the structural class should be accessible to chemical synthesis to allow derivatization within a reasonable period of time and at reasonable cost. Fifth, the properties of the drug determining absorption, distribution and metabolic stability in and excretion from the body should generally not be prohibitive to develop drugs with appropriate pharmacokinetic profiles. These ADME parameters are frequently expanded to include also toxicological profiling at early stages. For the pharmaceutical industry, a last, equally important requirement is novelty to ensure patentability of the resulting chemical structure. The necessity in pharmaceutical research to develop solid intellectual property positions is the major driving force for investing in lead discovery.

The value of a lead structure associated with a disease-relevant molecular target has frequently been underestimated. The success stories of the statins (inhibitors of HMG–CoA reductase; Fig. 2.2), sartans (blockers of the angiotensin II receptor) and opioids are examples which demonstrate that a single lead structure can potentially generate a multi-billion dollar market. Successful leads typically are followed by many companies in parallel competing optimization programs. For the statins and sartans, a number of pharmaceutical companies have generated patented drugs available for therapy today. However, advances in medicinal chemistry have enabled researchers to expand SARs far more rapidly and to cover larger chemical space around lead structures. In addition, patent strategies have advanced in parallel to disclose molecules effectively with suitable properties in strategically timed and carefully tailored structure claims, posing increasingly difficult challenges to "me-too" approaches from competitors. Thus discovering novel lead structures first provides research organizations with a clear competitive advantage.

Structure-based drug design efforts have thus far fallen short in the prediction of novel lead structures. Moreover, a large fraction of relevant drug targets are membrane proteins for which highly resolved structural data are frequently not available. Hence experimental HTS through collections of small molecules is the most relevant technology for the identification of novel lead structures. Corporate compound libraries have grown by incorporating compounds synthesized in the course of past medicinal chemistry programs. These molecules share common motifs and properties since they have been employed to explore the SAR around specific lead structures. Also, combinatorial chemistry can make a significant contribution to SAR exploration and expansion by providing "targeted" or "focused libraries" purposefully constructed around a known lead candidate [12]. A large fraction of a typical screening collection, if not the majority of compounds,

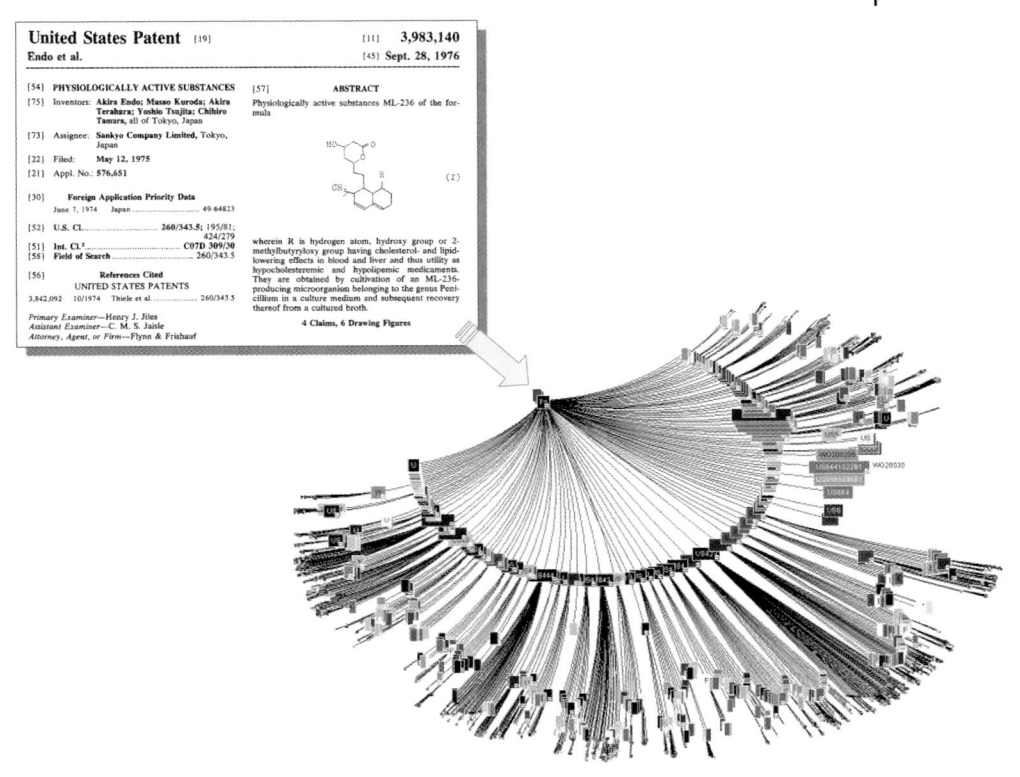

Figure 2.2 The potential impact of a novel lead structure. In 1976, Endo's group at Sankyo published the structure of ML-236 isolated from *Penicillium sp.* and described its lipid- and cholesterol-lowering effects [47]. ML-236B is now known as mevastatin. The citation tree illustrates the interrelationship of a large number of follow-up patents in at least four generations derived from the original patent by Endo's group. A large number of "me-too" projects in competing companies have resulted in four "blockbuster" drug products currently on the market. The citation tree was generously provided by Dr W. Thielemann of Bayer HealthCare.

originates from unbiased synthesis programs in combinatorial chemistry. The purpose of these "discovery libraries" is to provide potential novel lead structures through random screening.

2.3
Challenges to High-throughput Screening

The comprehensive list of attributes required from a lead structure explains why the discovery of novel molecules is expected to be a rare event. In most cases, HTS-derived lead candidates display liabilities regarding one or more of the above-mentioned properties and significant medicinal chemistry efforts are needed to assess completely the potential of a novel compound class. The importance of this "hit-to-lead" phase has attracted increasing attention, when analyzing frequent reasons for failure of HTS-driven projects [13]. One important lesson from the past decade of HTS was that HTS campaigns should ideally deliver more than one candidate compound to provide a selection of molecules for further profiling. To achieve this aim, the primary test should have a high sensitivity sufficient to detect also weakly active molecules within the screening collection. In addition, the major pharmaceutical companies invested significantly in combinatorial chemistry programs to built up comprehensive compound libraries close to or exceeding 1 million compounds. The overall productivity of this approach is difficult to quantify because of a significant lag between HTS and publication of a lead structure, the difficult assignment of a lead structure to a distinct discovery technology and the large fraction of information which remains unpublished. From the reports published there has been considerable success using HTS of chemical collections derived from classical synthesis or combinatorial chemistry in the recent past (e.g. [14, 15]). On the other hand, the early promise of HTS and combinatorial chemistry expanded libraries to "deliver more leads faster" [16–18] has remained largely unfulfilled.

In the recent past, two major issues have been raised to explain some of the key factors affecting success rates in HTS-driven projects (for a recent review, see [5]). One issue relates to the probability for a small molecule screening program to identify chemical matter for a given target or target–ligand combination. This probability, frequently referred to as "druggability" is a measure associated with a defined molecular target. It is determined by aspects of the molecular recognition process to be inhibited or mimicked by the candidate compound in addition to past experience with a target or target class. The basic concept postulates that because of the complementary requirements for a ligand to bind to a defined molecular receptor site, only certain binding sites on drug targets are compatible with the structural and physicochemical properties of drug candidate compounds [19, 20]. The second critical issue relates to the properties of the compounds in the screening collections. For example, from a retrospective analysis of databases containing marketed drugs, rules have been deduced predictive for a higher oral bioavailability of compounds (e.g. the "Rule of Five" [21]). Such rules of "drug-likeness" can be expanded to included many more properties readily described by computational descriptors, e.g. the molecule flexibility quantitated by the number of rotatable bonds, substructure filters to remove unwanted motifs or undesirable groups and positive selections of pharmacophore topologies (for a review, see [22]). Whereas in many organizations probing the drug-like properties of "virtual compounds" using chemoinformatic tools is an integral part of the design process prior to the synthesis of combinatorial libraries, these rules are typically handled with more pragmatism when applied to compounds already in the screening library. Property catalogues such as the "Rule of Five" are not relevant for the detection of active molecules in *in vitro* assays. In addition, the property prediction is certainly not accurate enough to discard non-compliant molecules from further evaluation right away. The experienced molecular pharmacologist or chemist can still extract meaningful information on structural features required for activity or possible SARs even from data based on non-drug-like compounds.

2.4
Assay Technologies for High-throughput Screening

The assay technologies used in today's high-throughput screening can be traced back to three different origins (Fig. 2.3). Cellular growth and proliferation assays have been employed to search for novel chemotherapeutics in the fields of anti-infectives and cancer (Fig. 2.3A). Bacterial and parasite growth assays were among the first tests to be employed in the systematic screening through chemical libraries for novel drug candidates. The ease of use, the potential for adaptation to miniaturized high-throughput formats and its immediate relevance to the disease pathology are reasons for their widespread use also in modern drug discovery paradigms. Growth and viability assays are prototypes of "phenotypic assays". For active compounds, the molecular target is not known and for many compounds the process of target identifications ends without conclusive results. The simple experimental realization of this test format is balanced by the difficulty of discriminating pharmacologically meaningful mechanisms from cytotoxic principles. This lack of discriminatory power of many phenotypic assay formats to differentiate between pharmacological target mechanisms and an unspecific mode of action is also one of the major pitfalls of "chemical genetic" approaches relying on similar assay technologies. As a consequence, drug discovery programs based on this strategy bear an inherently higher risk of failing toxicological testing even in late stages. Phenotypic assays are not restricted to cellular proliferation as readout. They have in common a complex cellular pathway providing, in theory, multiple targets for drug interaction. Another common feature of phenotypic assays is that the functional cellular response takes several minutes, hours or even days to become apparent. In addition, many readout formats, e.g. growth, are critically dependent on the metabolic integrity of the cell. Consequently, compounds interfering with the cell energy metabolism provide a potential source of unspecifically acting hits.

A cellular growth/proliferation

B tissue functional response to pharmacological stimuli
smooth muscle strip

C enzyme test

➢ multiple targets in signalling pathway
„Chemotherapy"

➢ specificity towards target mechanism or receptor
„Receptor Pharmacology"

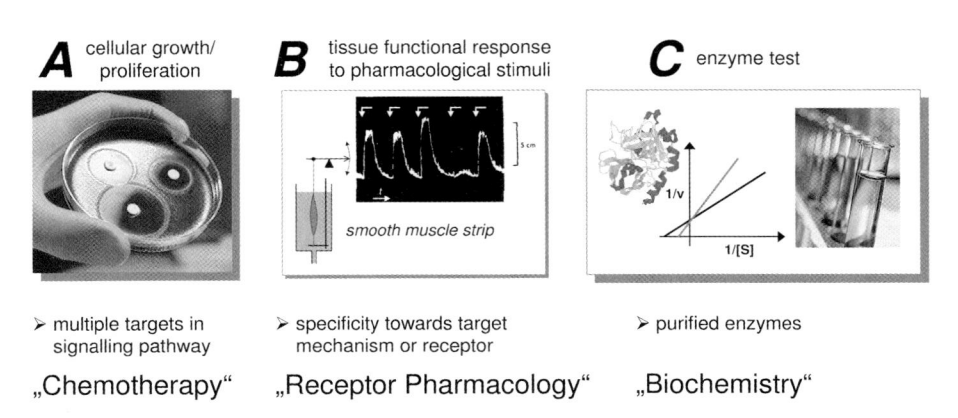

➢ purified enzymes
„Biochemistry"

Figure 2.3 Origin of assay technologies used for high-throughput screening.

Assays using a phenotypic readout can also be biased towards a molecular target by transforming a cell with, for example, a growth-promoting (proto-)oncogene. According to this strategy, active compounds inhibiting growth of the transformant but not the parental cell should be specific for the recombinant gene product or the affected pathway. Some recent reports, however, document that the specificity claim of this methodology is over-simplifying the complex underlying biology. For example, Fantin et al. [23] described a compound with "specific" growth-inhibitory action on cells transformed with the EGF family proto-oncogene neu^T but not the corresponding parental cell line. Further characterization of its effect revealed that the compound acts by compromising mitochondrial function rather than interacting with the EGF (neu^T) receptor or its downstream signaling partners. In a similar approach testing a small library of compounds with annotated pharmacological activity for killing of a variety of tumor cells versus their corresponding primary cells, ionophores were found to kill "specifically" the tumor cells [24]. From these examples, it is apparent that the complexity of cell-based assays with a "phenotypic" readout does not allow a meaningful specificity test by a simple comparison between two cell lines differing theoretically only in one trait or gene.

A different experimental approach can be traced back to classical pharmacological bioassays (Fig. 2.3B) introduced as tool for quantitative analytics and to study compound pharmacodynamic effects (e.g. [25–27]; for a comparison for NPY receptors of traditional bioassays with recombinant receptor expression system, see [28]). Classical bioassays frequently employed, for example, smooth muscle strips from different origins which respond to stimulation of cell surface receptors with readily measurable changes in contractile state. It was recognized early that sensitivity and specificity towards a defined type of receptor were crucial for a meaningful bioassay (e.g. [29]). Recently, molecular biology has revolutionized this type of assay format by providing recombinant cell lines, tailored to monitor the functional activity of a target of interest using a readout technology amenable to high-throughput formats, e.g. fluorescent Ca^{2+} indicators or reporter genes. It is obvious that the phenotypic assay and the targeted functional cell-based assay share many similarities, with no clearly defined transition between the two formats. However, the latter format can be differentiated from phenotypic assays by the degree of specificity and sensitivity towards the predefined target and the complexity of the signal transduction pathway coupling receptor stimulation to the readout signal. The time-scale relevant to this assay type ranges from seconds, e.g. hormone-stimulated intracellular Ca^{2+} signals, to a few hours for reporter gene assays. Knowledge of the target identity also allows for the establishment of additional tests to probe the specificity of compound effects. Therefore, this type of assay can be considered as "targeted" towards a specific biomolecule of interest. The functional readout of this assay format allows one to monitor all possible drug–receptor interactions, including allosteric modulation. Also agonist and inhibitor can be readily distinguished.

For many enzymes and receptors, the biomolecular target of interest can be studied in isolation using a biochemical test format (Fig. 2.3C) in which the activity of a purified enzyme, for example, is monitored directly. Moreover, by providing sensitive binding assays, receptor biochemistry has greatly catalyzed the introduction of automated parallel screening of compound libraries in modern drug research [30]. Today, a great variety of different binding assay formats relying on competition with labeled ligands or label-free formats are available. In addition to being readily formatted for use in a HTS environment, biochemical tests are sensitive, can be run at high compound/solvent concentrations and provide great experimental freedom for varying the test conditions. Again, functional biochemical tests can readily pick up allosteric modulators and discriminate inhibitors from stimulators. Biochemical assays require the protein expression in suitable host organisms and subsequent reconstitution of biological activity in an artificial test system (see Chapter 4).

Recently, "fragment screening", which also relies on biochemical assay technologies, has attracted increasing attention as an alternative route to lead discovery [31, 32]. In this approach, structural fragments or scaffolds, i.e. compounds of low molecular weight (<200 Da) and limited functionalities, are screened for typically low-affinity binding in the high micromolar to even the millimolar range at isolated target receptors. Binding to the target's active site and the detailed binding mode are resolved by NMR spectroscopy [33, 34] or X-ray crystallography [35]. Alternatively, Erlanson et al. [32] introduced a tethering strategy using disulfide-containing compounds and a modified target carrying a cysteine in close proximity to the binding site of interest to capture and subsequently identify low-affinity binders. Fragments (or scaffolds) identified are subsequently grown, linked and decorated to form larger compounds with greatly enhanced affinity. This fragment-to-lead process is often supported by structure-guided design. Because the readout technologies allow one to resolve the molecular binding site directly, these approaches are truly targeted strategies. Protagonists of this technology claim that smaller libraries can cover a significantly larger chemical space, thus compensating for the limited throughput of fragment-based screening techniques. From the few reports in the literature, however, the productivity of this novel fragment-based approach cannot be fully evaluated. The process of linking possible fragments and elaborating the most suitable decoration to achieve "drug-like" molecules producing activity in a meaningful concentration range is time and resource intensive and likely to have a high attrition rate. The minimal structural requirements needed to produce low-affinity binding might result in a high percentage of "dead end" fragments which cannot subsequently be converted into meaningful leads. In addition, because of the sophisticated readout technologies, NMR or X-ray crystallography, fragment-based screening can be applied only to targets accessible by these technologies. In summary, fragment screening requires not only a different assay expertise but also different approaches in medicinal chemistry since most of the fragments employed in these tests contain only few or no functionalities to serve as starting material for chemical derivatization. "Conventional" bioassay technologies used in HTS are not capable of resolving low-affinity interactions (>20 µM). For most receptors, a minimal structural complexity is required for the candidate compound to produce a measurable effect in the test [4].

2.5
Laboratory Automation

Laboratory automation has been a major driver of HTS. The introduction of the microtiter plate as assay carrier together with liquid handling automats and appropriate plate readers mark the beginning of HTS in the late 1980 s. Consequent miniaturization and further parallelization from 96- to 384- and finally 1536-well plates and the concomitant adaptations in instrumentation reflect the main route of HTS development. Generally, assays are either run semi-manually in the batch mode (Fig. 2.4) or on robotic systems integrating the different plate processing automats.

Robotic automation of all assay procedures not only frees laboratory staff to focus on optimizing the assay quality rather than tedious and repetitive processing of assay plates, but also contributes significantly to data quality (Fig. 2.4). Underlying every assay there are processes displaying a characteristic time course. Obvious signal kinetics for functional, non-equilibrium assays or hidden processes such as adhesion of assay reagents to the well surfaces impact the final assay reading. In semi-automated batch processing of microtiter plates, small differences in the timing of individual assay steps cannot be avoided. Depending on the kinetics of the signal-determining step, these differences result in an additional source of assay noise ("jitter"), which is largely reduced when robotic systems are used which ensure constant cycles times for each plate. In addition, constant cycle times for all assay plates further facilitate the identification of systematic trends and errors affecting the assay results which can be corrected for in primary data analysis.

The transfer of an assay protocol from manual handling in the laboratory to a robotic system to handle 150–200 plates in a row requires additional adaptations and optimization to ensure a stable assay signal and sensitivity during the course of an HTS run. Each task on a liquid handling automat, e.g. dispenser or pipetter,

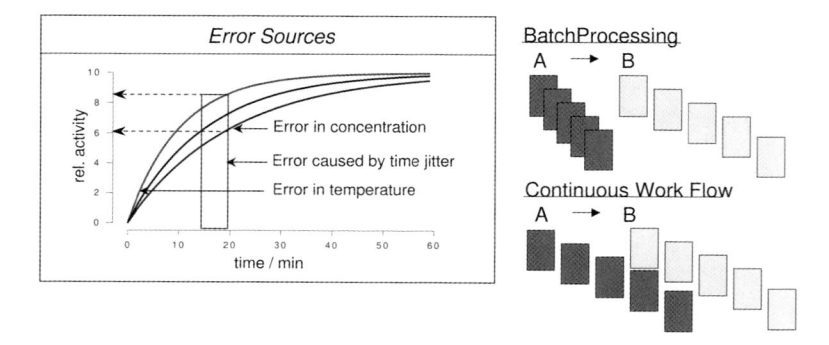

Batch Processing
- increases errors in temperature
- increases errors by ‚time jitter'
- easy to use for manual screening

Continuous Work Flow
- reduces errors in temperature
- reduces errors by time jitter
- needs sophisticated scheduling algorithms

Figure 2.4 Batch processing versus continuous flow.

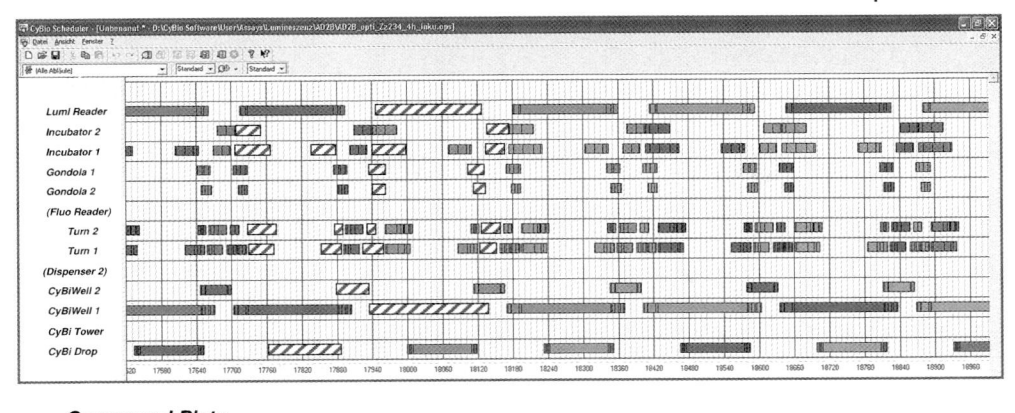

2.6 From Target Selection to Confirmed Hits – the HTS Workflow and its Vocabulary

Compound Plate
Get from Incubator 1
Turn 1
Fill w/ Buffer @ CyBi Drop
Turn 1
Transfer compounds @ CyBi Well 1
Turn 1
Store into Incubator 1

Test Plate
Get from Incubator 2
Turn 2
Gondola 1
Transfer compounds @ CyBi Well 1
Gondola 1
Turn 2
Store into Incubator 2
Get from Incubator 2
Turn 2
Aspirate Buffer @ CyBi Well 2
Turn 2
Gondola 2
Measure Signal in Lumi Reader
Discard Test Plate

Figure 2.5 Robotic realization of a reporter gene assay.

plate reader or plate transport device incorporated into the robotic systems, is programmed after careful optimization of each individual process. The scheduling software executes the individual device programs in a previously defined sequence, i.e. schedule. Figure 2.5 shows a "Gantt Chart" illustrating the realization of a simple reporter gene type assay on a robotic system (CyBi ScreenMachine®, CyBio, Jena, Germany). The chart illustrates the tight packing of processes once the system has reached "steady-state" conditions during which multiple plates are handled simultaneously at different stations of the robotic system. With a cycle time of 230 s within this test protocol, during one overnight robotic run of 16 h 250 microtiter plates or 384 000 wells (using the 1536-well plate) can be tested.

2.6
From Target Selection to Confirmed Hits – the HTS Workflow and its Vocabulary

Figure 2.6 summarizes the workflow underlying a typical target-based discovery program. In the complex target selection process, medical and strategic considerations have to be balanced with the scientific rationale of a disease hypothesis linking a gene (product) to a disease state [36]. This rationale can be based on a variety of experimental data from genetic studies, information on gene expression and regulation to protein, small-molecule or other tools employed in functional studies. Such disease-linked genes or proteins are commonly referred to as targets. Frequently cited examples for successful target approaches to drug discovery are the inhibitors of angiotensin-converting enzyme (ACE) [37] and HIV protease [38]. With the completion of the human genome project, a large number of potential drug targets have been catalogued. However, only very few have delivered successful drug discovery programs thus far. The degree of validation of a target or the underlying disease hypothesis can vary substantially. *In vitro–in vivo* disconnects, i.e. failures to demonstrate compound efficacy in a relevant secondary test or animal disease model, represent a major reason for termination of early discovery programs. Flawed disease hypotheses contribute significantly to failures in lead discovery with the corresponding candidate compounds being dropped in most cases.

To screen a chemical library for possible modulators of the target, an appropriate assay has to be designed. For many targets a variety of technically feasible assay systems are available (e.g. [39–42]). The choice of an assay technology depends on many factors, including already existing expertise in a target or assay class and the compatibility with existing HTS hardware, e.g. signal detection instrumentation. It is also driven by aspects related to assay sensitivity, the desired profile of the modulators and the susceptibility of the test system to possible assay artifacts and unspecific interference. Finally, the costs for assay realization can also play a major role. Assay development aims at delivering a test system to detect target modulation reproducibly. Signal scattering and high signal-to-noise ratios which frequently are described quantitatively by the Z-factor [43] represent only one aspect for assay quality. More importantly, it is the sensitivity of the assay towards detecting compounds with a desired molecular mechanism that primarily determines the test quality.

In primary HTS, the small-molecule library is tested in the assay. Tests are typically performed in microtiter plates with 96 (100 µL), 384 (10–50 µL) or 1536 (5–10 µL) wells per plate. One main factor determining the choice of the plate format for HTS is the size of the corporate compound library. Compound pooling and testing of mixtures necessitate complex deconvolution processes and limit the

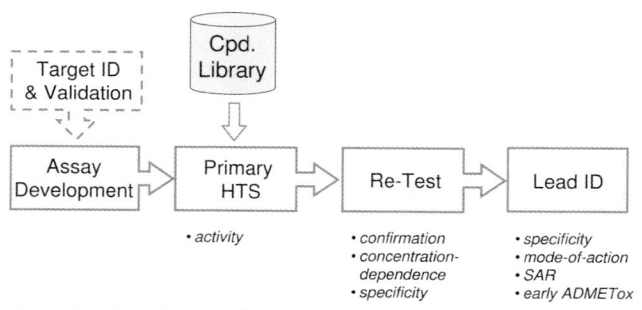

Figure 2.6 The HTS workflow.

concentration of compounds introduced into the test. Hence the apparent benefit of increased throughput is counterbalanced by an increased risk of false negatives and, consequently, the global industry trend is towards testing single compounds (HighTech Business Decisions, 2004). In large pharmaceutical companies the size of the compound collections averages around 800 000–1 000 000 compounds. However, only in rare cases is this library also routinely tested in all screening programs. Nevertheless, most large pharmaceutical companies have or intend to establish screening platforms allowing to test comprehensive libraries with throughputs exceeding 100 000 compounds per day (ultra-high-throughput screening; uHTS). Individual compounds are typically tested in a given assay only once at a concentration between 2 and 20 µM. However, the final assay concentration cannot be controlled for precisely. Because of these uncertainties in compound concentration, signal scattering of the biological test and random technical failures resulting, for example, from liquid handling errors, false negatives cannot be excluded in HTS. False negatives are truly active compounds not detected in a given assay. This problem is extremely difficult to quantitate. It can only be approximated in target assays for which the screening file already contains a number of known actives. Realistic estimates range between 5 and 15 % probability of false negatives, depending on, among other things, potency of the compound and the assay format. Typical chemical libraries, most prominently those derived from combinatorial chemistry, include compounds structurally related to each other. If the similarity within a class is high enough that a given assay produces clusters of multiple actives, the risk of missing potential lead candidates because of random errors (statistically false negatives) is greatly reduced. For example, with two structurally similar and active compounds and a false negative rate of 10 %, one has a 99 % chance of positively identifying this compound class.

Technical failures occurring randomly throughout the course of the assay, frequently associated with the different liquid handling steps in an assay protocol, cross-contaminations of active compounds into wells containing inactive compounds and data scattering of the biological test system, are the major sources of false positives, i.e. non-reproducible actives in primary HTS. The rate of false positives caused by assay noise can be approximated conservatively from the scattering of data points of controls, i.e. assay wells with no compound but only solvent added. Obviously, such "fake hits" can be identified readily through repetitive testing. The rate of false positives and therefore the confirmation rate is highly dependent on the assay format and the robustness of its experimental realization. Retesting of primary HTS hits under identical test conditions is a mandatory step in any HTS workflow to narrow down effectively a primary hit set to those compounds generating a true and reproducible effect in the assay.

In Fig. 2.7, results from a recent in-house 1536-uHTS campaign targeting a serine protease are shown illustrating a typical data set from a screening project. In this project, 1.5 million compounds were tested as singles in a biochemical assay monitoring enzyme activity directly using a fluorogenic substrate. The entire campaign was completed in 2 weeks with nine fully-automated robotic runs averaging 170 000 compounds tested per day. All activity data were normalized to the data values obtained for "neutral controls" on the same microtiter plate, i.e. in the absence of any test substance. When trying to identify also weakly active compounds using very sensitive test conditions and defining the hit threshold to the minimal confidence level, i.e. inhibition of roughly more than three times the standard deviation (SD), significant capacity has to be allocated for "hit (or cherry) picking" and retesting. For the example protease assay, the hit threshold was defined at 20% inhibition (>6 × SD) at a final concentration of test compounds around 7 µM. Roughly 2800 hits were picked from the library and reformatted on to new compound plates for retesting. Compounds were tested at four different concentrations (not shown) in quadruplets under conditions identical with primary HTS. The correlation between retest and primary HTS is shown in Fig. 2.7B. In this scatter plot, compounds can be readily separated into confirmed hits populating the bisecting line with comparable activity values in both tests and false

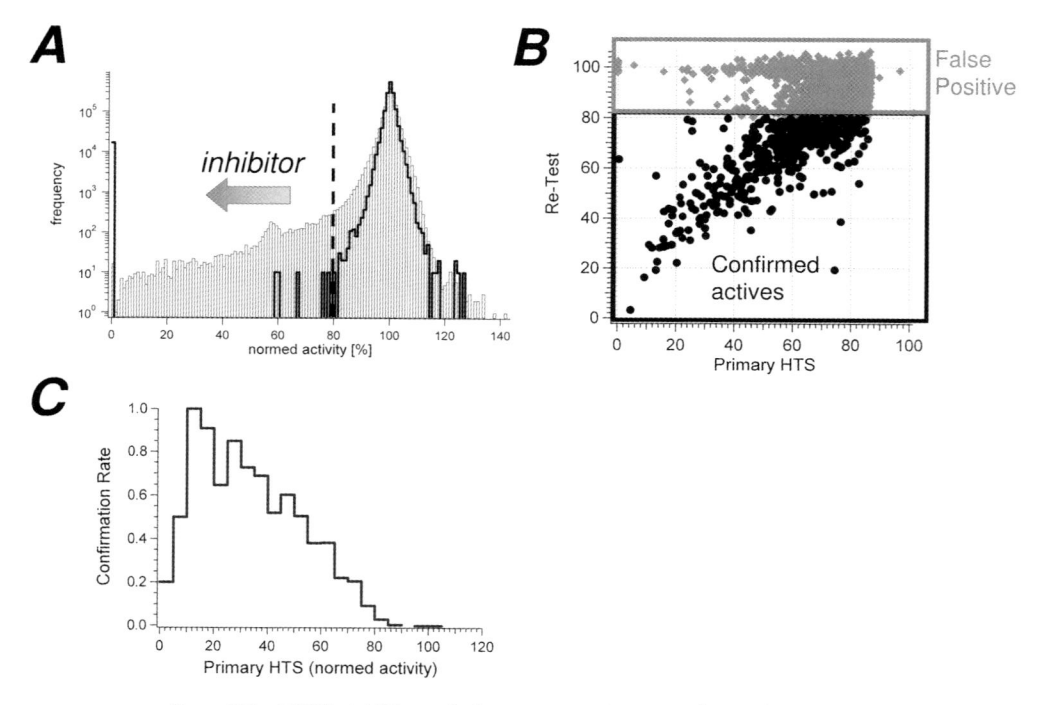

Figure 2.7 uHTS for inhibitors of a Ser protease. (A) Frequency distribution histogram of activity values from solvent controls (dark trace) and compounds (light bars) summarizing the data from a recent in-house campaign. More than 1.5 million compounds were tested as single compounds for their activity to inhibit enzyme activity measured by a fluorogenic substrate. Activity data were normalized to solvent controls on the same microtiter plate. The standard deviation of controls was ±3%. Compounds with >20% inhibition were considered primary hits and, after activity profiling on the HTS database, 2800 primary hits were followed up in a robotic retest. (B) Scatter plot comparing the activity in retest and primary HTS. Confirmed hits displaying similar activity in both tests can readily be distinguished from irreproducible false positives. (C) The confirmation rate clearly correlated with the potency of the primary hit.

positives scattering around 100% for the retest activity value. With a hit threshold of 20% inhibition for the retest, about 48% of the primary hits could be confirmed. The correlation between efficacy in the primary test and the rate of confirmation (Fig. 2.7C) revealed the expected behavior that the confirmation rate correlates positively with activity.

Confirmed actives identified in HTS need further characterization to ensure target specificity, a meaningful mode of action and finally a non-prohibitive ADMETox profile. Therefore, in most organizations, HTS is followed by a lead identification phase (Fig. 2.6) to address these issues. In addition, for many compounds in this early discovery phase, also synthesis of analogs is needed to probe for a possible initial SAR around a candidate molecule.

2.7
Separating Specific Modulators from Off-Target Effects

In HTS, the small number of lead candidate compounds acting through a meaningful target interaction typically are obscured by a large number of unreproducible hits and compounds acting through unspecific mechanisms. Statistically, false positives resulting from random errors within the assay realization can be readily identified through repetitive testing. A more difficult problem originates from compounds acting through unspecific mechanisms to produce the same effect on the assay readout as the desired target modulators. The probability of "off-target" interference by test compounds increases with increasing complexity of the assay. As discussed earlier, this problem complicates the interpretation of cell-based assays providing multiple sources of unspecific interactions. For unbiased, phenotypic assays, no clear rationale for an experimental strategy to separate pharmacologically relevant target-mediated effects from simple "assay artifacts" or toxic compounds is available until the target is identified. In contrast, for targeted assays such strategies can be readily developed. Typically, historical assay data from previous screening campaigns provide useful information to identify compounds with promiscuous activity in tests based on, for example, similar readout technologies. Because unspecific effects detected across many different cell-based target assays might be caused by compound interactions with endogenous receptors, comprehensive databases containing HTS-derived activity data can be mined for such activities to serve as starting points for "chemical genetic" approaches (see Chapter 10). In addition, confirmed active compounds typically are subject to testing in a series of reference assays probing for specificity and selectivity. Ideally, specificity tests should rely on readout technologies not affected by the same mechanisms of unspecific interference as the test used for primary HTS. For example, for a functional cell-based assay probing the activity of a membrane receptor, a binding assay can provide such an "orthogonal" test system. However, it is desirable, especially for testing large compound libraries, to gain information on the specificity of compound action already in primary testing. By filtering down the number of unwanted hits early in the process, downstream capacity for confirmation and specificity testing can be focused on a smaller number of hits. Early filters can be introduced into the primary HTS assay by implementation of reference signals. In a simple case such an assay integral reference signal might be the detection of compound autofluorescence to exclude artifacts in tests with fluorescence readouts.

Figure 2.8 summarizes a more complex strategy utilizing a functional cell-based assay for detecting inhibitors of a G-protein-coupled receptor (GPCR). Receptor activity is monitored by the ligand's ability to initiate Ca^{2+} release from intracellular stores through formation of the intracellular messenger inositol-1,4,5-trisphosphate. Changes in intracellular free Ca^{2+} concentration were visualized by means of a Ca^{2+}-sensitive photoprotein (e.g. [44]) targeted to the mitochondrial matrix compartment [45]. Stimulation with the ligand for the recombinant target receptor

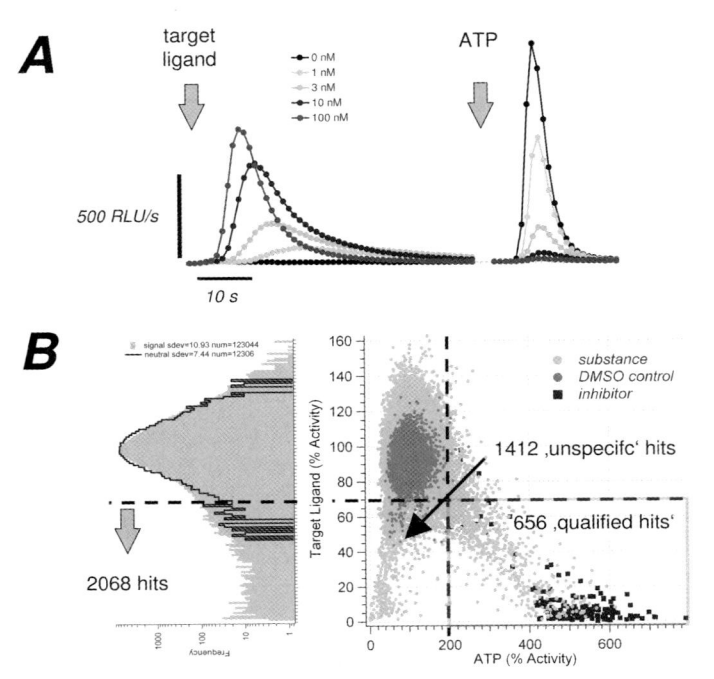

Figure 2.8 Discrimination of target modulators from unspecifically acting compounds by integrated assay reference signals. A Ca^{2+} release assay monitoring the function of a recombinant GPCR. (A) Kinetic reading obtained from recombinant screening cells seeded in 1536-well plates in response to consecutive stimulation of the recombinant target receptor (first challenge; target ligand) and the endogenous P2Y receptors (second signal; ATP). Changes in intracellular Ca^{2+} concentration were followed by means of a recombinantly expressed Ca^{2+}-sensitive photoprotein. Inhibitors of the targeted cell surface receptor were expected to inhibit the first signal, but enhance the light emission upon stimulation with ATP. (B) Frequency distribution histogram summarizing data from one robotic run (~120 000 compounds) for the first signal revealed an expected number of 2068 hits inhibiting the target receptor signal. Additional 2D analysis using also the reference signal (scatter plot; right) visualized two populations of hits and discriminated >60% of hits as unspecific.

resulted in luminescence signals resembling the time course of the underlying intracellular Ca^{2+} transient. The amplitude of the Ca^{2+} or light transient increased and the time-to-peak decreased with increasing concentration of ligand injected into the well. The photoprotein is consumed in the reaction with Ca^{2+} and intracellular Ca^{2+} stores become depleted because of a net loss of cellular Ca^{2+} in the continuous presence of receptor agonist. Both consumption of photoprotein and cellular Ca^{2+} depletion contributed to the desensitization towards a second challenge of the cells with ATP which activated endogenous purinergic receptors coupling to the same biochemical signaling molecules to elicit Ca^{2+} release (Fig. 2.8A). The stronger the cellular response to the target receptor stimulation, the smaller was the signal output upon subsequent stimulation with ATP. Thus, in the presence of a receptor antagonist specifically interfering with the targeted receptor, the first signal should be reduced while the response to the ATP reference signal is expected to increase. This behavior is used to differentiate between two populations of hits.

Figure 2.8B summarizes the data from a single robotic uHTS run performed in 1536-well plates testing ~120 000 compounds. For each well, the cells were challenged first with the agonist of the target receptor before ATP was injected to stimulate endogenous P2Y receptors (Fig. 2.8A). Temporally resolved luminescence signals were captured simultaneously using a custom-made plate reader. All activity values were normalized to the DMSO controls on the same assay plate. From the 120 000 compounds tested 2068 compounds inhibited the response to target receptor stimulation by more than 30%. The total library tested with ~1 000 000 compounds revealed around 10 000 hits (not shown), far too many for follow-up confirmation and specificity testing. Taking advantage of the reference ATP signal in the 2D scatter plot (Fig. 2.8B), the initial hit pool can be differentiated into two discrete populations. Large fractions of this compound set could also be labeled "unspecific" and excluded from further analysis because of their broad activity in other assays using the same readout (not shown). In contrast, compounds inhibiting the target signal and enhancing the response to ATP ("qualified hits") displayed the desired profile expected from inhibitors of the target receptor. Positive controls, i.e. known inhibitors of the target receptor included in the test at various concentrations, also distributed with the qualified hits. By use of this reference signal, the number of compounds to be included in further testing was reduced to 656 hits (or 20% of the total hits) displaying the activity profile expected from receptor antagonist acting specifically on the target receptor. The majority of compounds inhibited both signals. A large fraction of these compounds could be flagged as unspecific based also on activity profiling with historical assay data. Promiscuous inhibitors included, among others, ionophores, detergents and compounds severely impacting cellular energy metabolism like uncouplers of oxidative phosphorylation. Focusing early in the process on compounds with a higher probability of being specific modulators of the target of interest is particularly helpful when using cell-based assay formats combining significant data scattering and intrinsically higher hit rates. Follow-up confirmation and specificity testing is typically limited in throughput and therefore excluding compounds acting unspecifically through off-target mechanisms allows resources to be used more effectively.

Reference signals to exclude prominent sources of systematic assay "artifacts" can be integrated into a large variety of test formats. Integration of reference signals provides additional challenges for the realization in HTS because they require one to integrate additional signal detection and liquid handling steps into the assay protocols. However, the example illustrated in Fig. 2.8 demonstrates that even in uHTS settings using 1536-well microtiter plates such complex protocols can be realized.

2.8
Data Analysis and Screening Results

Data analysis in HTS combines biological activity data with chemoinformatic tools to consolidate and interpret experimental observations and to extract additional information sometimes hidden in noisy data or otherwise not readily detectable. Grouping active compounds into structural classes, e.g. through hierarchical clustering, and considering the structural clusters rather than single compounds greatly enhance the resolution looking at scattering HTS data. Detecting structurally related compounds with a desired activity might provide early clues for possible SARs guiding early optimization. Another frequent application of this method is to discriminate compounds in functional cell-based tests which are likely to act through off-target mechanisms (Fig. 2.9). This method is based on the assumption

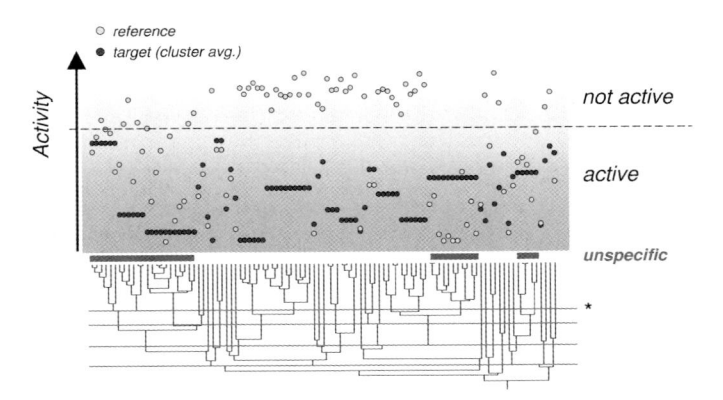

Figure 2.9 Advanced specificity testing using cluster analysis. Compounds with confirmed activity in a target assay were tested on a second related test used to inspect the specificity of compound action. Compounds were clustered hierarchically based on 2D fingerprints. The structural relationship between active compounds is coded in the dendrogram. Slicing the dendrogram at different levels revealed groups of related molecules, i.e. clusters. Alignment of the compound activity in the target (dark points) and reference test (light points) with the structure dendrogram readily visualized clusters with promiscuous representatives (marked unspecific) and compound clusters active only in a defined target context.

2.8 Data Analysis and Screening Results

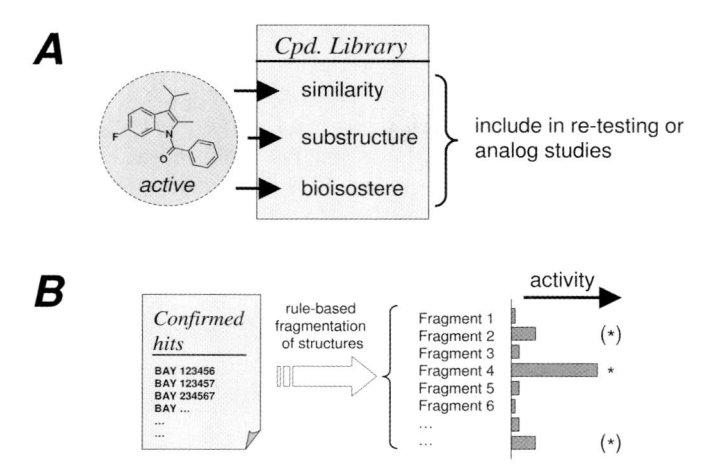

Figure 2.10 Chemoinformatic tools supporting HTS data analysis. (A) Similarity and substructure searches are frequently utilized for analysis of HTS data. On the basis of known actives, the compound set for follow-up studies is enlarged by similar structures, structures sharing common substructures and structures with known bio-isosteric replacements. Within such extended compound sets, false negatives in the primary test and compounds with borderline activity can be identified. (B) The association of activity to defined substructural motifs within a hit set is aimed at provide insights into possible fragments correlating with overall compound activity.

that structurally similar compounds act through the same biological mechanism. Examination of the activity of a cluster of related compounds in a relevant reference or specificity test frequently identifies clusters with unspecifically acting representatives. Because the reference test might not reliably detect this unspecific interference with sufficient sensitivity, some compounds might pass this filter when not analyzed in the context of their structural neighborhood. Elimination of unspecific compounds from hit sets is only one typical application of chemoinformatics in HTS. Broadening of possible SARs around identified actives and detection of possible false negatives in primary HTS is a second important field. To this end, active compounds are used as templates to search the library for compounds with an overall similarity, a shared substructure or known bio-isosteric replacements (Fig. 2.10A). The compounds yielded by this search are either included in the retest following primary screening or in the subsequent lead identification process to complete the data set on all physically available compounds.

In the recent past, computer tools have been introduced which aim at identifying molecular fragment contributing to biological activity. For example, Lewell et al. [46] developed a retrosynthetic analysis approach (RECAP) to "select building block fragments rich in biologically recognized elements and privileged motifs" from larger sets of pharmacologically active compounds. These fragments subsequently serve as building blocks to synthesize combinatorial chemistry libraries enriched in relevant motifs. The RECAP analysis can be performed on large databases such as the World Drug Index. Alternatively, it can be applied to sets of confirmed actives from HTS. Although significantly smaller in number, such hit sets contain potentially novel and proprietary fragments biased towards a target class or a cellular pathway of interest. The most advanced fragment research tool to analyze HTS data is provided by TRIPOS. "Structural unit analysis" can be applied to sets of several hundred compounds found to be active in HTS. The principle is illustrated in Fig. 2.10B. Compounds are dissected *in silico* into fragments by cutting along rotatable bonds. After removing trivials and very rare fragments, all remaining fragments are correlated with their biological activity measured in the assay and based on statistical models rules are extracted to identify structural units which positive correlate with activity.

Fragment analysis of compound sets identified in HTS provides means for extracting information hidden in screening data and not readily detectable by clustering methods. The identification of novel and unexpected fragments in screening hits which can be combined with known scaffolds opens up new possibilities in expanding existing structure–activity relationships. Detection of such fragments can therefore be a meaningful and productive outcome of screening.

2.9
Conclusions

Experimental testing of candidate drug compounds remains the major route for lead discovery. HTS using targeted assays together with the design of chemical libraries have matured over the last decade to provide a technology platform for lead discovery. Still critical to experimental compound testing is the early discrimination of compound effects exerted through a pharmacologically relevant target interaction from unspecific "off-target" mechanisms. Building large databases containing relevant reference data and implementation of reference signals into the assay itself are effective means to filter out these compounds within the early screening process. Further proof of a specific compound–target interaction, particularly for cell-based tests, requires additional tests with orthogonal readouts to exclude also subtle assay artifacts. Converting the knowledge of the target mechanism and underlying molecular recognition principles into robust and sensitive assays is a prerequisite of successful screening. Assay results have to be viewed not only from the affinity or potency perspective. Ideally, HTS should deliver more than one candidate compound and subsequent prioritization includes early testing of ADME-Tox properties.

References

1 Koehn, F. E., Carter, G. T. *Nat. Rev. Drug Discov.* **2005**, *4*, 206–220
2 Reinbauer, R., Vetter, I., Waldmann, H. *Angew. Chem. Int. Ed.* **2002**, *41*, 2878–2890.
3 Schreiber, S. L. *Bioorg. Med. Chem.* **1998**, *6*, 1127–1152.
4 Hann, M. M., Leach, A. R., Harper G. *J. Chem. Inf. Comput. Sci.* **2001**, *41*, 856–864.
5 Lipinski, C., Hopkins, A. *Nature* **2004**, *432*, 855–861.
6 Harper, G., Pickett, S. D., Green, D. V. S. *Comb. Chem. High Throughput Screen.* **2004**, *7*, 63–70.
7 Wintner, E. A., Moallemi, C. C. *J. Med. Chem.* **2000**, *43*, 1993–2006.
8 Roth, H.-J. *Curr. Opin. Chem. Biol.* **2005**, *9*, 293–295.
9 Valler, M. J., Green, D. *Drug Discov. Today* **2000**, *5*, 287–293
10 Böhm, H.-J., Schneider, G. (Eds.) *Virtual Screening for Bioactive Compounds.* Wiley-VCh, Weinheim, **2000**.
11 Müller, G. *Drug Discov. Today* **2003**, *8*, 681–691.
12 Gordon, E. M., Gallopp, M. A., Patel, D. V. *Acc. Chem. Res.* **1996**, *29*, 144–154.
13 Bleicher, K. H., Böhm, H. J., Müller, K., Alanine, A. I. *Nat. Rev. Drug Discov.* **2003**, *2*, 369–378.
14 Golebiowski, A., Klopfenstein, S. R., Portlock, D. E. *Curr. Opin. Chem. Biol.* **2001**, *5*, 273–284.
15 Golebiowski, A., Klopfenstein, S. R., Portlock, D. E. *Curr. Opin. Chem. Biol.* **2003**, *7*, 308–325.
16 Gordon, E. M., Barrett, R. W., Dower, W. J., Fodor, S. P. A., Gallop, M. A., *J. Med. Chem.* **1994**, *37*, 1385–1401.
17 Bevan, P., Ryder, H., Shaw, I. *Trends Biotechnol.* **1995**, *13*, 115–121.
18 Williard, X., Pop, I., Horvarth, D., Baudelle, R., Melnyk, P., Deprez, B., Tartar, A. *Eur. J. Med. Chem.* **1996**, *31*, 87–98.
19 Hopkins, A. L., Groom, C. R. *Nat. Rev. Drug Discov.* **2002**, *1*, 727–730.
20 Hopkins, A. L., Groom, C. R. *Ernst Schering Research Foundation Workshop* **2003**, *42*, 11–17.
21 Lipinski, C. A., Lambardo, F., Dominy, B. W., Feeney, P. J., *Adv. Drug Deliv. Rev.* **1997**, *23*, 3–25.
22 Mügge, I. *Med. Res. Rev.* **2003**, *23*, 302–321.
23 Fantin, V. R., Berardi, M. J., Scorrano, L., Korsmeyer, S. J., Leder, P. *Cancer Cell* **2002**, *2*, 29–42.
24 Root, D. E., Flaherty, S. P. Kelley, B. P., Stockwell, B. R. *Chem. Biol.* **2003**, *10*, 881–892.
25 Vane, J. R. *Br. J. Pharmacol.* **1983**, *79*, 821–838.
26 Black, J. W. *Science* **1989**, *245*, 486–493.
27 Black, J. W., in J. C. Forman, T. Johansen (Eds.), *Textbook of Receptor Pharmacology.* CRC Press, Boca Raton, FL, **1996**, pp. 277–285.
28 Pheng, L. H., Regoli, D. *Regul. Pept.* **1998**, *75*, 79–87.
29 Cuthbert, A. W. *Trends Pharm.Sci.* **1979**, *1*, 1–3.
30 Snyder, S. H. *J. Med. Chem.* **1983**, *26*, 1667–1674.
31 Carr R., Jhoti, H. *Drug Discov. Today* **2002**, *7*, 522–527.
32 Erlanson, D. A., McDowell, R. S., O'Brien, T. *J. Med. Chem.* **2004**, *47*, 3463–3482.
33 Shuker, S. B., Hajduk, P. J., Meadows, R. P., Fesik, S. W. *Science* **1996**, *274*, 1531–1534
34 Hajduk, P. J., Gerfin, T., Boehlen, J. M., Haberli, M., Marek, D., Fesik, S. W. *J. Med. Chem.* **1999**, *42*, 2315–2317.
35 Card, G. L., Blasdel, L., England, B. P., Zhang, C., Suzuki, Y., Gillette, S., Fomg, D., Ibrahim, P. N., Artis, D. R., Bollag, G., Milburn, M. V., Kim, S.-H., Schlessinger J., Zhang, K. Y. J. *Nat. Biotechnol.* **2005**, *23*, 201–207.
36 Knowles, J., Gromo, G. *Nat. Rev. Drug Discov.* **2003**, *2*, 63–69.
37 Vane, J. R. *J. Physiol. Pharmacol.* **1999**, *50*, 489–498.
38 Caporale, L. H. *Proc. Natl. Acad. Sci. USA* **1995**, *92*, 75–82.
39 Walters, W. P., Namchuk, M. *Nat. Rev. Drug Discov.* **2003**, *2*, 259–266.
40 Landro, J. A., Taylor, I. C. A., Stirtan, W. G., Osterman, D. G., Kristie, J.,

Hunnicutt, E. J., Rae, P. M. M., Swettnam, P. M. *J. Pharmacol. Toxicol. Methods* **2000**, *44*, 273–289.
41 Sundberg, S. A. *Curr. Opin. Biotechnol.* **2000**, *1*, 47–53.
42 Gonzales, J. E., Negulescu, P. A. *Curr. Opin. Biotechnol.* **1998**, *6*, 624–631.
43 Zhang, J.-H., Chung, T. D. Y., Oldenburg K. *J. Biomol. Screen.* **1999**, *4*, 67–73.
44 Stables, J., Green, A., Marshall, F., Fraser, N., Knight, E., Sautel, M., Milligan, G., Lee, M., Rees, S. *Anal. Biochem.* **1997**, *252*, 115–126.
45 Rizzuto, M., Simpson, A. W., Brini, M., Pozzan, T. *Nature* **1992**, *358*, 325–327.
46 Lewell, X.-Q., Judd, D. B., Watson, S. P., Hann, M. M. *J. Chem. Inf. Comput. Sci.* **1998**, *38*, 511–522.
47 Endo, A. *Atherosclerosis Suppl.* **2004**, *5*, 67–80.

Part II
Automation Technologies

3
Tools and Technologies that Facilitate Automated Screening
John Comley

3.1
Introduction – the Necessity to Automate

3.1.1
Compound Libraries

The mean size of a compound library in a large pharmaceutical company was recently (2005) estimated to be 1.05 million compounds, and this is expected to grow to 1.80 million by 2008 [1]. In contrast, small pharmaceutical/biotech companies are estimated to have 0.73 million compounds today (2005), and this is expected to grow to 1.24 million by 2008 (Fig. 3.1). On average, around 90% of the compounds in a library are accessible for screening [1].

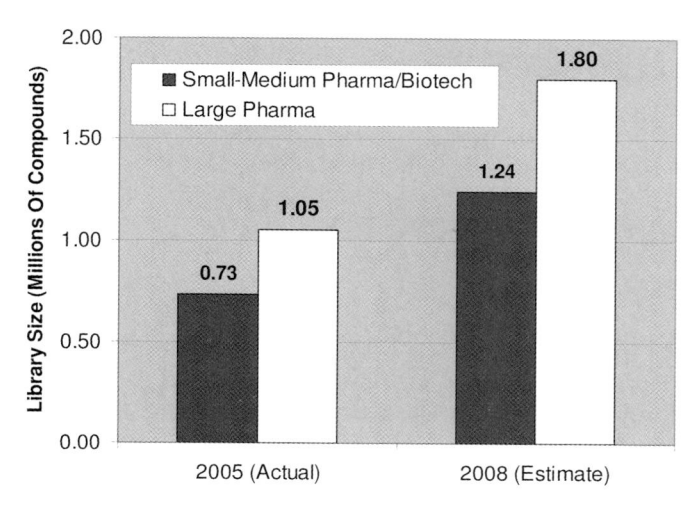

Figure 3.1 The average size and estimated growth of compound libraries.

High-Throughput Screening in Drug Discovery. Edited by Jörg Hüser
Copyright © 2006 WILEY-VCH Verlag GmbH & Co. KGaA, Weinheim
ISBN: 3-527-31283-8

3.1.2
Targets and Data Points

High-throughput screening (HTS) groups will screen on average 24.2 targets per laboratory in 2005, but there may be multiple screening laboratories or sites per company [2]. The majority of these targets are drawn from three target classes in 2005: G-Protein Coupled Receptors (GPCRs), kinases and ion channels, representing 33, 19 and 19% of all targets screened, respectively (Fig. 3.2). An average of 792 000 wells (data points generated) will be screened against each of these targets in 2005. The total number of wells screened per year (data points generated) against all targets is 19 008 000 per laboratory per year in 2005, increasing by about 20% annually [2].

3.1.3
Main Issues Facing HTS Groups Today

The greatest pressure on HTS groups today is to improve data quality and to achieve higher success rates. This is set against a background of resource constraints – most screening groups operate on fixed or reduced budgets (i.e. less head count and operating costs) relative to several years previously. In addition, there is pressure for greater workflow capacity from existing screening systems. Furthermore, groups are increasingly adopting a more cautious approach when implementing new technologies, there is less Capex (Capital Expenditure) money available and there is much greater emphasis on return of investment and accessing fully validated technologies. Current interest in acquiring new technologies is therefore confined mainly to those that: (1) allow access to new ways of screening [e.g. high-content screening or high-throughput electrophysiology (patch clamping)]; (2) confer significant advantage and add value (e.g. method simplification, potential cost benefits when used as a generic platform) over existing methodologies; or (3) overcome perceived limitations of existing technologies and formats (e.g. allosteric inhibition has put into question the use of β-lactamase reporter systems or Fluorescence Lifetime (FLT) may overcome many assay interferences)

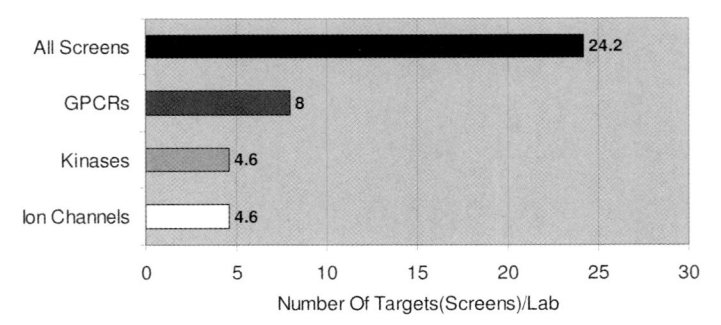

Figure 3.2 The average number of targets screened per HTS laboratory in 2005.

or facilitate access to new or difficult target classes (e.g. electrochemiluminescence). Cell-based assays are on the increase, they now represent greater than 50% of all assays screened and this also creates resourcing issues (see Section 3.5.7). Further issues arise as groups struggle to maintain the flexibility (both hardware and in-house staff expertise) needed to screen many diverse assay technologies. As screening operations become more like factories, there is a desire to focus screening personnel away from the repetitive tasks towards assay development and data analysis. The need to speed up and simplify the assay transfer process to HTS is also a problem for some groups.

3.1.4
Benefits of Miniaturization

Miniaturization is perceived as one of the key mechanisms to making better use of declining budgets: it facilitates reduced reagent consumption (one of the few costs over which screeners have direct control), a reduced amount of target (protein or cells) is needed and compound consumption is minimized (saves precious libraries). In addition, miniaturization necessitates the greater use of laboratory robotics and automation: it is not possible to screen in 1536-well plates without some degree of automation. Parallel processing leads to improved efficiency, higher throughput (speed) and the potential for increased capacity (facilitating the screening of larger compound libraries). Miniaturization also provides for improved logistics by minimizing the number of plates handled and by reducing waste stream (especially plastics) generated and requires less inventory/warehouse storage space.

3.1.5
Benefits of Automated HTS

Apart from the obvious benefits of automation in allowing for unattended operation and freeing up labor resources by enabling redeployment of staff to less repetitive tasks, there are other important paybacks. These particularly include improvements in data reproducibility and quality achieved through higher consistency in plate processing (particularly liquid handling) and reduced inter-plate variations. Without doubt the higher levels of throughput achieved today would not have been possible without automation. Associated with automation are audit trails, sample tracking and inventory control, all aspects which are increasingly viewed as highly important and necessary in today's sophisticated drug discovery laboratories. There are also important biochemical, radiochemical and biological safety benefits to be derived from the reduced exposure to potentially toxic or harmful substances or pathogens that contained automated processing can offer. Automated systems are highly suited to environmental control (e.g. AlphaScreen beads are light sensitive and it is advantageous to work in a subdued green light environment (~100 lux) as exposure of beads to light reduces the signal; this is a lot easier to achieve in an enclosed automated system than on the laboratory bench where the impact of subdued light on humans needs to be considered), and this aspect has been particularly exploited to improve sample integrity in compound management (see Section 3.5.8.3).

3.1.6
Screening Strategies

There are two main screening strategies practised today. The first is diversity or random library screening, where one screens everything or the majority of what one has in the liquid compound library against a target as rapidly as possible. The second is focused or targeted library screening, where computational methods are used to cluster or group compounds into smaller subsets representative of the entire compound library, and one then screens only the subsets thought to be appropriate for the target under investigation. Focused screening relies on the ability rapidly to pick a library subset from the compound store, and limited screening resources have been one the main reasons that have necessitated a focused approach. The majority of companies still rely on full diversity screening, but also still employ focused libraries as an additional screening strategy/tool, particularly downstream of primary screening. Few companies rely exclusively on focused libraries. A third strategy, virtual screening, is also increasingly employed alongside the other approaches to help prioritize the design of compounds in order to focus resources on those hits most likely to generate biologically tractable leads. Virtual screening aims to predict which compounds will have the highest affinity for the target by using (1) 3D coordinates of a target, (2) a virtual collection of compounds to test, (3) software that docks putative ligands to the target site and (4) methods that rank the affinity of test compounds for the target (must be fast, accurate and able to handle a broad range of target molecules). A further complication is that some companies use a pooling strategy where multiple compounds are tested in the same well, typically in range 5–20 compounds/well. The initial justification for pooling was to increase throughput, reduce costs, reduce radioactivity, consolidate laborious assays and decrease screening time and in many cases this had its origins in combinatorial chemical synthesis. Increasingly pooling is losing favor as it necessitates a deconvolution strategy, compounds need to be clustered and it is associated with higher rates of false positives and negatives, interference/cytotoxicity to cells and, more importantly poor data quality. Singlet compound data results are now preferred and <15% of HTS in 2005 is based on pooling [3].

3.1.7
Ultra HTS (UHTS)

HTS was initially used to describe those laboratories with a capability of processing 10 000 samples/day, i.e. equivalent to 100 × 96-well plates per day or 25 × 384-well plate per day. Typically this level of throughput can be achieved with batch processing using workstations and does not necessarily require full robotics. As technology developments and aspiration to achieve higher throughput increased, the term ultra HTS (UHTS) came into usage around 1998 and was associated with the advent of higher density microplates, especially 1536-well plates. UHTS is used to describe those laboratories with a capability of processing ≥100 000 samples per day, i.e. equivalent to 260 × 384-well plates per day or 66 × 1536-well plates per day. In most cases, this level of throughput can only be achieved using fully automated screening systems often referred to as UHTS systems (e.g. EVOscreen, Zeiss and Kalypsys). In all cases, the number of samples screened is usually equivalent to the number of compounds tested or the number of wells processed or data points generated.

3.2
Sample Carriers

3.2.1
A Brief History of the Microplate

The microplate is variously known as a microtiter plate (MTP) or multiwell plate. The name refers to any flat plate, tray, panel or dish with multiple wells. The first microplate originated in 1951 as a micro-machined 8 × 12 array in acrylic. The first molded 96-well plate was made in 1963. Microplates entered common usage in the late 1960 s. By the 1990 s, microplates were available in a diverse range of formats, well shapes and polymers. In 1996, the Society for Biomolecular Screening (SBS) proposed standardization of microplate dimensions with a center-to-center well spacing of 9 mm in 96 wells, an external footprint of 127.76 × 85.47 mm and a plate height for standard wells of 14.35 mm. In 2003, the SBS Standards gained ANSI approval. The microplate is the main platform for bioanalytical research, clinical diagnostic testing and HTS today [4].

3.2.2
Microplate Usage Today

The main microplate formats used in automated screening are compared in Table 3.1.

Table 3.1 The main microplate formats, dimensions and working volumes.

Microplate type	Format	Center spacing (mm)	Multiple of 96	Working volume (μL)
96	8 × 12	9.0	1 × 96	100–300
384 regular volume	16 × 24	4.5	4 × 96	5–100
384 low volume	16 × 24	4.5	4 × 96	1–10
1536	32 × 48	2.25	16 × 96	1–10
3456	48 × 72	1.5	36 × 96	0.2–3

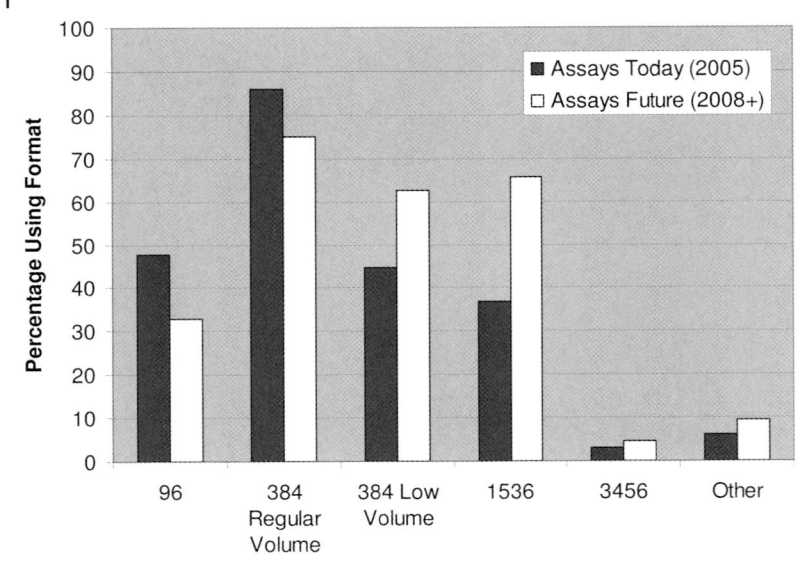

Figure 3.3 Comparative use of different microplate formats in automated screening assays.

The 384-well (regular volume) format is the most popular plate format used today (2005) for screening and is expected to remain so up to 2008. However, by 2008 the use of 1536- and low-volume 384-well plates is expected to be roughly comparable and both will exceed that of 96-well plates (Fig. 3.3). Interestingly, by 2008, more than six out of 10 users will be using a combination of 384- (regular volume), 384- (low volume) and 1536-well formats for their assays. Only 5% will use 3456-well plates and about 10% other formats (mainly microfluidic systems) [5].

3.2.3
Microplate Arrays

Microplates (usually a 96-well plate) which contain an arrays of analytes, spots or tests per well, typically in the range 16 (4 × 4) or 25 (5 × 5) spots, are referred to as microplate arrays. The array (e.g. Pierce's Searchlight Cytokine Panel) is usually identical in each microplate well. Microplate arrays allow the parallel testing of 96 compounds simultaneously against all the spots in the panel. They make use of existing 96-well liquid handling and the required robotics processing is less demanding to automate. Most microplate plate arrays are imaged with a charge-coupled device (CCD) plate imager (see Section 3.4.3) to quantify the signal of interest (usually fluorescence or luminescence). The HTAPlate (Griener Bio-One) has been specially designed for microplate arrays; it has square wells to enlarge the printable area and a low well rim to minimize Z-travel during arraying. The current focus for microplate arrays is secondary screening/compound profiling.

3.2.4
Non-microplate Alternatives

3.2.4.1 **Labchips**

CaliperLS LabChip® technology is a chip-based platform using highly precise etched microchannels in glass or quartz with on-chip electrokinetic (electroosmotic) flow or pressure-driven microfluidics. Chips use a sipper capillary to bring compounds (~1 nL) from microplate wells on to the microchip and this allows for serial continuous flow assays inside a microchannel of a multi-sipper chip. The LabChip 3000 Screening System provides an easy-to-use platform for kinase selectivity screening with standardized reaction conditions for a broad range of kinases, which allows direct data comparison within a panel. Assay development is simple on the LabChip 3000 system with a common set of buffers being used over a broad range of kinase assays, all of which operate in a fixed ATP/K_m ratio. Plug-and-play assay development software is provided to define further the optimal assay conditions for a particular enzyme and substrate. The microfluidic-based separation assay has extremely low sensitivity to compound-associated false positives and negatives and gives highly reproducible quality data with Z' values >0.95 and compound inhibition of <20% can be identified with high confidence. This potentially allows for structure–activity relationship generation directly from a primary screen, thereby avoiding the expense of secondary screening reagents and accelerating the overall lead development time. Other targets classes and assay types are possible, including ion channels and GPCR activation.

3.2.4.2 **LabCDs**

Tecan's LabCD-ADMET™ system is a miniaturized turnkey system for the full automation of ADMET assays. The key attributes of the LabCD™ are that it can perform fluidic functions (unlike microplates); it integrates a wide range of assays types on a single disk; it permits assay miniaturization (nanoliter volumes); it offers highly parallel processing with simultaneous initiation of assays; and it has sealed wells (i.e. no evaporation or meniscus effects). The LabCD utilizes centrifugal forces to move liquids and spinning (rpm), capillary forces and design to gate samples through the CD, with on-disK microstructures providing on-board dilution and reagent distribution. Validated ADMET applications include a family of CYP 450 inhibition assays for measuring drug–drug interactions and a family of serum binding assays for measuring the binding characteristics of drug compounds at various protein binding sites. The LabCD translates the microplate format directly into the circular format on the LabCD, using the eight channels of the Tecan Genesis pipetting head. Dispensing precision is not critical as it is the CD's microfluidics that meter volume.

Gyros's alternative lab-on-a-CD system is based around their GyroLab™ workstation, which shares many of the benefits of the LabCD. The GyroLab is a flexible benchtop instrument platform that automates almost every assay step from sample application to detection on the Gyros CDs. Application-specific methods control a high-precision robotic arm that transfers samples and reagents from microplates or vials to the CD positioned on a spinning station; the speed of spinning generates the exact flow rate required for each step of the application. Gyros-supported applications include a CD designed to prepare digests optimally for peptide mapping or sequencing by matrix-assisted laser desorption/ionization mass spectrometry (MALDI-MS) and a bioaffinity CD for protein quantification on a nanoliter scale, in effect the equivalent of a streptavidin-coated microplate but in a CD format allowing the transfer of established heterogeneous sandwich immunoassays to a CD.

3.2.4.3 LabBrick

SpinX Technologies' LabBrick is the latest microfluidics system that aims to automate all steps involved in optimizing and running assays, using 500 nL final volumes in a closed system. Unique among microfluidic approaches, one type of chip can be configured via software for any assay protocol under any set of conditions. SpinX's programmable microfluidics are based on a novel valving system, the Virtual Laser Valve (VLV), which uses a short pulse of focused laser light to perforate a thin foil separating two microfluidic structures. The two core functions of the VLV are metering of sub-microliter liquid volumes with high precision and directing liquid volumes from any liquid container on the chip to another liquid container on the chip. By alternating these two functions, any assay protocol can be expressed in a sequence of valving operations, performed while the liquids are present in the microfluidic chip. The user defines the assay strategy and protocol and dedicated software translates these instructions into the appropriate sequence of valving operations. Compatibility with conventional liquid handling instruments is ensured by SpinX's innovative interface: input wells on the edge of the microfluidic chip with dimensions identical with those of the wells in 1536-well plates. Each chip corresponds to one row of a 1536-well plate and chips are stacked together to form "bricks" with the footprint of conventional plates. SpinX's first product is a system for 500 nL kinase profiling assays, with fully integrated readout based on fluorescence polarization.

3.2.4.4 Arrayed Compound Screening

Arrayed compound screening (ARCS) is a novel, very high density, "well-less" microplate-sized screening format developed, validated and used in UHTS at Abbott Laboratories. Discovery Partners International has licensed the technology for use in their drug discovery operations. The system consists of 4608 compounds (in duplicate) arrayed on a plastic sheet of microplate dimensions (the ChemCard). Assays are assembled by overlaying layers of agarose gel containing the assay reagents or cells on to the ChemCard, eliminating need for robotic liquid handling. ARCS is applicable to most assay types done in 96-well plates and utilizes imaging detection modes. Images are processed and hits identified by the deconvolution of duplicate spots. ChemCards facilitate the simplified storage/distribution of compound libraries in an assay-ready format. Throughputs of >100 000 compounds per day can be achieved by manual processing without the need for screening automation.

3.3
Liquid Handling Tools

3.3.1
Main Microplate Dispense Mechanisms

There are probably seven main dispensing mechanisms in use for automated screening applications today. These are as follows.

3.3.1.1 Pin Tools
Pin tools rely on transferring the drop of liquid that adheres to tip of the pin or nail. The surface area of the tip determines the volume of the drop, which roughly equates to the volume transferred. Pin tools represent a method of contact dispensing, that requires surface tension (touch off on the destination plate) to remove the drop from the tip of the pin. In addition to the solid pins, there are other varieties (e.g. split pins, grooved, slots, quills, tweezers and pins with a hollow capillary core) that hold the liquid being dispensed. Dispensing variation between pins has long been an issue and it is generally recognized that the precision [coefficient of variation (CV)] of pin dispensing is on the high side. Various modifications to the pin (e.g. surface coatings to tip and pin shaft), the dispensing head (e.g. pins can be made to float independently within the array, so uniformity in the substrate is less important) and the dispensing process (tight control over the speed and heights used during the pickup and subsequent dispensing) are employed to improve the precision. The advantages of pins are that they are relatively cheap, can be arrayed at high density (e.g. 384 and 1536) and offer access to nanoliter dispense volumes. Pin tools are not suited to volumes greater than 500 nL and the minimum dispense volume is around 5 nL. Some bolt-type pins transfer liquid in their threads are able to transfer larger (microliter) volumes. Pin tool heads compatible with a range of third-party dispensing systems are offered by V&P Scientific.

3.3.1.2 Air and Positive Displacement
Air and positive displacement is the most commonly used dispense mechanism on pipetting devices. These systems rely on a plunger or piston rod operating within the confines of a tightly fitting syringe, cylinder or a solid dispense head block with multiple cores. Movement of the plunger out of the syringe causes liquid to be drawn in (aspirated) and reversal of the plunger direction causes liquid to be expelled (dispensed). In an air-displacement system the liquid never touches the

plunger, as there is an air gap between the two. In a positive placement system there may be no air gap between the plunger and the liquid being dispensed, alternatively the entire air gap or the majority of it may be replaced with a systems fluid that is immiscible with the liquid being dispensed. Air displacement is the dispense mechanism used in the majority of the 96- and 384-channel pipetters that are used and sold today. Most air-displacement pipetters are used with disposable plastic tips. In contrast, most positive displacement heads typically have fixed tips or stainless-steel cannulas. Air-displacement systems operate (with acceptable CV) mainly within the dispense range 250 nL–1 mL (e.g. CyBio CyBiWell and Tecan Te-MO™ multipipetter); some positive displacement systems can, however, dispense lower volumes. TTP LabTech's Mosquito is an example of a disposable positive displacement device capable of dispensing between 50 and 500 nL. The Mosquito utilizes a continuous reel of miniature precision pipettes mounted on a bandolier to combine the performance of fixed-head dispensers with the convenience of disposable tips. One recent option offered on air and positive displacement systems is the interchangeable dispense head, such that it possible to change rapidly the head format (i.e. 96, 384 or even 1536 channels) or piston volume capacity. The fluidics head on the Molecular Devices FLIPR Tetra is an example of a interchangeable 1536-channel dispense head. The head operates by air displacement and is composed of 1536 individual piston-driven cannulas. The novel aspect of the device is a proprietary technology that replaces individual seal gaskets with a single elastomeric gasket for the entire array of tips. This contact-based dispensing system is designed for routine liquid transfers in the range 0.5–3 µL.

During the use of an air-displacement system, it is not uncommon to pull up an air gap prior to aspirating the liquid and then to dispense the liquid followed by the air gap to ensure complete dispensing. In addition, piston rods or plungers on air-displacement heads can be made to dispense with an over-stroke, i.e. to move out beyond the position they were at prior to the start of the aspirate–dispense cycle. Because air is compressible, it is important to keep the air gap between plunger and liquid as small as possible to reduce the dispensing error. While dispensing using an air-displacement multi-channel head, is important to be aware of cylinder air tightness and cylinder-plunger linear motion properties, because these can restrict the performance of pipetting. There are two sources of air tightness problems, namely the joint between plunger seal and cylinder and the seal between cylinder and disposable tip. In linear motion, critical moves must be avoided during the change (reversal) in plunger direction otherwise backlash will result. There are three sources of backlash: seal flexion, motor screw tolerance in changing the movement direction and pressure equalization within the cylinder. In practice, many air-displacement systems offer a backlash correction in their software.

3.3.1.3 Peristaltic

Peristaltic pump mechanisms involve tubing that is compressible being clamped between a rotating cylinder with paddles and a fixed surface. As the cylinder rotates, the paddles move the point of compression along the tube and in doing so drive liquid along the tube. A peristaltic pump typically takes liquid from a bottle or reservoir, although it is usually possible to reverse the flow to collect any unused liquid that may remain the system tubing (dead volume). The best known example of a dispenser using the peristaltic mechanism is the Thermo Labsystems Multidrop. This system has a different fluid path (compressible tubing) for each of its eight or 16 channels. Tubing for the Multidrop is provided as part of an easy-to-install disposable cassette that can also be obtained presterilized. The pump on the MultiDrop forces liquid along the tubing and by use of a nozzle at the end of the tube; the liquid is ejected into the dispensing wells from above the plate. This is referred to as non-contact dispensing and it relies on force to eject drops from the dispensing element; it does not involve direct contact with the destination plate. The current minimum volume on the MultiDrop Micro is about 1 µL.

3.3.1.4 Solenoid-syringe

In a solenoid-syringe dispenser, a syringe is used to aspirate the sample and to provide a pressure source applied against a closed high-speed micro-solenoid value. Aspiration occurs through the open valve. A nozzle or disposable tip is fitted to the valve to direct the flow and regulate the droplet size. Opening the valve under pressure causes nanoliter-sized liquid droplets to be ejected from the tip in a noncontact fashion. Precise control of the valve opening time allows for regulation of the volume dispensed. Dispensing is limited by the volumetric capacity of the syringe and for continuous operation it is necessary to re-aspirate during the dispense cycle. Typically 1–8 solenoid-syringe fluid paths are offered in a single system. Each solenoid syringe is capable of independently dispensing a different fluid and/or volume. Dispensing and XY movement above the plate are synchronized such that is possible to dispense very rapidly on-the-fly (e.g. Deerac Equator and the GSI SynQuad dispensers). Solenoid-syringe dispensers are mainly used in the dispense range 5 nL–50 µL, but the exact volume range offered depends on the syringe capacity and orifice size of the nozzles. In one version of the technology (e.g. CaliperLS Sciclone iNL-10), an in-line MEMS flow sensor is placed between the valve and the syringe and 96 fluid paths are arrayed in 12 banks of eight at microplate (9 mm) spacing. The iNL-10 is the first instrument to offer 100% dispense verification in real time and clog detection. In addition, each channel has a built-in liquid level detection capability and a temperature sensor that enables it to compensate automatically for changes in viscosity.

3.3.1.5 Solenoid-pressure bottle

Solenoid-pressure bottle devices are a variant of the solenoid-syringe dispenser which in one form (e.g. PerkinElmer FlexDrop, Aurora Instruments FRD) has replaced each syringe by a single pressurized bottle of varying capacities. This means that these systems have in effect sacrificed their aspirate capability and are used entirely by backfilling the bottle. This can be advantageous for continuous use or where it is necessary to control the temperature or stir the reagent or suspension in the bottle. Earlier systems had a large dead volume; this has now been addressed in most systems. In some designs, the individual valves (one per channel) have been replaced by a common valve, which leads to a multi-way manifold, from which arise between eight and 32 nozzles (e.g. Genetix AliQuot). Some solenoid-pressure bottle devices can dispense as little as 50 nL, but in most cases the dynamic range is between 500 nL and 300 μL.

3.3.1.6 Capillary Sipper

GSI's Hummingbird is the only commercialized example of a capillary sipper dispenser. This non-contact technology transfers sample by dipping an array of 96 or 384 narrow-bore glass polyamide capillaries into a source plate, filling the capillaries by capillary action and dispensing (ejection) into the destination plate by applying pressure to the back side of the capillaries. The transfer volume is determined by the volume of the capillary and the technology has been proven down to 20 nL. The appeal of this technology is its simplicity; there are no moving parts involved as in syringe pumps and there is no dilution or sample loss during transfer. Hummingbird's capillary dispense heads (cassettes) are easily exchangeable and come in a range of fixed volume dispenses (25, 50, 100, 250, 500 and 1000 nL volumes).

3.3.1.7 Piezoelectric

Piezoelectric dispensers typically consist of a glass or quartz capillary tube to which is attached a piezoelectric collar or crystal, usually within a plastic casing. One end of the capillary is open and drawn out into a fine tip and the other end is attached to the system fluidics (syringe pump or reservoir). When the piezoelectric mechanism is activated via a shaped electrical pulse, the piezo crystal contracts and induces a compression wave that causes sub-nanoliter-sized droplets to be ejected from the tip. Specifying a discrete voltage allows accurate and adjustable dispensing of sub-nanoliter volumes. Aurora Discovery currently offers two piezo-based dispensers, the Multitip Piezo Dispenser(MPD), which has a 96-channel piezo head, and the PicoRAPTR, which has eight individually addressable nano piezo tips. Both systems provide non-contact dispensing in 500 pL increments down to 1 nL. In each case the entire fluid pathway is 100% DMSO compatible. GeSim piezo tips manufactured from silicon, glass and ceramics have a drop on demand capability down to 100 pL. Tecan's ActiveTip is a nano-pipetting option/upgrade for the Freedom EVO liquid handlers; it is a piezoelectric dispensing tip that is capable of handling anywhere from 0.5 nL to 10 μL volumes.

3.3.1.8 Acoustic Transducer

Dispensing devices based on an acoustic transducer (e.g. Labcyte's Echo 550) support true non-contact dispensing (i.e. the ejection of liquid drops from a source without entry of a dispensing element into the source or contact with the destination substrate).This technology enables single liquid drops in the volume range from 0.1 pL to over 1 mL to be ejected from starting volumes as low as 1 nL. The advantages of the contactless drop-ejection mechanism is no plugging, no cleaning, no cross-contamination and no waste, together with compatibility over broad viscosity and liquid vapor pressure ranges. The process involves coupling the transducer to the bottom of the source plate with water and, in the case of a destination microplate, mounting it inverted on an X–Y stage, directly above the source plate. Surface tension forces dominate over gravitational forces for liquids in these small wells and the liquid is held in place even though the plate is inverted. Labcyte's Echo 550 is focused on supporting compound reformatting applications (see Section 3.3.2.2), but acoustic transducers have the potential not only to add value through eradicating wash steps and eliminating tip utilization, but also in acoustic auditing (see Section 3.5.8.3).

3.3.2
HTS Liquid Handling Applications and Dispensing Technologies Used

3.3.2.1 Bulk Reagent and Cell Addition

Bulk reagent additions are required when the 2–5 components that make up a typical biochemical or cell-based assay are assembled together. The assay volume range and volumes of additions for the main microplate formats are given in Table 3.2. Usually all wells receive the same component; however, dispensing into selected rows or columns may be omitted for controls. A cell or bead suspension may be one the bulk additions made.

Table 3.2 Total assay volumes and volume of additions for the main microplate formats.

Well plates	Total assay volume (μL)	Volumes of additions (μL)
96	25–300	5–100
384 regular volume	10–100	5–50
384 low volume	2–20	0.5–5
1536	2–10	0.5–3

Existing bulk reagent dispensers fall into two categories: (1) 96- and 384-channel head and (2) 8-, 16- and 24-channel head types.

The first type are typically based on air-displacement dispense heads, which have an aspirate and dispense capability and fill all wells of a plate simultaneously. They may be fitted with disposable plastic tips or stainless-steel fixed needles/cannulas. If the source of the aspirate is a plate (rather than a common reservoir), it is possible to dispense a specific pattern of reagents across the plate.

Typically the 8-, 16- and 24-channel head type are based on peristaltic, syringe-solenoid or solenoid-pressure bottle dispensing mechanisms. They are usually backfilled from a reservoir or bottle and the dispense heads have fixed nozzles whose diameter is matched to the dispense volume range desired. It may be necessary to change the head/nozzles for a different (e.g. lower) volume capability. Typically these dispensers fill a column or row at a time and move across or down the plate. Multiple heads may be grouped together in parallel dispense mode, with a separate dispense head/fluid path for each different reagent or separate assay component. It is usually possible to sterilize or autoclave the fluid path, which is required for the most cell-based applications. In some cases the dispense head and fluid path are configured as a pre-sterilized disposable dispensing cartridge, suitable for a fixed number of dispensing cycle before deterioration of dispense quality. Most devices operate standalone and are sufficiently compact to fit within a sterile flow cabinet.

3.3.2.2 Compound Reformatting and Nanoliter Dispensing

Compound reformatting is the transfer of aliquots of compound (sample) from a library stock plate or array of microtubes into an assay plate. Stocks, dissolved in 100% DMSO, are stored in 96- or 384-deep-well plates (or plated arrays of microtubes) that are usually referred to as "mother plates". The resulting assay or intermediate dilution plates are usually referred to as "daughter plates". Reformatting can take place between plates of the same density, e.g. 96 to 96 or 384 to 384, or between plates of different densities, e.g. 4×94 to 1×384, 16×96 to 1×1536, 4×384 to 1×1536 or even 36×96 to 1×3456. Low-volume (<50 nL) dispensing is essential for compound reformatting for the following reasons: (1) the industry is currently focused on using less compound per assay data point; (2) most laboratories want to transfer samples from neat DMSO stocks to low-volume assays (<5 µL); (3) the DMSO concentration needs to be kept low (usually <1% is desirable for cell-based assays); (4) in most cases it is advantageous to keep the sample concentration as high as possible; (5) uncertainty exists over sample integrity if pre-diluted in aqueous buffer; and (6) there is a preference to prepare assay plates by direct dilution, i.e. saving on compounds, dilution steps, money and time.

The ideal requirements for a compound reformatting device are as follows: (1) whole-plate parallel transfers (i.e. all wells at a time, mainly to save time); (2) 96- but ideally 384- or even 1536-channel dispensing; (3) DMSO-compatible components; (4) dispense volume range 10–500 nL; (5) minimal wastage of aspirated volume; (6) non-contact dispensing is preferred; (7) good precision (ideal CV < 5%, acceptable CV < 10%); (8) rapid cycle time (<1 min for 384 to 384) and (9) effective, fast washing with a carryover of <0.1%. For biochemical assays, it is usually sufficient for compound to be added to the assay plate before assay components are assembled and the assay started. However, with the growing number of cell-based assays relying on adherent cell lines, it is increasingly important to deliver nanoliter amounts of compound to wells already partly filled with media and a cell layer. In those cases where compound addition is no longer the first step in the screening process, reformatting needs to be fast, fully integrated and appropriately scheduled to ensure uniform incubation times and to avoid manual plate handing steps.

When transferring compound solutions in the low nanoliter volume range the influence of physical effects such as surface tension, diffusion, evaporation and surface-to-volume ratio become very important. In addition, many instrument-related effects become more apparent, for example: (1) mechanical tolerances – which affects the positional accuracy of the dispense; (2) relative dead volume – the total volume needed for instrument set-up tends to increase as the volume actually transferred decreases, killing the advantages of miniaturization; (3) susceptibility to clogging – which increases dispense time; (4) tip orifice uniformity – which influences droplet formation/quality; (5) liquid viscosity – which requires temperature correction; and (6) instrument stability – calibration needs to be adequately demonstrated.

Most conventional low-volume dispensers show a significant degradation in CV as the dispensed volume decreases into the low nanoliter range [6]. The fundamental advantage of acoustic transfers performed using Labcyte's Echo 550 over more conventional low-volume liquid handling is that acoustics scale drop volumes for transfers by adjusting the wavelength of sound in a liquid and all other technologies change something which is tied to the physical dimensions associated with the transfer. Sources of variability for "solid-based" liquid transfer methods arise from geometric nonuniformity between the dispensing elements (such as the orifice size or pin diameter) and surface characteristics of those transfer devices (roughness, coatings, contamination, etc.). As volumes decrease, the variations in these factors are further amplified by the growing impact of surface tension and electrostatic forces. The only "solid" having a role in acoustic transfers is the well plate bottom and one can measure its impact on the process by using the transducer as an acoustic microscope. The Echo 550 also takes an active role in reducing electrostatic fields by deionizing both source and destination microplates. Hence the remaining source of variability in the process rests in the well fluid samples themselves and that is why the composition and depth detection process are important in reducing the CVs. The viscosity of DMSO increases from ~2 to ~3.5 cP as it hydrates from a few percent to 30% water in a normal laboratory atmosphere. The better the Echo 550 can know how much energy to deliver (from well fluid composition detection) and at what height to deliver it (from the fluid speed of sound and echo traverse time in the fluid), the more accurate the transfer volume will be. Without composition detection, the Echo 550 would still retain its ability to dispense the required drop volume, but would see a significant degradation in CVs for all transfer volumes [6].

3.3.2.3 Cherry Picking and Serial Dilution

Cherry picking is the ability to aspirate from a selected plate well and to dispense the aspirate to a predefined well in another plate. Cherry (or hit) picking is used to pick HTS actives (hits) from a stock mother plate and to transfer them to a retest plate during the hit reconfirmation process. Hit picking may also involve picking from a store a microtube containing a fresh (unthawed) aliquot of the compound of interest (see Section 3.5.8.2).

Serial dilution is an integral step in the pharmacological proofing of compounds and involves the ability to make repeated dispense, mix and aspirate cycles in successive wells across columns or down rows of a plate, to prepare a dilution series. Serial dilution can also be done in a series of microtubes arrayed in a rack at microplate spacing. In either case, it usually involves a secondary step of transferring the compound dilutions into the assay plate or plate, if multiple copies are made. The resulting drug titrations or dose–response curves are used to determine the relative potency of HTS actives (IC_{50} concentration). The number of dilutions, the dilution ratio or step (e.g. 1 in 10) and number of replicates vary widely between laboratories.

Existing dispensers (e.g. Tecan Genesis) capable of cherry picking from plates or microtubes and subsequent serial dilution typically are based on 1–8 multi-channel robot sample processors (RPS). Each channel is a separate syringe or Cavro pump, with independent control of channel volume and the Z (aspirate or dispense) height. It is highly desirable for tips or probes to adapt to a range of plate or tube formats (i.e. 96 and 384 and possibly 1536 wells). This is made possible by a variable span capability, i.e. the ability of a multi-channel device to adjust the center-to-center spacing of the probes, e.g. from 9 mm for 96-well dispensing to 4.5 mm for 384-well dispensing. Some devices are fitted with liquid level sensing, typically based on conductance using black-carbon-filled plastic tips. Most offer liquid surface tracking, based on software calculation of the volume aspirated and source container dimensions. Systems are offered with disposable and fixed tip (PTFE-coated) options. Most of these systems use a combination of tip washing via backfilled reservoir and/or aspirate plus dispense cycles in a circulating bath or custom reservoir with individual receiving cones to direct wash fluid flow around the dispense tips.

An emerging alternative to serial dilution within a plate is direct dilution, i.e. making dose–response curves (e.g. IC_{50}s) on a volume dispensed basis from a single aspirate. To undertake such a task requires a low volume dispense capability with a wide dynamic range, over several logs volume, ideally descending into the picoliter volume range. Few instruments currently possess this capability; examples include Labcyte's Echo 550 and Aurora Discovery's PicoRAPTR. The latter has a dispense range from 500 pL to 100 µL, using piezoelectric dispensing for volumes below 1 µL and a positive displacement syringe for volumes above 1 µL. During the conventional serial dilution process, compounds (held as stocks in 100% DMSO) frequently come out of solution (precipitate) immediately when diluted with aqueous buffers, with potential impact on, usually a reduction in, the true potency of a compound. Direct dilution from stock DMSO into assay buffer should add quality to this process by limiting compound loss during the dilution process.

3.3.2.4 Microplate Washing

Microplate washers are used for a variety of applications in screening: (1) heterogeneous immunoassay processing (e.g. ELISA and DELFIA®); (2) to reduce background in radioactive binding assays (e.g. Flashplate®); (3) to remove unbound (free) fluorescent label in cell-based assays (e.g. FLIPR® assays); and (4) during routine maintenance of adherent cell lines (e.g. media removal and replacement). Microplate washing devices may be based on strip washing, in which 8, 12, 16 or 32 channels (corresponding to a 96-, 384- or 1536-well plate column) move along a plate and perform a series of defined aspirate and fill commands at selected heights in each column before progressing to the next column (e.g. Molecular Devices Aquamax DW4). Alternatively, plate washing devices may be based on entire plate washing, where all plate wells are processed simultaneously, with separate aspirate and dispense heads with 96- or 384-channel manifolds offered (e.g. Molecular Devices Embla 384, Tecan PowerWash, Bio-Tek ELx405HT) Plate washers are capable of much greater throughput that strip washers.

3.4
Detection Technologies

3.4.1
Main Detection Modalities Used in HTS

The relative importance of four main detection modalities used in automated screening today is shown in Figure 3.4 [2]. Fluorescence (52% of all screening assays) is the most important HTS detection technology and includes the following fluorescent readouts in approximate order of importance: FLIPR® ion flux (mainly Ca^{2+} activation); FI (fluorescence intensity); TR-FRET (time-resolved FRET, e.g. LANCE and HTRF); FP (fluorescence polarization); HCS CCD imaging; FLIPR® voltage (mainly membrane potential); FRET (fluorescence resonance energy transfer); TRF (time-resolved fluorescence, e.g. DELFIA), FLT (fluorescence lifetime); VIPR® (FRET); HCS laser scanning; and FCS (fluorescence correlation spectroscopy). Luminescence (19% of all screening assays) is the second most important HTS detection technology and includes the following luminescent readouts in approximate order of importance: Glow; AlphaScreen™; Flash; BRET (bioluminescence resonance energy transfer); and ECL (electrochemiluminescence). Radio-

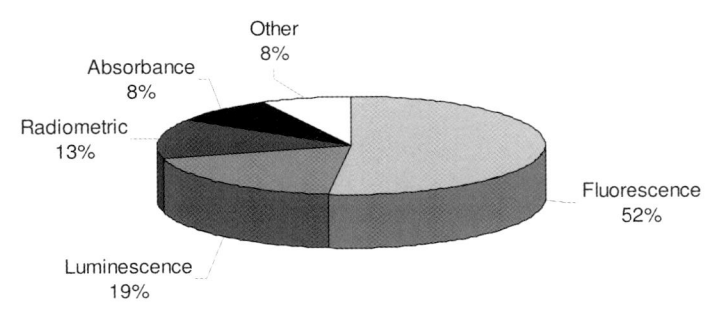

Figure 3.4 Main detection technologies used in automated screening.

metric (13% of all screening assays) is the third most important HTS detection technology and includes the following assay readouts in approximate order of importance: SPA (scintillation proximity assay); LEADseeker™; Filter Binding; and FlashPlate®. Absorbance (photometric or colorimetric) is the fourth most important detection technology contributing to around 8% of all screening assays. Other detection modalities (8% of all screening assays) include the following assay readouts in approximate order of importance: Automated Patch Clamping; Rubidium Flux; and Label Free.

3.4.2
Plate Readers

The majority of plate readers are based on single point detection using photomultiplier tubes (PMTs). Typically the plate moves under the detector, with the latter located in a fixed position. Most readers use a white light source such as a tungsten lamp or more commonly a xenon flash lamp. Where extra sensitivity is required for an application (e.g. Cis-Bio HTRF® or PerkinElmer AlphaScreen), lasers may be used; typically only one or two spectral laser lines (e.g. 488 or 630 nm) are offered. Some new readers are now using light-emitting diodes (LEDs) to provide a cheap source of intense light at a specific wavelength. Most readers offer fixed-wavelength excitation and emission using bandpass filters optimally chosen for specific fluorescent probes (e.g. PerkinElmer EnVision, BMG PheraStar). A few are available with variable-wavelength selection, using either single or double monochromators; in such systems the emission and in some cases also the excitation wavelengths are tunable, giving flexibility to use almost any fluor (e.g. Tecan Safire2). Other systems have combined fluorometric and photometric capabilities allowing for spectral scanning and unlimited wavelength selection for assay optimization (e.g. Thermo Varioskan). However, at the wavelength extremities of the monchromator the detection sensitivity can be less efficient than with a fixed-wavelength system.

PMT-based readers generally offer higher sensitivity, a wider dynamic range and lower cost relative to plate imagers. They are more compact, have less crosstalk and more even illumination. The disadvantages of PMTs are that they are generally slower than imagers in reading higher plate densities, normalization may be required if multiple detectors are used, kinetic sampling is limited to a few wells or unless a delay (interval) between samples is acceptable. PMT readers are also less suited to dispense and read applications, allowing only low-throughput processing as they lack a parallel read capability, and as such they tend to be used for assay development of applications requiring dispensing.

The majority of readers today are multimode, i.e. they are able to read all the detection modalities in common use (e.g. PerkinElmer EnVision, BMG PheraStar, Molecular Devices Analyst and SpectraMax M5). The advantages of multimode readers is that one device does all and saves on bench or robot space; they are cheaper than buying multiple instruments; they offer greater flexibility to change technologies; and only one device needs to be integrated into a robot. The downsides of multimode readers are that sensitivity may be compromised in that not all modes may have the highest possible sensitivity; some devices may be limited by the modes offered; and typically they do not include proprietary detection technologies (e.g. AlphaScreen).

Automated microplate radiometric detection is enabled by scintillation counting (e.g. PerkinElmer MicroBeta® TriLux and TopCount NXT). The MicroBeta has up to 12 detectors. Each detector consists of two PMTs, one positioned above the sample and the other below. These count the sample from the top and bottom at the same time (coincidence counting). The TopCount also has up to 12 detectors, but in this case they are single PMTs positioned to count the sample only from above. Both MicroBeta and TopCount are suitable for counting SPA, FlashPlate® and filtration assays in 24-, 96- and 384-well plate formats only. They can also be used for luminescence detection. Higher throughput radiometric detection of LEADseeker™ proximity assays or Image FlashPlate® is possible using plate imagers (e.g. PerkinElmer ViewLux, GE Healthcare LEADseeker) (see Section 3.4.3.1).

Increasingly, flow cytometers are being developed that support automated microplate- and tube-based acquisition for cell- and bead-based assays (e.g. Guava EasyCyte, Beckman FC 500 MPL). Such systems incorporate up to two lasers allowing for five-color and two-scatter analysis and mainly perform a support/ validation function in automated screening.

Nephelometers (e.g. BMG NEPHELOstar and Thermo Nepheloskan Ascent) are available to support the automated measurement of light scattering from particles and the detection of precipitates in microplates. In such systems, a highly focused light beam is produced by a light source below the microplate and an optical unit filters and directs the light through the sample. Another optical unit only allows the scattered light (about 30°) to pass towards the detector, which is a photomultiplier tube above the microplate. Applications of nephelometry in automated screening include drug solubility testing and the assessment of bacterial and fungal growth (turbidity).

3.4.3
Plate Imagers

Plate imagers can be divided into macro and micro imagers based on the criteria set out in Table 3.3.

Table 3.3 Image-based screening

Criterion	Macro-imaging	Micro-imaging
Application	Primary screening	High-content screening
Area per image	Entire microplate	Part of one well
Useful resolution	~1 mm	~1 μm

Criterion	Macro-imaging	Micro-imaging
Specimen format	Solutions or cells in plates	Discrete cells in wells
Object size	2–8 mm (a well)	2–50 μm (a cell or organelle)
Imaging time per plate	Seconds to minutes	Tens of minutes
Throughput (assays/day)	100 000+	1000 s to 10 000s
Detection modes	Fluorescence, luminescence, absorbance, radiometric	Mainly fluorescence and some brightfield

3.4.3.1 Macro-imaging

Macro (whole plate) imagers mainly use CCDs that are cooled (sometimes as low as −100 °C) to enhance sensitivity. To avoid parallax issues (i.e. the loss of signal at the edges of the plate, which arises from the reduced light collection efficiency of a normal lens for emitted light, where the plate is not positioned directly beneath the center of the lens), the CCD is usually coupled directly to either a telecentric lens (e.g. PerkinElmer ViewLux, GE Healthcare LEADseeker, MesoScale Discovery Sector HTS) or a 96-multi-lens array (e.g. Zeiss Plate Vision) or to 96 paired optical fibers (e.g. PerkinElmer ImageTrak). Other systems use a 2D image intensifier with transfer optics to an uncooled CCD that permits single photon counting (e.g. Hamamatsu FDSS and CyBio Cybi-Lumax). The input window of the intensifier is a photocathode which converts photons to electrons just like a PMT. The difference is that the intensifier also gives area information. The location of the photon event is transferred to the output of the intensifier so that the readout camera can not only measure the intensity but also record the geography. The specialty of this technology is that single photons are amplified with an extremely low background, keeping the 2D positional information of the detected photons, thereby intensifying the image. PerkinElmer's LumiTrak™ use a single fiber-optic taper that is coupled directly to a cooled (−100 °C) back-thinned CCD. The fiber-optic taper allows "contact imaging" – a patented means of highly efficient light gathering without parallax, no reflection, minimal cross-talk or the focusing requirements of a normal lens. Both types of imager read either all plate wells simultaneously or a proportion of a plate (e.g. a 96 array or tile or sector), thus making them suited to kinetic measurements. Both cooled CCD cameras and photon counting (image intensifier) cameras require integration. The difference is where they integrate. A cooled CCD integrates images on the CCD chip before read-out. Hence the cooled CCD requires longer exposure times at low light levels (i.e. slow frame rate). On the other hand, the photon counting camera integrates the images on computer memory after read-out. The frames are read out quickly (i.e. fast frame rate), but needs a certain integration time after read-out. Comparing frame rates can be misleading; a more useful measure is the potential throughput that can be achieved using either system.

The main advantages of plate imagers are that they are very fast; they simultaneously read all wells or tiles (sectors) of a plate; it is possible to dispense and image all wells at same time; the quantum efficiency of the CCD is better in the red region of the spectrum (i.e. 500 nm upwards) and as such is optimally matched to red-shifted reagent chemistries (e.g. GE Healthcare LEADseeker beads); imagers can read free formats; they can result in better data quality and improved inter-plate precision; and they offer a visual (image) readout.

The main disadvantages of plate imagers are their relatively large size; image processing necessitates applying many corrections to raw data [flatfield, parallax, pixel binning, shading, vignetting (light falling off at the corner of some images, a characteristic of the lens), background subtraction, etc.]; images are subject to interference from dust; additional services may be needed (e.g. cooling for the -100 °C cameras); and imagers can typically cost a order of magnitude more than most PMT readers.

3.4.3.2 Micro-imaging

In screening, micro-imaging is now is mainly associated with high-content screening.

High-content screening (HCS) There is a lack of uniformity over the definition of HCS, but most understand it to include multi-color fluorescence imaging of intact single (including live) cells using a combination of fluorophores, antibodies and biosensors. HCS permits the evaluation of compound effects on multiple independent or interacting targets or pathways. It provides detailed information on subcellular, temporal and molecular events and allows for the discrimination/analysis of different cell populations. Typically, multiple images are collected per microplate well at different magnifications and processed with or without pre-optimized algorithms (these are the software routines that analyze images, recognize patterns and extract measurements relevant to the biological application, allowing the automated quantitative comparison and ranking of compound effects) to derive numerical data on multiple parameters. This allows for the quantification of detailed cellular measurements that underlie the phenotype observed. HCS utilizes automated fluorescence microscope imaging systems, some with confocal optics (for optical sectioning or enhanced resolution), fluorescence-based reagents [e.g. GE Healthcare/BioImage green fluorescent protein (gfp) or BD reef coral fluorescent proteins (Living Colors)] and advanced bioinformatics tools. The advantages of HCS are that it allows walk-away automation of HTS targets previously impossible to screen by other assay methods. Manual image evaluation is far too repetitive, time consuming and subjective for such comparisons. HCS reduces the cost of cell-based assays by minimizing cell line development and validation. It simplifies the design of cell-based assays. It allows cross-correlation of potency, specificity and toxicity in a single assay. HCS identifies cell sub-populations and sub-cellular compartments and helps in the understanding of the mechanism of action of drug candidates. As with all other assay methodologies, false positives are a major problem in HCS. However, increased understanding of the common failure modes (e.g. compound auto-fluorescence, cytotoxicity, over-confluence of cells and confounding phenotypic changes) means that it is now possible with HCS to understand false positives better and to devise fully automated strategies to flag outliers rapidly.

Currently in HCS there is still a need to standardize the file format of the images generated and to move away from proprietary image formats, as these limit assay development and algorithm transfer between different brands of HCS reader. Lack of an industry standard data format causes inconvenience as image analysis and data manipulation often involve multiple data mining applications. There is also a need for open-source software that handles HCS images and derived data and interfaces with other software. As most current HCS users currently plan to store raw images for at least 6 years, attention must be given early on to the data storage and management solution needed to store, transfer, share and mine terabytes of data [5].

HCS detection There are two types of instruments used for automated HCS detection: (1) laser scanners and (2) CCD imagers. (1) Laser scanners have their origins in fluorescence cell sorting. Some systems have confocal optics, all have laser excitation (at one or more fixed wavelengths) with simultaneous multicolor emission (via multiple PMTs). Laser scanning results in point detection at varying resolutions (depending on linewidth and sample interval) and produces pseudo-images (i.e. reconstituted objects). It is particularly suited for whole cell and bead analysis (differentiating free from bound or internalized label) and optimized for high speed (numbers). Examples of laser scanners include the TTP Labtech Acumen Explorer and Compucyte iCtye. (2) CCD imagers have their origins in fluorescence microscopy and in their simplest forms are automated microscope plate readers (e.g. MAIA Scientific MIAS-2 and Molecular Devices Pathway 1). More complex systems typically have confocal optics or switchable optical modes, with autofocus and either use white light (variable wavelength) or laser (fixed wavelength) excitation sources. A single CCD maybe used sequentially to collect light emitted at different emission wavelengths or multiple CCDs (up to four) may simultaneous collect multicolor emissions in parallel. CCD imaging results in area detection and the ability to drill down (zoom in) to very high resolution and is optimized for maximum information (content). Some CCD imager examples are the Cellomics ArrayScan Vti, Evotec Technologies OPERA QD, GE Healthcare IN Cell Analyzer 3000, Molecular Devices ImageXpress and Beckman Cell Lab IC 100. Some CCD imagers with live cell imaging/kinetic capability (i.e. include a single-well liquid handling/dispense and image capability) include the Cellomics Kinetics Scan, GE Healthcare IN Cell Analyzer 1000 and Becton Dickinson Atto Pathway HT. For high-resolution sub-cellular analysis, confocal CCD imaging is essential. The basis for the division of high content into analysis and screening is summarized in Table 3.4.

Table 3.4 High-content analysis versus screening.

Parameter	High-content analysis (HCA) – emphasis on content	High content-screening (HCS) – emphasis on numbers
Instrument type	CCD imager	Laser scanner
Main application area	Therapeutic areas, target validation assay development, secondary screening, lead optimization (H2L)	Primary screening, increasingly compound profiling
No. of compounds tested	100s–1000s	10 000s–100 000s
Throughput needed	Low/medium throughput acceptable	Needs higher throughput
No. of cells required	Algorithms typically report on 1000, requires high plating densities	All the cells in a well are processed, requires lower plating densities
Information needed	Want all the information one can get	Only want to know what happened, outcome drilled down to a single number, 99.9% of the wells do nothing
Assay requirements	Need well-validated/characterized system with fully understood biology, flexibility to permit live cell or kinetic assay with environmental control	Need robust assay with maximum window, performance criteria to address Z'. etc., typically fixed end-point assays to permit batch processing of large numbers of plates
Analysis	High-resolution analysis of small sample area	Ideally whole well analysis to account for patch effects, autofluorescence, cytotoxicity, etc.
Data storage	Large file size to store all high-content information for detailed analysis	Small file size to cope with increased throughput – no re-analysis required

In practice, once a high-content assay has been fully developed and algorithms validated, it may be possible for some types of assays (e.g. kinase activation) to transfer the screening of that assay from a CCD imager to a laser scanner. By use of both types of instrument it is possible to derive the benefits of throughput and reduced file size (data storage) that laser scanning affords, but to make use of the resolution and content of the CCD imager to investigate further the events identified initially by primary screening using a laser scanner.

3.4.4
Dispense and Read Devices

Dispense and read devices are essential for the measurement of events that occur immediately after liquid addition (e.g. cell-based assays for GPCR activation or ion channel flux). They are also needed for time-dependent reactions and assay kinetics. Liquid handling is usually integrated directly into the detection system. Low-throughput devices are PMT-based plate readers which may have between one and eight injectors (e.g. PerkinElmer EnVision or Molecular Devices Flex Station). Higher throughput devices are based on CCD imaging systems with multichannel heads that can dispense 96, 384 and 1536 wells simultaneously. Most liquid additions are based on dilutions in the range from 1 in 10 to 1 in 3. Systems typically require integrated tip washing, in some cases tip loading, stirred reagent reservoirs and cell suspension devices, all within a temperature-controlled, light-tight environment. Systems are available optimized mainly for fluorescence detection (e.g. Molecular Devices Flipr TETRA, PerkinElmer ImageTrak); for flash luminescence detection [e.g. CyBio Cybi-Lumax (eight-channel dispensing only) and PerkinElmer LumiTrak]; or for both fluorescence and flash luminescence detection in the same instrument (e.g. Hamamatsu FDSS).

3.4.5
Other Detection Technologies

Several other detection technologies are noteworthy from an automated screening perspective, namely rubidium flux, high-throughput electrophysiology [sometimes called automate patch clamping (APC)] and label free. Rubidium (Rb^+) efflux is usually monitored by atomic absorption spectrometry (AAS) and, prior to the advent of APC, was regarded as a higher throughput alternative to electrophysiology in the pharmacological assessment of ion channels. A multi-detector systems with automated sample transfer capable of unattended plate processing is available (Aurora Biomed ICR 12000). The recent availability of higher throughput APC turnkey workstations [e.g. Molecular Devices IonWorks® Quattro™ system with Population Patch Clamp™ (PPC) technology and Sophion Q Patch] is expected to have a major impact on patch clamping, also simultaneously driving down the per compound assay costs. By the use such APC systems, it will be possible to automate processes that previously were extremely labor intensive, with throughput gains of between 100- and 1000-fold over manual alternatives and to permit electrophysiology measurements to be made much earlier during the discovery process, including primary screening in some companies, although the main application of APC systems is expected to be non-compliant hERG testing. Label-free detection encompasses a broad range of nonlabeled methodologies; perhaps the best known systems are those based on surface plasmon resonance (SPR) offered by Biacore. To date, SPR-based detection systems have rarely been applied in HTS; the main reasons for this have been lack of commercial systems that had adequate throughput or were cheap enough per data point to be applied *en masse*

[7]. The next generation of label-free tools (e.g. SRU Biosystems BIND, Corning Life Sciences Epic™, MDS Sciex CDS, ACEA Biosciences RT-CES™) will be predominantly microplate-based sensors, which for the first time may facilitate wider interest in automated label-free detection, particularly from those involved in primary screening.

3.4.6
Automation of Detection Technologies

Nearly all detection technologies aimed at primary screening today are microplate compatible and for most that compatibility extends from 96- to 1536-well plates. All of these detection devices have a microplate holder or carrier or cassette, which is usually part of the plate loading mechanism. Most plates are loaded in landscape orientation. Typically two types of automation can be enabled: (1) integrated plate stacking or (2) robotic loading and interface with the plate carrier. Robotic loading is the most common as it gives the user the option to source standalone plate stacking options (see Section 3.5.4). Robotic loading requires that the plate carrier extends out beyond the detector box, so that the robot can access it, and that the plate carrier is robot friendly, i.e. able to present a loosely held plate for robot gripper pickup and able to move the plate to a locked, fixed position after robot release. In some cases, particularly where the detection has specific processing requirements, there may be integrated liquid handling or dedicated sample processing (work-up) within the system. Some instruments (e.g. PerkinElmer MicroBeta® TriLux and ViewLux) utilize a plate cassette support system into which plates must be individually loaded before they are placed on a shelf within the instruments stacker. Cassette support had it origins in providing a practical generic way to allow counting in a diverse range of sample types, e.g. flexible microplates, complete filtermats or 4 mL LSC vials. The cassette/shelf approach permits random access and facilitates detection, e.g. in the MicroBeta by applying a mask to reduce crosstalk or in ViewLux by providing orientation reference points used to apply the well grid after imaging. In both MicroBeta and ViewLux, robot arms are able to load/unload plates into the cassette when presented in the instrument's robot loading bay.

3.4.7
Potential Sources of Reading Error

Users need to be aware that there are a large number of potential sources of reading error. These include crosstalk (signal spillover) between wells, particularly important in radiometric assays; meniscus shape and volumes effects; trapped air bubbles; light scattering from bright specs of dust, fluffs, scratches (particularly important for imaging) and increasingly particulates in solution (insoluble compounds); auto-fluorescence; inner filter effect (i.e. absorption of light, sometimes incorrectly referred to as color quenching); dark quenchers; photo-bleaching; inadequate signal or reader sensitivity; and corrections applied to imaged data (see Section 3.4.3.1). In addition, infrared emitters (i.e. anything not black) can be a problem in low-light chemiluminescent imaging. Solid white microplates typically employed in SPA to enhance the signal are strongly phosphorescent in the red (>700 nm). This phosphorescence is derived from the titanium dioxide filler used in white plates and, unless controlled for by either dark adaptation overnight or by the select use of a bandpass filter, would make CCD detection of low-light chemiluminescence impossible.

3.5
Laboratory Robotics

Three alternative approaches to laboratory robotics are compared and contrasted in Table 3.5.

Table 3.5 Comparison of the different approaches to screening automation.

Feature	Approach		
	Traditional workstation	Fully integrated robotic system	Turnkey workstation
System size	Typically bench-top	Large, often requires dedicated room and services or fully enclosed robotic cell	Compact, small footprint, bench-top or custom table
Number of tasks performed	One or specialized part of a process	Multiple and diverse, fully automated process	Several including reading, specific validated processes, walkaway automation
System flexibility	Moderate	High	Limited
Walk-away plate processing capability	<100 plates	<1000 plates	<250 plates
Plate movement	From manually loaded stacker	Fully articulated robot arm, usually on a track or conveyor, other proprietary plate moving systems	Typically plate gripper or plate mover, limited axis pick and place robot or instrument-specific mechanism
Scheduling software	None	Essential	Partial requirement
Task/application complexity	Specialized component(s)	Less specialized, but greater diversity	Specialized and/or fully optimized

3.5 Laboratory Robotics

Feature	Approach		
	Traditional workstation	Fully integrated robotic system	Turnkey workstation
Staffing resource	Minimal, supervisor	Several dedicated, highly trained personnel	Key user(s)
Staff training needed	Moderate	Extensive	Limited
Vendor-supported applications optimized to instrument	Confined to specific workstation functions only	Partially supported, depends on source of system	Fully, with significant gains in precision, reproducibility and throughput
Source of components/system integration	Single supplier	Multiple suppliers/ typically made to customer order by third-party integrator	Single supplier/emphasis on single supplier integrated solution
Promotes new technologies	Partially	Limited, expensive to reconfigure, depends on upgrade of peripherals	Yes, but focus may be limited by suppliers' ability to access assay technology
Key strength	Task focus	Fully automated, better suited to larger screening campaigns	Pre-validated applications, enhanced processing speed, better service
Main area of use	Batch processing, typically plate reading or sample preparation	High-volume HTS and UHTS	Focused libraries, secondary screening, assay development, ADMET, low–medium-throughput HTS
Ease of implementation/time frame	Simple/immediate	Difficult/1–6 months to agree specification, 6–12 months from placing order	Simple to install/ rapid, no need to develop interfaces, no application validation required
Price range ($)	50 000–200 000[a]	500 000–>1 million	125 000–300 000
Examples	PerkinElmer TopCount (reader), Tomtec Quadra (liquid handler)	RTS LifeScience Assay Platform	CyBio CyB-Lumax, Beckman Biomek FX ADMEtox (see Section 3.5.6)

a Some specialized plate imager workstations can cost up to $500 000, some CCD HCS imagers nearly $1 million.

3.5.1
Traditional Workstations

These are independent systems that are highly specialized to perform a single function or task as efficiently as possible, e.g. a plate reader or liquid handling device. In a workstation (semi-automated) screening environment, it is the human operator who moves the plates between workstations. Most workstations have plate stackers to allow walk-away processing of small batches of plates (typically <50, although they can be up to 100).

3.5.2
Robotic Sample Processors

Most laboratory robotics originated from the simple robotic sample processors (RSP). The very first RSP consisted of a single pipetter mounted on an XYZ platform and performed a limited range of automated liquid handling tasks. Subsequently four- and eight-channel pipetting devices were developed. Plate moving capabilities, within and to and from the XY table, were then added. From these devices evolved today's liquid handling workstations, many with multiple liquid handling capabilities including interchangeable 96- and 384-channel heads, automated loading of disposable tips, eight-channel pipetters, bulk reagent dispensers, plate movers/grippers and plate stackers (e.g. Tecan Freedom EVO and Beckman Biomek FX).

3.5.3
Plate Storage Devices

There are two main types of plate storage devices: open shelves (or hotels) and plate stackers.

Open shelves (or hotels) are relatively basic examples of plate storage, but from these can be a configured high-density rotating carousels. All these systems allow for full random access to any of the plate storage positions. They may also be used to store microplate-compatible consumables, e.g. pipette tips and lids. Some plate carousels or hotels have been encased to provide incubators with full environmental control (temperature, humidity and atmosphere). The more advanced incubators and plate stores have small letterbox-type robotic loading positions to minimize the impact of plate loading/withdrawal on the plates within the enclosed environment. Optimized capacity to a compact footprint and fast plate retrieval with barcode reading are important attributes of the larger plate stores (e.g. Thermo Molbank, RTS Life Science SmaRTStore).

Plate stackers consist of vertical stacks of microplates, held in alignment by a metal frame. Stackers allow only for ordered access, i.e. first in–first out. Stackers are usually configured in pairs, an in-stack, to store plates prior to processing, and an out-stack, to store processed plates. Reprocessing or access to specific plates requires stacked plates to be shuffled between in and out stacks. Typically integrated stackers are an option on many plate readers and liquid handlers with decks that are designed for batch processing. The destacking/stacking mechanism may limit compatibility to certain types of microplate, mainly owing to the skirt dimensions, and software may restrict the interleaving of different plate types. Some plate stackers are capable of handling lidded plates and the automated removal and retention of plate lids during plate processing. Although most microplates now conform to the ANSI standards, there is no such thing as a standard microplate stack and, unless one's detection, liquid handling and automation systems come from a single source or supplier, it is usually not possible to interchange the stacking component between different instruments.

3.5.4
Plate Moving Devices

The simplest plate moving devices are those which were designed to provide a generic add-on plate stacking for those instruments that lack an integrated stacking capability (e.g. Thermo RapidStak, Bio-Tek Bio-Stack). Some stand-alone stackers are associated with a limited axes (two or three degrees of freedom) robotic arm and plate gripper, e.g. CaliperLS Twister II and Hudson Plate Crane, which have a fixed, centrally positioned robotic arm, or the Hamilton Swap, which has a robotic arm on a small linear track. The robotic arm is able to address the stackers and those instruments positioned within the work envelope or reach of the arm, depending on whether the gripper also has a rotational capability. Thermo's CataLyst Express uses a compact five-axis robotic arm in a small bench-top configuration that includes a 45-shelf random-access plate hotel and can be used to create small workstations with peripherals. In comparison, larger, fully articulated robotic arms (e.g. Staubli six-axis TX40 robot) are much more versatile, they offer greater flexibility in terms of reach, angle and positioning, greater speed, lower noise and what they can manipulate and handle as payload. Such arms may be located in a fixed position at the center of a workcell table or move along a linear track or even hang from a ceiling or wall. Articulated arms are to be found at the heart of most fully integrated automated systems and typically provide a more industrial solution, but at greater cost. Thermo's high-speed distributed motion technology (HSDM) uses an alternative approach to plate moving. The HSDM has two components, the Flip Mover, which permits rapid, single-axis plate motion to peripherals located on either side of a linear track, without disturbing the well contents, and the Linear Plate Transport ,which provides high-speed bi-directional plate motion and optimized acceleration for individual plate types.

3.5.5
Fully Integrated Robotic Systems

These are a collection of diverse peripheral instruments (such as workstations), which are served by an articulated robot arm or conveyor to move plates between devices. Such systems are controlled by scheduling software and capable of performing many different procedures (assays). Earlier systems were room sized and were configured to perform a relatively limited range of HTS assay methodologies. From such systems evolved modular components or workcells capable of performing specific applications separately or as part of larger systems (Tecan, Beckman Coulter, Caliper LS, Velocity 11, Thermo CRS, Protedyne). Later additions included specialized screening factories specifically optimized for UHTS (e.g. Zeiss plate::explorer™ uHTS system, Evotec Technologies EVOscreen, TAP Asset, RTS Life Science Assay Platform, Kalypsys). Most recently, RTS Life Science have combined their UHTS factory with a compound library store to give RTS Symphony, a closed-loop screening system in which hits automatically generate secondary screening requests, which are then screened immediately.

3.5.6
Turnkey Workstations

These are small footprint systems that on delivery are immediately capable of performing an entire automated process (e.g. a specific assay or generic assay method) including the detection. They come complete with pre-validated protocols for reagent/assay applications and application-specific software [e.g. CyBio CyBi-Lumax automating EuroScreen Aequorin assays or Beckman Biomek FX ADME-Tox workstation automating the pION method for PAMPA (parallel artificial membrane permeability analysis).

3.5.7
Automated Cell Culture Systems

These are specialized, fully integrated systems designed to perform aspects of automated cell culture. With increases in the proportion of assays that are cell based, increases in the size of compound libraries and screening throughput, the demands for cells by screening groups has escalated in recent years. It has been estimated that a screening robot processing only 100×384 well plates per day typically requires one FTE (Full Time Equivalent) to produce enough cells. However, screens must run at least 5 days per week to justify the investment and to achieve the throughput required and many cell-based assays require cells to adhere to plates at least 24 h prior to screening. This either means weekend work or only 3 or 4 days of screening per week. Cell-based assays are notoriously variable, with much of the variation attributed to person to person differences in cell processing and counting. Automated cell culture offers a way of improving assay quality, of meeting the demands for cells and of facilitating work outside normal hours.

Large, fully automated robotic cell maintenance and culture systems able to support the needs of a department or screening facility are available from TAP (Cellmate, SelecT and Cello), RTS International (acCELLerator), Velocity 11 and Protedyne. Smaller batch processing robotic workstations to support cell-based assay screening more suited to use by a group or laboratory are available from Beckman, CaliperLS, Hamiliton, Genetix, PerkinElmer, Tecan and others. Systems for the rapid automated production of recombinant proteins are available from TAP (Piccolo).

3.5.8
Compound Management Systems

A variety of different commercial automated compound storage options are available today. Most stores are able to accommodate all types of container, i.e. microplates, microtubes and dram vials. They differ depending on whether the focus is solid compounds or DMSO solutions. Stores vary widely in size and the flexibility of the automated solution enabled. The earliest systems were based on industrial paternoster equipment combined with robotic sample picking [8]. Paternoster stores (e.g. TAP Haystack) have multiple shelves mounted in a vertical chain mechanism such that the shelves remain horizontal while they are rotated up and down. Shelves are pulled from the paternoster to pick the samples (e.g. vials of powder) or microtubes of interest. The main benefit of the paternoster is that the storage density is high. The disadvantages are their height, often necessitating customized (tall) buildings; in addition, environmental control is difficult, and as a result they are now used mainly for ambient dry compound storage. A more recent alternative to the paternoster is the narrow isle-based systems which use two parallel rows of static shelving accessed by one or more robots running down the aisle between them (e.g. TAP Homebase, REMP AAS and MSS, RTS International Sample Store™). Labware (plates or racks of tubes) are typically stored in carrier trays that are located on the shelving. Selected trays are pulled by the robot and requested labware is picked either on that robot or moved to a dedicated picking tool within the system. The main advantage of narrow isle approach is that it can be configured to fit within existing laboratory space, it can be modular and it is less expensive and quicker to install. Most liquid stores have the ability to control fully the storage environment [temperature –20 to +20 °C, humidity and atmosphere (nitrogen or argon)], which also extends to the picking area. Newer systems are designed to be configurable (modular) to meet the capacity required, flexible to allow for rapid adjustment of processes and offer easy integration into changing corporate software environments (e.g. Matrical MatriStore and Discovery Partners International Universal Store). Some of the latest systems are highly compact (e.g. RTS International SmaRTStore) and are designed to offer competitively priced standardized solutions particularly suited to smaller satellite laboratories or for use by biotech or university screening groups where limited capacity is required. In some cases these compact stores are mobile (e.g. TekCel's TubeStore/TubeServer system) or capable of being relocated after installation (TTP Labtech comPOUND®).

3.5.8.1 Current Practice in Compound Management
Most companies start off making *ad hoc* manual use of laboratory freezers, but find that they outgrow these within a few years and need to consider an automated

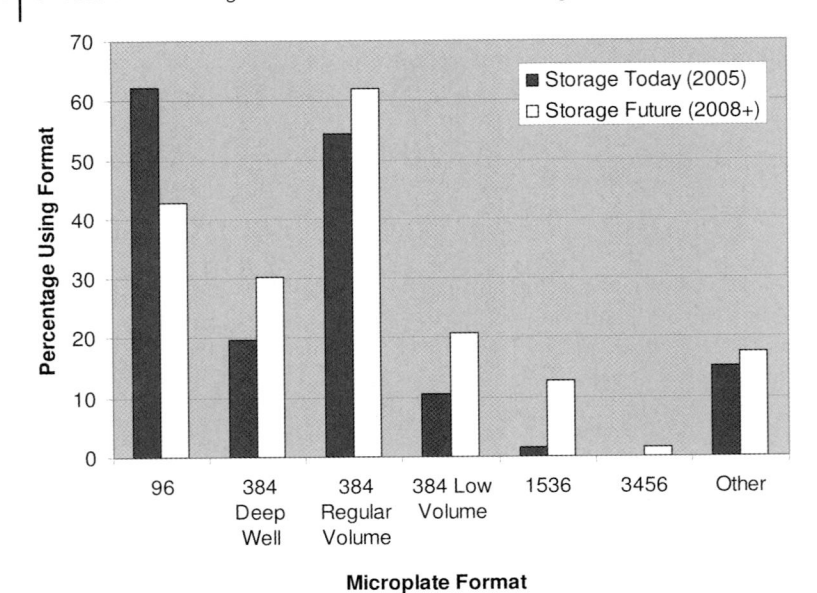

Figure 3.5 Comparative use of different microplate formats in automated compound storage.

solution. The key factors in justifying the acquisition of an automated compound management (CM) system are their cherry-picking (retrieval) requirement, the size of their library and lifetime costs [1]. A recent survey [1] of CM groups revealed that about 60% of all screening operations have access to some form of automated liquid CM store. The remainder still use an *ad hoc* arrangement of laboratory freezers as the main component of their CM store. The majority (55%) use microplates and microtubes for storage in part of the same CM facility. Only 20% use microplates alone and the remaining 25% use only microtubes. The majority (67%) store multiple replicates, the remainder (33%) only single replicates. Most companies (60%) store their master stocks at a single concentration, the remainder (40%) store at multiple concentrations. Most (63%) store variable volumes, only 37% have standardized on one or more fixed volumes. Nearly all (89%) storage facilities used 100% DMSO (assumed starting concentration) for their stock solutions: 6% use 90% DMSO, 2% use 75% DMSO and 4% do not use DMSO. The most common storage volume today (2005) was 0.1–0.5 mL of stock solution. The 96-well type is still the main plate storage format today (2005), although this is expect to change to predominantly 384-well in 2008 (Figure 3.5).

3.5.8.2 Plate-based versus Tube-based Liquid Compound Storage

The big debate in CM as to whether plates or tubes offer greater flexibility and minimal adverse effect on sample integrity for liquid DMSO stocks has largely been won by tubes, if current vendor promotion of tube-based systems is an indication. However, most pharmaceutical companies today make use of a mix of both storage containers for their liquid library.

Plate-based storage provides samples pre-arrayed in the plate format of choice, typically 96 or 384, but today there is increasing interest in 1536-well plate storage. Samples are frequently sorted into sub-sets represented by a collection of plates. Plate storage particularly facilitates the preparation of replicate copies of those stocks, using a multichannel pipetter and the shipping of liquid stocks to different laboratories/satellite stores. However, plates need to be delidded or desealed or matts removed to gain access to the samples contained in the wells, and in many cases this is a manual process. Although it is technically possible to pierce individual wells, in practice this is not common and removal of one sample from a plate typically involves exposing all samples to the operating environment. In addition, it is currently not possible to freeze–thaw individual wells without thawing the entire plate. Furthermore, cherry-picking can be time constrained if the samples of interest are spread over a large number of different plates. Cherry-picking (see Section 3.3.2.3) involves removal of an aliquot from a well and picked plates may end up with some wells devoid of compound whereas others are untouched. Although plate storage has been successfully automated, it is also readily amenable to manual processing and as such is widely used by groups (e.g. small biotech and academic laboratories) who have not invested in automated compound storage and retrieval systems.

Tube-based storage is associated with a variety of formats. Tubes of various sizes and storage volumes (ranging from 1 mL to 50 µL) may be stored in racks, in racks (or inserts) on store trays and direct in special store trays. Minitubes are generally stored in special receiving microplates (96- and 384-well formats only). Racks and store trays may take the form of a microplate array or custom close packing array (e.g. TTP Labtech comPOUND). Tubes are usually picked into a microplate format so that they can be processed (e.g. diluted) or replicated *en masse* like a microplate. Most tubes are identified by a bar-code e.g. a 2D code on the tube base. Some types of tubes can be sealed with either a piercable septum or a removal cap, facilitating access multiple times to the tube. Alternatively, DMSO-resistant seals (e.g. plastic-coated aluminum foil) are applied that can be pierced only once. Usually these seals are applied to an entire plate of tubes and are then cut into individual (singlicate) tubes 96 or 384 at a time. The main advantage of tube-storage is the ability to pick a single tube from within a controlled dry environment without compromising the integrity of the other tubes. The trend today is towards single-shot (use) and single freeze–thaw of foil-sealed minitubes that contain a minimal stored volume (<25 µL) of DMSO stock solution. Reasonably fast random access cherry-picking can be accomplished by *XYZ* robots that physically pick (remove) tubes with a gripper from a storage tray and place them in a receiving tray or plate (e.g. TekCel, Matrical). Some minitubes are pushed upwards out of the source plate into a gripper that can hold multiple (12) tubes [e.g. The Automation Partnership (TAP)], which reduces travel and gives higher pick speeds. The picking tolerances of 384 mini-tubes are, however, tiny, so for higher speed and accuracy some companies advocate avoiding *XYZ* robot manipulation. Such mini-tubes may be picked by punching out tubes, arrayed in the microplate, through the base into a receiving plate positioned below (e.g. REMP). Faster picking rates are achieved if this can be done *en masse* or in parallel (e.g. RTS Life Science), rather than punched singly. The pick rate of comPOUND (TTP Labtech), which uses pneumatic picking, is 5 s per tube irrespective of the tube location in the store; however, with multiple comPOUND stores one could get up to 12 tubes to arrive in parallel every 5 s (i.e. 0.4 s per picked tube). The latest tube stores are designed to permit high-speed picking, with advertised pick rates of between 1000 and 5000 tubes per hour. The effective picking rate of an automated compound store not only depends on the cycle time of the picking robots, but also is typically limited by a range of factors including the order size distribution (many small orders are typically much less efficient than one large order), whether samples are retrieved as entire racks or individual cherry-picked items, the number of items on a storage tray, etc., relative to the time taken to fetch it, whether required items can be co-located on the basis of likely retrieval patterns, whether alternative vessels are available which can be selected dynamically to maximize efficiency, whether sample identity is verified during picking and whether this represents a time overhead on the picking process, and, especially at higher rates of access, how efficiently the store logic can deal with loading samples back into the store at the same time as needing to service retrieval orders. Unless careful attention is paid to all of these factors, the performance of an automated system may be disappointing, compared with what might be theoretically possible based on the robot cycle times.

3.5.8.3 Associated Automated Instrumentation

Screening automation, in particular compound management, has driven the development of a range of automated peripheral devices that are compatible with microplates. These include bar-code labelers, bar-code scanners, shakers, sealers (based on heat seals, adhesive seals and laser welding), seal piercers, seal cutters and sonicators.

3.5.8.4 Sample Integrity and QC Testing

Pharmaceutical screening operations today are increasingly focusing on achieving and maintaining quality. In compound management groups, this is reflected by their desire to ensure sample integrity, with respect to both new compounds entering the collection and the handling and storage of existing compounds. Sample integrity is generally regarded as a measure of the quality of the material in a container (e.g. vial, tube, microplate well) and relates to: (1) the degree to which the chemical identity (structure) of a compound in a collection is what it is supposed it to be; frequently the starting material used to prepare the solution is not what is on the label; (2) the purity of that sample and whether any impurities or salts are present; these may arise, for example, as a consequence of impurities introduced (or not eliminated) during the original synthesis or new species arising as a consequence of compound decomposition; and (3) the actual concentration of a liquid sample in DMSO, how it deviates from the intended concentration at the time of solubilization and including any post-solubilization dilution processes. Here the influences of precipitation, water uptake and solvent evaporation are all significant. Making sure that compounds received into a library at the start are in fact of high purity, in addition to being the compound they are supposed to be, is therefore highly important.

Water uptake (ingress) and repeat freeze–thaw cycles were seen as the main reasons for sample integrity being compromised in DMSO liquid stores [1]. Implementation of quality control (QC) measures to assess sample integrity has rapidly become a major priority for compound management groups. Of the QC measures being applied to preselected and to randomly selected compounds, high-performance liquid chromatographic separation followed by mass spectrometric detection (LC–MS) is widely used in QC analysis to determine concentration (and purity) [1]. Currently the industry is focused on finding a universal detection scheme and the evaporative light-scattering detector (ELSD), the chemiluminescent nitrogen detector (CLND) and NMR spectroscopy have proved useful [9]. Although the throughput of these technologies is adequate for a one-off assessment, they cannot routinely be applied today to all liquid samples in corporate compound collections at the point of delivery to the customer (i.e. to measure the actual sample concentration of the sample diluted in the assay buffer in parallel with the screen), where the number of analyses required would be considerably larger.

Of the QC measures now being considered, the greatest interest is in the hydration status of the DMSO solvent and solubility (precipitate) detection [1]. Acoustic auditing (e.g. Labcyte's Echo 380) is regarded as the most promising new tool for the rapid, noninvasive monitoring of the "health" of a compound library. In acoustic auditing, the acoustic transducer (see Section 3.2.2) is operated as a small-scale sonar device that is positioned underneath the source plate well. By measuring the time-of-flight of sound energy (the echo return time) to the surface of liquid in the well, it is possible to calculate the depth of the well fluid. The amplitude of well fluid bottom reflection, which varies with the fluidic impedance, is then used to determine the fluid composition and the speed of sound. Echoes can then be processed to calculate the DMSO/water ratio and fluid volume. This auditing process can occur even if the wells are sealed or lidded, can be extended to mini- and microtubes and has great potential in quantifying the degree of hydration of compound libraries, originally dissolved on 100% DMSO, in addition to calculating the well volume.

Compound precipitation from DMSO stock solutions is increasingly viewed as a more serious problem than chemical degradation. Current opinion is that uncontrolled water uptake into DMSO stocks in synergy with freeze–thaw cycles is the primary cause of precipitation [10]. Once compounds precipitate from wet DMSO, redissolution is very difficult and simple methods are not very effective. Opinion as to the best approach to deal with compound precipitation is divided between those who advocate trying to prevent the problem in the first place, i.e. the use of a single shot/single freeze–thaw tube approach under a controlled environment in the compound storage area, and those who accept that the problem exists in their collection and place more emphasis on measuring solubility when the compound reaches the assay. Of the other approaches, the use a microplate sonication device (Matrical's SonicMan) to resolubilize compounds is perhaps the most amenable to implementation. By sonicating samples with SonicMan just prior to liquid handling, those compounds which are capable of being dissolved will be driven back into solution, at least to the limit of their thermodynamic solubility. Resolubilization just prior to screening has the potential to eliminate a major source of false negatives and is expected to have a noticeable impact on the quality of hits by facilitating better, more consistent data.

Disclaimer

The author has attempted to balance the mention of specific instruments, products and brands. However, the reader should be aware that space does not permit the mention of all possible alternatives. Please note the use of specific examples to illustrate certain technologies does not represent endorsement of those products or brands by the author.

Addendum

This chapter was written in March 2005, by the time of publication some of the technologies referred to in the text will inevitably have changed, been withdrawn or even replaced by newer alternatives. In particular, the author wishes to draw the reader's attention to the adaptive focussed acoustic (AFA) devices recently developed by Covaris/KBioscience. AFA offers a non-invasive approach to mixing and homogenisation that looks set to have a dramatic impact on quality in both compound management, (by minimising sample precipitation during primary dissolution, freeze thaw and aliquoting) and screening (by enhancing compound solubility and accelerating mixing thereby decreasing assay variability and reducing false positives).

List of instrument photographs (see Appendix)

References

1 J. Comley, *Compound Management Trends 2004*, HTStec, Cambridge, **2004**.
2 J. Comley, *Detection Trends 2003*, HTStec, Cambridge, **2003**.
3 S. Fox, *High-throughput Screening 2003: Improving Strategies, Technologies and Productivity*, HighTech Business Decisions, Moraga, CA, **2003**.
4 http://www.microplate.org/history/det_hist.htm.
5 J. Comley, *High Content Screening Trends 2005*, HTStec, Cambridge, **2005**.
6 J. Comley, Continued miniaturisation of assay technologies drives market for nanoliter dispensing, *Drug Discov. World* **2004**, 5, (2) 44–54.
7 J. Comley, *Label Free Assays Trends 2004*, HTStec, Cambridge, **2004**.
8 J.R. Archer, History, evolution and trends in compound management for high throughput screening, *Assay Drug Dev. Technol.* **2004**, 2, 675–681.
9 J. Comley, Compound management – in pursuit of sample integrity, *Drug Discov. World* **2005**, 6, 59–78.
10 C. Lipinski, Compound precipitation from DMSO and the synergy between water uptake and freeze/thaw cycles, oral presentation at LRIG meeting, **2005**.

Part III
Assay Technologies

4
Functional Cell-based Assays for Targeted Lead Discovery in High-throughput Screening

Jörg Hüser, Bernd Kalthof, and Jochen Strayle

4.1
Introduction

Functional bioassays employing isolated organs to measure and quantify the response, e.g. muscle contraction, to hormone stimulation have made great contributions to pharmacological science and early drug discovery. These classical pharmacological tests still provide important links in "screening cascades" aiming to transfer *in vitro* results to *in vivo* effects. With the growth of corporate compound libraries, they have been abandoned as test formats for primary drug screening because of their very limited throughput, high costs and consumption of animal material. Today, functional cell-based assays employing recombinant cell lines continue the pharmacological bioassay tradition but circumvent the outlined shortcomings. They provide cost-efficient access to large amounts of material, are amenable to miniaturization in microplate formats and contribute to minimizing the use of laboratory animals for experimental lead discovery. Small-molecule drugs and pharmacological bioassays have greatly facilitated the functional characterization and molecular identification of drug receptors (for reviews, see [1, 2]). During the 1980 s, functional cloning started to unravel the molecular nature of the most relevant classes of drug receptors, including G protein-coupled receptors (GPCRs), ligand- and voltage-gated ion channels and nuclear hormone receptors and many others. In a second wave, degenerative primers and other techniques have been used to clone additional members of these receptor classes. The identification of large gene families soon extended the catalogue of classical pharmacology and provided a rich source of receptors as potential future drug targets for therapeutic interventions.

The introduction of recombinant DNA technology into routines in pharmaceutical research laboratories at the beginning of the 1990 s allowed the study of a defined drug receptor in host cells without any appreciable background. The use of recombinant expression systems allowed one to screen chemical libraries against human receptors. Moreover, receptor cloning and heterologous expression have facilitated experimental strategies to identify and discriminate receptor subtype-specific ligands to deliver in the end drugs with improved selectivity and less side-effects [3, 4]. In addition, advances in optical readout technologies visualizing subtle cellular functions have rapidly replaced crude phenotypic readouts such as cell growth. Early robust high-throughput screening (HTS)-compatible assay techniques which could be readily adopted to the microtiter plate format utilized reporter gene technologies to follow receptor activation through the cellular transcriptional response [5]. The introduction of Ca^{2+}-sensitive fluorescent indicators, e.g. Fluo–3, developed in Tsien's laboratory [6, 7], further advanced functional cell-based tests for HTS. Its successful implementation in the HTS environment was largely due to the advent of fluorescence microplate readers with integrated liquid handling. Transfer of test compounds or receptor agonists in the measuring position allows the online recording of the resulting intracellular Ca^{2+} transients with fast temporal resolution. Recently, the introduction of improved fluorescent probes for the detection of changes in membrane potential have expanded the experimental repertoire of functional HTS techniques to include also electrogenic transporters and ion channels.

Functional cell-based assays targeted towards a defined receptor provide a number of advantages over binding assays based on competition with labeled ligands. Binding assays typically provide only a limited dynamic range of the recorded signal whereas functional tests frequently utilize amplifying cellular signal transduction cascades. In particular, the identification of receptor agonists benefits from the use of cell-based systems. Through the combination of artificially high receptor densities and strongly amplifying signal transduction cascades, weak receptor activators with a low affinity or minimal efficacy can also be resolved. Measuring temporally resolved cellular signals, e.g. hormone-induced intracellular Ca^{2+} signals, provides kinetic data bearing additional information on the possible mode of action. Generally, the functional test can differentiate between agonist and inhibitor. In fact, the assay design and even the choice of the assay format will be determined by whether the focus will be on stimulators or inhibitors. Moreover, functional tests also offer the potential to detect "allosteric modulators", which by definition act on molecular sites distant from the "orthosteric site" occupied by the endogenous ligand. As a consequence, allosteric modulators are not necessarily detected in competition binding experiments. The increasing awareness of receptors as allosteric proteins, in particular for G protein-coupled receptors [8], has further underlined the relevance of functional tests but also challenged experimental protocols frequently employed in HTS [9–11]. The frequent use of radiometric assay formats in competition binding experiments, e.g. scintillation proximity assays, also poses technical hurdles in testing large compound libraries. Because of safety issues in the handling of radioactive reagents, the use of fully automated robotic systems is frequently abandoned in favor of semi-manual batch processing in combination with stand-alone instrumentation. In addition, radiometric formats frequently cannot be adopted to microtiter plate formats beyond 384-well plates. As a consequence of the resulting throughput constraints, both issues limit the use of radiometric formats in screening large compound collections. A frequent strategy to compensate for throughput limitations is to introduce compound mixtures (typically 5–40 compounds tested per well) in the test. "Compression format" testing in many cases limits the concentration at which the compounds are introduced into the test and necessitates an error-prone deconvolution step. These limitations clearly reduce the resolution of the test with respect to identifying also weakly active target modulators. As a consequence, with the increasing relevance of functional cell-based tests in HTS, compression format testing has been abandoned by many groups.

Compared with biochemical tests, cell-based assays are generally burdened by higher levels of signal scattering. Even among clonal cell populations, the response of individual cells to chemical stimuli can vary significantly. Inter-cell heterogeneity in the signal quantified by the assay, variations in the number of cells per well, together with variations introduced by liquid handling steps during the assay (e.g. addition of receptor ligand) account for the signal scatter in cell-based assays. In addition, cell-based assays provide for test compounds a plethora of "off-target" mechanisms by which effectiveness in the test can be mimicked. Strategies have to be devised to discriminate such systematically unspecific hits early in the screening process (compare Chapter 2). True target hits are rare events. Thus, assay formats that also pick up large numbers of unspecifically acting compounds will dilute target modulators in a large number of hits acting through off-target mechanisms. For example, using a reporter gene test coupling the activity of a cell surface receptor to induction of a reporter gene expression to screen for inhibitors of the said receptor will deliver hits that interfere with ligand–receptor interaction. In addition, the assay will detect an even larger number of compounds interfering in downstream signaling events, transcription and translation or even cellular energy metabolism or structural integrity (compare Fig. 4.1). Thus, to allow for experimental follow-up studies also on weakly active compounds, HTS assays or the

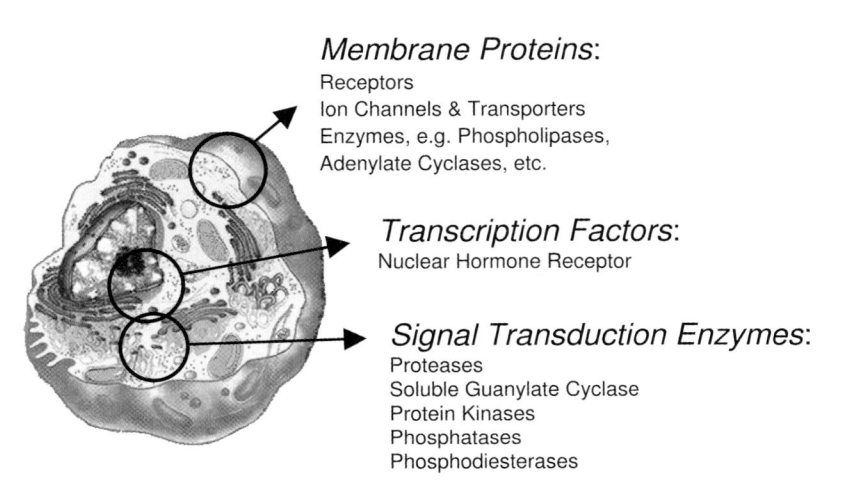

Membrane Proteins:
Receptors
Ion Channels & Transporters
Enzymes, e.g. Phospholipases,
Adenylate Cyclases, etc.

Transcription Factors:
Nuclear Hormone Receptor

Signal Transduction Enzymes:
Proteases
Soluble Guanylate Cyclase
Protein Kinases
Phosphatases
Phosphodiesterases

Figure 4.1 Pharmacological targets frequently addressed in cell-based HTS assays.

subsequent analysis process need to discriminate early in the process between specific and unspecific hits.

Pharmacological targets that are effectively probed through cell-based assays in most cases are integral surface membrane proteins involved in membrane transport processes or transmembrane signal transduction. Assay systems designed to identify target inhibitors are typically based on a "competition" scenario in which the candidate small molecule competes with a known target ligand, e.g. the natural receptor agonist or transport substrate. Presentation of the receptor site towards the extracellular space allows experimental control of the ligand concentration during the experiment and thus adjustment of the assay sensitivity towards this molecular receptor site based on the law of mass action. Intracellular ligand concentrations, in most cases, cannot be controlled experimentally. As a consequence, assay sensitivity towards molecular receptor sites not accessible from the extracellular space is far from optimal, thus frequently excluding the detection of weak inhibitors competitive with an intracellular target ligand. In addition, binding to intracellular receptor sites always requires sufficient membrane penetration from test compounds to be effectively detected in the assay.

In the remainder of this chapter, we will briefly summarize some of the main aspects of the three most relevant cell-based assay techniques used for targeted drug discovery: reporter gene tests, potentiometric fluorescence probes to follow cellular membrane potential changes and fluorescent and luminescent indicators for intracellular Ca^{2+} ions.

4.2
Reporter Gene Technologies

Reporter gene assays utilize the capacity of many cell surface receptors to stimulate changes in gene expression through intracellular signaling cascades ultimatively acting on cis regulatory DNA elements controlling the transcription of target genes (Fig. 4.2). (For a brief review, see [12].) A number of identified response elements have been used as starting points to design artificial promoters responsive to specific intracellular signals. Much of this knowledge has been derived from the promoter analysis of immediate early genes, e.g. *c-fos*. Palindromic arrangement of response elements and other maneuvers have been used to optimize "designer promoters" to function as molecular switches regulating transcription of downstream genes. Ideally, the constitutive activity of the promoter should be very low to minimize the background signal in the absence of stimulation, while upon stimulation the promoter should drive transcription at significant rates. "Leaky" promoters generating a significant background signal in the absence of receptor stimulation will clearly limit the dynamic range and thus resolution of the assay.

The function of the "gene switch" is quantified by transcription of a downstream gene, i.e. the reporter gene. Ideal reporters confer to the host cell a readily detectable and quantifiable phenotype clearly discernible over the cellular background. In addition to its technical detectability, sensitivity of detection is another essential property of a reporter gene. In many cases, reporter genes encode enzymes for which quantification is amplified by product accumulation. Stability of the reporter gene-coded protein in the host cell is another relevant parameter affecting the utility of the system. Accumulation of stable reporter protein in the cell through constitutive background activity of the promoter can limit its usefulness. For typical applications, the ideal reporter should display only a relatively short lifetime in the cellular environment.

Historically, the *E. coli* transposable element Tn9 consisting of the chloramphenicol acetyltransferase gene (CAT), which confers resistance to the antibiotic chloramphenicol, was the first to be used for this purpose [13]. Various analytical methods have been developed to detect the activity of the reporter gene-encoded protein, including radiometry, immunoassay, fluorescence and colorimetry. Other reporter genes commonly used today include β-galactosidase, firefly and *Renilla* luciferases, green fluorescent protein and β-lactamase. For use in HTS, luminescent, e.g. luciferases, and fluorescent readout techniques, e.g. β-lactamase, are dominant because of their sensitivity and applicability in miniaturized assay formats.

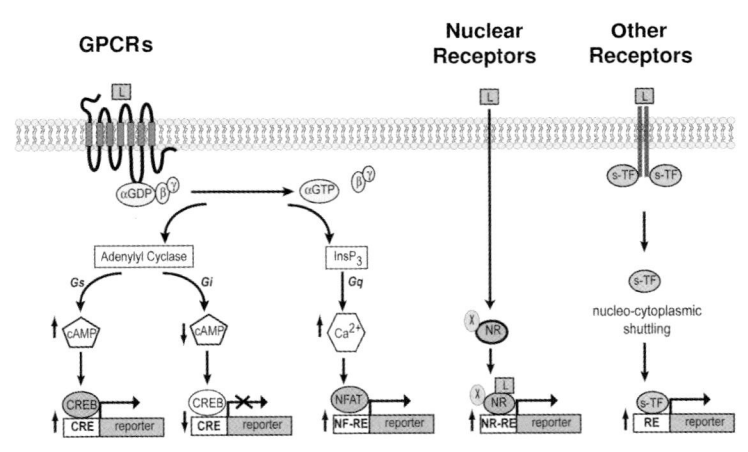

Figure 4.2 Reporter gene assays utilize the capability of many receptor signaling mechanisms to induce changes in gene expression. Assays for HTS employ suitable promoter constructs containing DNA response elements which bind transcription factors specifically activated upon stimulation of the targeted receptor. GPCRs typically couple through the second messenger cAMP, Ca^{2+} or activation of mitogen-activated protein kinases (MAPK) to regulate a number of transcription factors, e.g. CREB (= cAMP response element binding protein), NFAT (= nuclear factor of activated T cells), which activate transcription by binding to their corresponding DNA response elements (REs). Other REs frequently employed for HTS assays include the serum response element or DNA sequences recognizing activator protein 1. Reporter gene tests can also be engineered for any non-GPCR signaling pathway converging on any one of the second messenger Ca^{2+}, cAMP, diacylglycerol or MAPK activation. A variety of hormones and lipid intermediates bind and activate nuclear receptors (NRs), which function as ligand-regulated transcription factors. Upon ligand (L) binding of growth hormone and cytokine receptors, transcription factors are recruited and activated to translocate subsequently to the nucleus and activate transcription.

The most commonly used bioluminescent reporter genes are the firefly luciferase from *Photinus pyralis* and the *Renilla* luciferase derived from the sea pansy *Renilla reniformis*. Luciferases have relatively short half-lives (for firefly luciferase ca. 2–3 h in CHO cells; authors unpublished data) compared with other widely used reporter genes (e.g. GFP or β-galactosidase), contributing to a low background activity in the absence of stimulation. A clear advantage of luciferases as reporter genes in respect of sensitivity is that there are no endogenous mammalian enzymes known with comparable substrate usage or functional activity. The availability of membrane permeable luciferin esters also provides the possibility of non-invasive detection of luciferase expression. Alternative luciferases used as reporter genes include coelenterazine oxidizing photoproteins from different marine species, e.g. *Metridia*. The latter has been shown to be secreted by its host cell, thus allowing non-invasive observation of activity [14].

The Green Fluorescent Protein (GFP) derived from *Aequorea* sp. and its derivatives were introduced as reporter genes some years ago. However, the relatively slow folding and maturation into the functional fluorescent molecule together with its considerable stability in eukaryotic host cells has limited its application as a reporter gene. In order to overcome this problem, in one approach the cellular stability of GFP was reduced by incorporating a sequence into the protein which targets the construct to proteasomal degradation [15]. Alternatively, GFP and other fluorescent probes have been used in combination with engineered chimeric proteins as fluorescent labels to follow subcellular translocation events. Another GFP-based technique employs appropriate pairs of GFP-derived fluorophores attached to signaling molecules. In this approach, fluorescence resonance energy transfer (FRET) between the two GFP labels reveals information on protein–protein interaction or intramolecular conformational changes. Genetic engineering of suitable probes, their expression in cells together with fluorescence microscopy has facilitated the resolution of temporal and subcellular spatial aspects of signal transduction processes involving a variety of chemical signaling species, e.g. Ca^{2+}, cAMP, $InsP_3$ and many others [16, 17]. Although this technology has opened up new avenues to study the function of many aspects of cell signaling on the microscope stage, the impact on HTS for targeted lead discovery has been limited by poor signal dynamics and major throughput constraints. Advances in the automation of fluorescence microscopy brought forward the development of "high-content screening" as a recent trend in experimental lead discovery (compare Chapter 6).

Another widely used reporter gene is β-lactamase, a bacterial enzyme which hydrolyzes and inactivates β-lactam antibiotics efficiently. The recent introduction by Tsien's group of a membrane permeant FRET substrate composed of a cephalosporin moiety coupled to a coumarin donor and fluorescein acceptor dye (CCF2) has brought a renaissance of this reporter. The technique provides a number of advantages. First, the protein has only a short lifetime ($t_{1/2} \approx 3.5$ h [18]) in suitable host cells. The dye is loaded as the acetoxymethyl ester into the cells, where it is cleaved and trapped by cellular esterases. UV excitation (350–360 nm) to excite the coumarin moiety results in efficient FRET to emit light from the fluorescein acceptor at ca. 520 nm with only limited output from the coumarin at 450 nm. In the presence of β-lactamase, substrate cleavage unmasks the blue coumarin fluorescence while reducing the green signal originating from fluorescein. Dual emission ratiometric recordings render the signal robust against variability in cell size, cell number, loading heterogeneity and optical path. Accumulation of the product accounts for a high sensitivity capable of detecting as few as 10 copies of β-lactamase per cell [19]. Taken together, these properties make β-lactamase/CCF2 a very well suited reporter gene with a proven track record in HTS [20, 21].

Reporter gene assays are frequently employed for the detection of modulators targeted towards cell surface or nuclear hormone receptors. Typically, recombinant cells have been generated by functional cloning, providing cell clones for which receptor stimulation is reliably translated into expression of the reporter. It is important to note that receptor–ligand interactions, second messenger signals and transcription of the reporter gene operate on different time-scales. In most cases, the overall time course of the signal output is determined by translation of reporter mRNA into the corresponding protein. Typical incubation periods for reporter gene assays employed in HTS range between 4 and 6 h. Frequently, the kinetic requirements of the receptor-stimulus and/or the intracellular messenger (e.g. cAMP or Ca^{2+}) for a given promoter and the cell clone used for testing are not known. A number of recent reports, however, indicate a rather complex kinetic behavior of "stimulus-transcription" coupling. For example, oscillations of intracellular free Ca^{2+} concentration elicited by submaximal agonist concentration are effectively translated into transcriptional activity by means of Ca^{2+}-dependent transcription factors such as NF-AT or NF-϶B [22, 23]. Despite such complex kinetic behavior underlying stimulus-transcription coupling at the single cell level, the temporal integration of the input signal renders the reporter gene assay a very sensitive tool for detecting receptor agonists.

Reporter gene tests have also been used to screen through chemical libraries for receptor antagonists (Fig. 4.3). This experimental strategy has several shortcomings. Signal output is critically dependent on the function of signal transduction downstream of the targeted receptor, transcription and translation. Moreover, all steps are dependent on the physical and metabolic integrity of the cell. As a consequence, the few meaningful receptor antagonists possibly contained in a randomly assembled chemical screening library will be hidden in a large number of compounds interfering with any of the downstream processes, cellular metabolism and/or integrity.

In many cases, the discrimination of target modulators from unspecifically acting compounds is not trivial. Reference assays or activity profiling using historical data facilitate the identification of off-target or toxic compounds. However, each cell clone and the specific assay conditions used can produce a unique set of reproducible, but unspecifically acting, hits not readily discriminated by a reference cell or assay. In addition, Baker, et al. [24] have reported differences in the antagonist affinity derived from Schild analysis when tested in a reporter gene assay against full or partial agonists of the $β_2$-adrenoceptor. This discrepancy was caused by receptor modification, most likely through phosphorylation, in response

A control

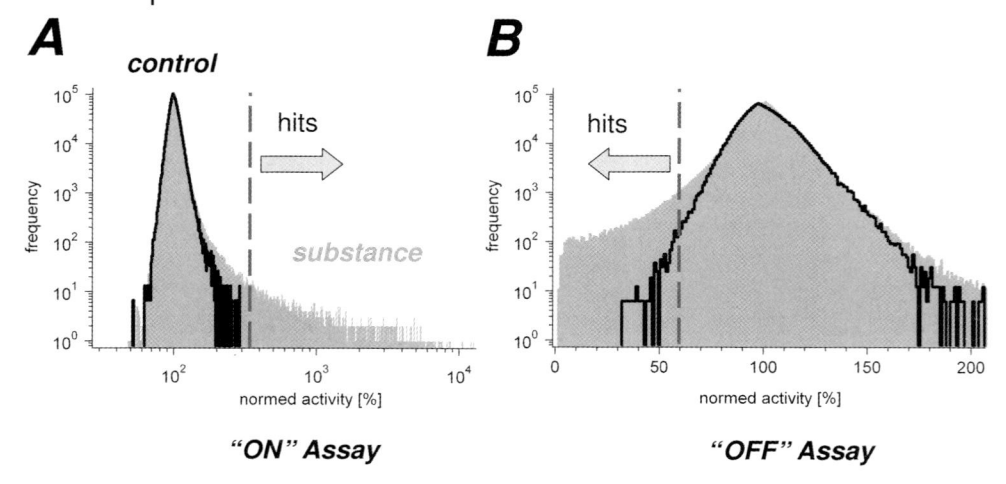

B

"ON" Assay

- receptor agonist <u>induces</u> reporter gene expression

1.55 Mio. compounds tested
3700 primary hits (0.2 % hit rate)

"OFF" Assay

- receptor antagonist <u>reduces</u> reporter gene expression induced by agonist

1.55 Mio. compounds tested
18700 primary hits (1.2 % hit rate)

Figure 4.3 Reporter gene "ON" and "OFF" assays. Frequency histograms summarizing HTS results from two firefly luciferase reporter gene tests. Activity data have been normalized to solvent controls (= 100%). Test (A) was seeking receptor agonists capable of inducing expression of luciferase. Test (B) employed the same reporter gene controlled by a different receptor. In this test, the receptor was half- maximally stimulated with agonist and candidate compounds were introduced to reveal compounds capable of inhibiting the agonist-induced luciferase signal. In addition to compounds acting on the targeted receptor, the test revealed a large number of unspecifically acting compounds contributing to significantly higher hit rate of the assay.

to highly efficacious agonists and a consequent reduction of antagonist affinity. These differences could not be resolved in the corresponding binding assays. Thus, receptor regulation in response to stimulation, e.g. phosphorylation, which remains unresolved in reporter gene tests measuring signals remote from ligand–receptor interactions can obscure experimental results.

4.3
Membrane Potential Indicators

In the recent past, ligand-gated ion channels have facilitated our understanding of receptor function at the molecular level. Single-channel analysis using the patch clamp technique provided detailed insights into "unitary" ligand–receptor transduction mechanisms for this receptor class [25]. The complex functional repertoire of ion channels is also reflected in the variety of possible drug–channel interactions

(Fig. 4.4). Drugs can not only compete with activating ligands for common binding sites or modulate ligand binding allosterically, but small molecules and peptides also have the potential to interfere with channel gating or ion permeation. For the voltage-gated sodium channel, for example, a minimum of six molecularly distinct binding sites have been described for small molecules or peptidic neurotoxins. Ligand binding to these sites results in a spectrum of functional consequences ranging from pore blocking to alterations of voltage-dependent gating affecting either activation or inactivation (for a review, see [26]). In addition, ion channels and most likely also other receptors exhibit a spectrum of functional states corresponding to discrete protein conformations. For ion channels, these conformational states have been introduced to explain the kinetic behavior of single channels in patch clamp experiments. Drugs might bind selectively to discrete receptor states or confirmations (e.g. [27]). As a consequence, small molecule accessibility of channel states or associated binding sites, respectively, will impact the sensitivity of the test systems towards certain mechanisms of modulation. In analogy, similar conformational and functional states with potentially distinct ligand binding properties are discussed for GPCRs (for a review, see [28]). Despite being model systems for receptor studies in basic science, ion channels have been difficult subjects for drug discovery. The frequently multimeric composition of native ion channels, the size of the proteins and the lack of appropriate readout technologies for Ca^{2+}-impermeable pores still pose significant hurdles for the development of functional assays using recombinant receptors.

Radioligand binding studies based on competition of labeled ligands, known drugs or neurotoxins [29] only reveal candidate compounds binding at the same ("orthosteric") site on the protein or affecting binding of the label through some allosteric mechanism. Moreover, biochemical tests employing membrane preparations with no electrical potential gradient present provide access to only a limited number of channel states. In addition, regulatory effects exerted, for example, through phosphorylation/dephosphorylation cycles of the protein cannot be captured using this technique. Assays detecting ion (or substrate) flux through channels or transporters provide an attractive alternative because they monitor channel or transporter function [30]. Measurements of radiotracer flux through membrane transporters typically require a step separating cells from bulk solution

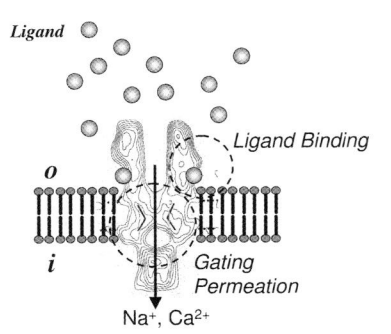

Figure 4.4 Ion channels as pharmacological drug receptors.

through filtration. To obtain sufficient signal-to-noise ratios, high expression levels of the studied transporter and temporal integration, i.e. accumulation of sufficient radiotracer inside the cell (or loss thereof), are needed. For a variety of potassium channels, permeant Rb$^+$ ions in combination with atomic absorption spectroscopy have been established as a methodology amenable to 384-well microtiter plates. However, for practical reasons, all the above-mentioned flux technologies are not feasible for high-density plate formats and therefore are limited with regard to throughput. The only exception is provided by ion channels permeable to Ca^{2+} ions, which can be monitored with high sensitivity and temporal resolution using either fluorescent or luminescent indicators.

The activity of electrogenic membrane transporters and channels, i.e. transporter generating an electrical membrane current, can also be monitored by changes in the transmembrane electrical potential gradient (ΔE_m) associated with transport activity. The availability of fluorescent potentiometric indicators has improved the experimental access to this target class (for a review, see [31]). Most assays reported to be HTS compatible employ "Nernstian" redistribution probes, e.g. DiBAC4 or the recently introduced probe from Molecular Devices (FMP = FLIPR Membrane Potential Dye). These charged fluorescent molecules can readily permeate the membrane. Their equilibrium distribution between extracellular solution and the cytosolic compartment is described by the Nernst equation:

$$\Delta E_m = -\frac{RT}{zF} \ln\left(\frac{[\text{dye}]_i}{[\text{dye}]_o}\right) \quad (1)$$

where R = gas constant = 8.315 J K^{-1} mol^{-1}, F = Faraday's constant = 9.65 × 10^4 C mol^{-1}, T = temperature (K) and z = charge of dye species.

For some dyes, e.g. DiBAC4, the membrane potential given by measuring cellular fluorescence is complicated by binding of the probes to intracellular proteins, which results in changes in the quantum yield. As an additional consequence, the fluorescence signal kinetics in response to changes in ΔE_m are further slowed by protein binding and/or unbinding.

The limited temporal resolution provided by this class of indicators is balanced by the readily resolvable signal amplitudes. In contrast, fast membrane staining dyes, e.g. di-8-ANEPPS, provide microsecond time resolution but suffer from poor signal [$\Delta(\Delta E_m)$] resolution with only ca. 10% fluorescence intensity changes per 100 mV change in ΔE_m. Because of this limited signal dynamics, these dyes have not been used for HTS assays. Gonzalez and co-workers [32, 33] introduced a fluorescent readout technique based on fluorescence–resonance energy transfer (FRET) from a immobilized coumarin-labeled phospholipid, CC2-DMPE, localized to the outer leaflet of the membrane bilayer and a mobile oxonol dye, DiSBAC$_n$. The dye combination is frequently referred to as "VIPR dye" because of its initial realization on the Voltage and Ion Probe Reader introduced by Aurora Bioscience [34]. Measuring only the emission of the FRET acceptor, i.e. the labeled lipid, renders this recording technique sensitive only to changes in oxonol dye concentration at the membrane interface towards the extracellular space. As a result, fluorescence changes rapidly follow changes in membrane potential with sub-second time resolution.

For a long time, measuring cellular fluorescence in microtiter plate experiments has been impaired by background fluorescence originating from the bulk solution. Directing excitation light in an angular fashion through the transparent bottom of the microtiter plate as realized, for example, in the FLIPR reader from Molecular Devices [35], and/or simple quenching of extracellular bulk fluorescence by addition of masking dyes [36] has greatly facilitated the recording of fluorescence signals from cell monolayers through a transparent microplate bottom.

Figure 4.5C illustrates some simplified principles of membrane potential measurements using a potentiometric fluorescent probe to resolve the activity of electrogenic membrane transport processes. A test to monitor the activity of a voltage-gated K channel should be designed. The channel will be recombinantly expressed in a suitable host cell with no significant endogenous K conductances. We assume that the membrane potential of the host cell is governed only by potassium and sodium conductances. Using a simple form of the Goldman–Hodgkin–Katz (GHK) equation [27], the membrane potential will depend on the ratio of K and Na permeabilities ($\alpha = P_{Na}/P_K$).

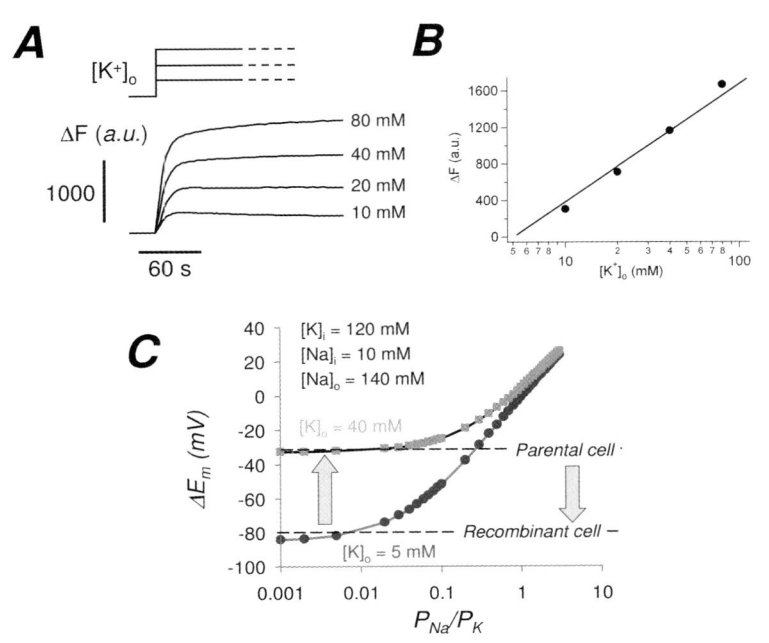

Figure 4.5 Recombinant K channel activity measured with FMP probe. (A), (B) Original FMP recordings (1536-well plate) in response to $[K]_o$ jumps from CHO cells expressing a recombinant K channel. (C) Dependence of cellular membrane potential on K channel expression (Goldman–Hodgkin–Katz equation; for details, see the text).

$$\Delta E_m = \frac{RT}{F} \ln\left(\frac{[K]_o + \alpha[Na]_o}{[K]_i + \alpha[Na]_i}\right) \qquad (2)$$

Whereas P_{Na} is assumed to stay constant, P_K will increase with increasing expression levels of the recombinant K channel. The contribution of recombinant K channel is determined by the product of "open probability" (p_0), unitary single channel current amplitude (i) and the number of functional channel proteins in the plasma membrane (N):

$$I = N p_0 i \qquad (3)$$

Hence also voltage-gated channels typically activating upon depolarization which have a low open probability at resting membrane potentials can contribute a current at negative membrane potentials depending on the number of channels expressed. In addition, their overall contribution to membrane potential depends on the relative contribution of endogenous conductances. As can be seen from the GHK relationship, increasing the expression level of the recombinant K channel is expected to shift the membrane potential towards the K equilibrium potential. Further increasing N (and thus P_K) will not result in additional hyperpolarization. This saturation behavior has important consequences for detecting K channel modulators by monitoring ΔE_m. In cells with high expression levels, a large fraction of K channels need to be blocked before a resulting depolarization can be resolved. For K channel stimulators, the resulting hyperpolarization is limited by E_K. Therefore, depending on the aim of the assay looking for stimulators or inhibitors, the expression level of the targeted K channel will determine the signal amplitude and assay sensitivity (Figure 4.5). In such a simplified model system, inhibition of the K channel activity would result in a depolarization of resting membrane potential and a reduction in the fluorescence change in response to KCl addition.

Fluorescence recordings using a potentiometric dye in a plate reader will not report absolute membrane potential values. Signals will be affected by the number of cells, the height of the cell monolayer and ΔE_m-independent background fluorescence adding a signal off-set. Hence only relative changes ($\Delta\Delta E_m$) can be resolved. Figure 4.6 illustrates data obtained from a recent in-house uHTS campaign seeking modulators of a K_{ATP} channel heterologously expressed in a CHO cell. Following compound transfer and a 10-min incubation period, the cells were challenged with a K$^+$ jump (5–40 mM). The assay was adjusted and optimized to detect both activators and inhibitors. As can be seen from the plot of the simplified GHK equation (Fig. 4.5C), activation of a K conductance should hyperpolarize the membrane potential, resulting in a signal reduction of the FMP signal measured upon transfer into the plate reader [$F(t = 0)$], whereas the amplitude of the signal change in response to the K$^+$ jump [$\Delta F(KCl)$] should increase. In turn, inhibitors should depolarize the cells and reduced the signal upon KCl addition. The scatterplot summarizes normalized data from a single overnight robotic run comprising >200 000 compounds tested. Values (as a percentage of solvent controls) below 100

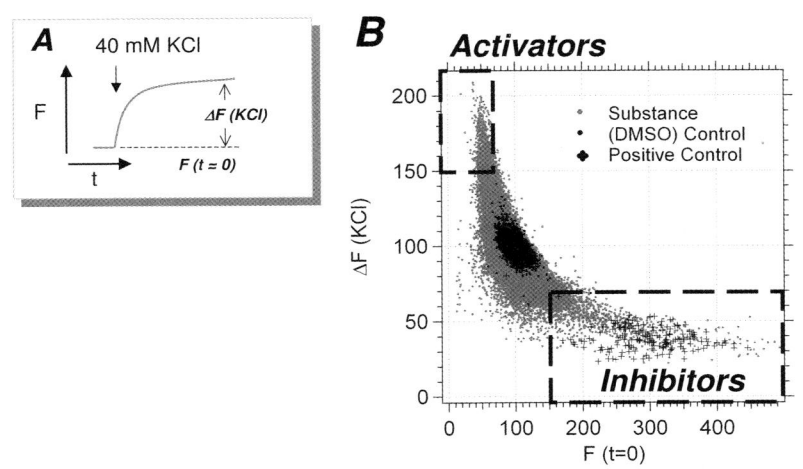

Figure 4.6 Screening for modulators of recombinantly expressed K(ATP) channels. (A) In this 1536-well uHTS assay, the response of the cells to a [K]$_0$ jump was recorded. For database import and analysis, the data were reduced to a background [$F(t = 0)$] and value [ΔF(KCl)] signal, both normalized to the corresponding solvent control values. Substance and control values for a single robotic run (16 h) comprising in excess of 200 000 compounds are summarized in the scatterplot (B). Sulfonamide K(ATP) channel blockers included in the test as "positive controls" revealed a significant contribution of K(ATP) activity to the resting membrane potential. Background fluorescence [$F(t = 0)$] was markedly elevated compared with control, indicative of a depolarization. Also, the response to the [K]$_0$ jump [ΔF(KCl)] was reduced. Test compounds which inhibited K(ATP) co-localize with the positive controls in the scatterplot. Compounds marked as "activators" which hyperpolarized the resting membrane potential and potentiated the response to [K]$_0$ jumps could also be identified in the assay. As expected, this functionally distinct hit pool included a large fraction of compounds known to compromise the cellular energy metabolism.

reflect a signal reduction (= hyperpolarization) and values greater that 100 correspond to a signal increase (= depolarization). The data revealed a large number of compounds displaying either significant inhibition [$F(t = 0) > 150$ and ΔF(KCl) < 70] or stimulation of a K conductance [$F(t = 0) < 70$ and ΔF(KCl) > 150], most likely K$_{ATP}$. The significant hit rates observed are representative of assays dependent on a membrane potential readout. Multiple pharmacological off-target effects together with a large number of robust assay artifacts converge on membrane potential changes. Activity profiling against assays employing the same readout technique provides a first means for filtering out unspecifically acting compounds. Known channel or transporter blockers or activators can be employed to characterize further the compound effect in microplate experiments. However, the lack of alternative high-throughput techniques measuring ion channel function aggravates the separation of meaningful target modulators from assay trivials and off-target acting compounds early in the HTS process.

4.4
Ca^{2+} Indicators

A large set of targets can be functionally studied via their ability to induce intracellular Ca^{2+} signals by increasing either surface membrane Ca^{2+} permeability or eliciting release of Ca^{2+} from intracellular stores, most prominently the endoplasmic reticulum (ER). The introduction by Tsien's group [5, 6] of fluorescent Ca^{2+} indicators derived from the Ca^{2+} chelator BAPTA or the dye fluorescein has greatly impacted cellular physiology and pharmacology research by providing membrane permeant ester forms readily accumulated by and trapped in living cells. The ease of use of this fluorescence technique has facilitated its wide application today. The UV dyes fura-2 and indo-1 provide the possibility of ratiometric fluorescence recordings on the excitation or emission side, respectively. Signal rationing allows one to quantify intracellular Ca^{2+} signals with less errors originating from heterogeneity in indicator concentration, number of cells or differences in optical path. The fluorescein derivative Fluo-3 [38] can be excited with 480-nm radiation from the widely used argon ion laser, provides a very high quantum yield when Ca^{2+}-bound and further simplified the measurement of cellular Ca^{2+} signals. With the advent of plate readers combining fluorescence imaging with integrated liquid handling, fluorescent Ca^{2+} recordings could be transferred to the microtiter plate. A plate reader such as the FLIPR [35] in combination with functional cell-based assays has found broad application in the HTS community.

Ca^{2+}-sensitive photoproteins, such as aequorin derived from the jellyfish *Aequorea victoria* [37], provide a technical alternative to fluorescent Ca^{2+} indicators. Transfection of suitable host cells with the cDNA coding the apo-photoprotein eliminates the necessity for microinjection of the protein [38]. Moreover, engineered fusion constructs combining the apo-photoprotein with recognition sequences necessary for import into cellular organelles allowed the targeted expression of the Ca^{2+}-sensing photoprotein in selected subcellular compartments [39]. Recombinant cells constitutively expressing the photoprotein and a receptor of interest, i.e. the target, can be employed to measure receptor-mediated Ca^{2+} entry or release using a luminescence plate reader. For the photoprotein to become functional, it needs reconstitution with the membrane permeant cofactor coelenterazine. Upon binding of Ca^{2+} ions, coeleterazine is oxidized, resulting in light emission. Because of the extremely slow dissociation of the oxidized cofactor, the photoprotein is practically consumed in the reaction. This consumption of photoprotein, the need for recombinant DNA techniques and the requirement for highly sensitive detection devices have limited the use of photoproteins for resolving cellular Ca^{2+} signals in the past. The absence of any appreciable background and the lack of interference with the detected signal from colored test compounds are clear benefits of this technique. The availability of improved plate readers with enhanced sensitivity has contributed to a broader applicability in the HTS community. Today, a variety of closely related photoproteins originating, for example, from the medusa *Obelia* sp. [40], the hydroid polyp *Clytia* sp. [41] and other hydrozoa are available as genetically encoded Ca^{2+} indicators. In addition, engineered Ca^{2+}-sensitive photoproteins such as Photina™ from Axxam [42] further facilitate their use for HTS assay development.

Intracellular Ca^{2+} signals can be used to monitor quantitatively the activity of a variety of therapeutically relevant drug targets directly or indirectly coupled to Ca^{2+} influx or release. Directly accessible targets include, among others, a variety of voltage- or ligand-gated plasma membrane Ca^{2+} channels, $G\alpha q$ protein-coupled receptors and receptor tyrosine kinases. In some cases, the molecular biology

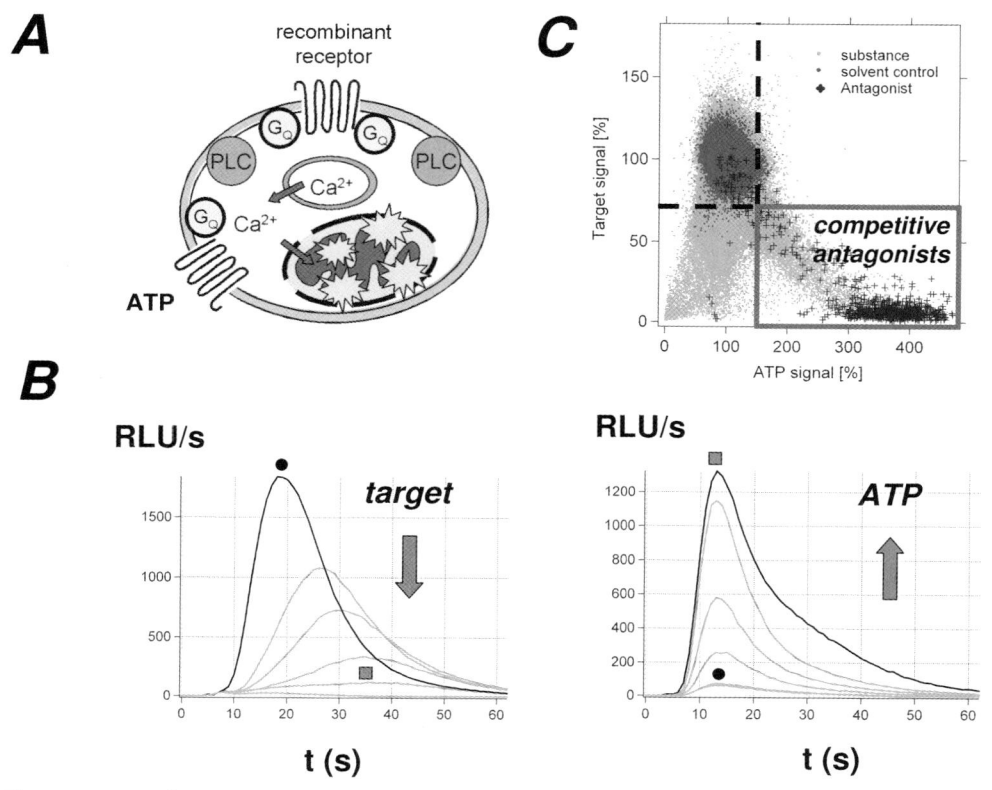

Figure 4.7 uHTS for GPCR antagonists monitoring receptor-induced intracellular Ca^{2+} release by a Ca^{2+}-sensitive photoprotein. The target receptor was introduced recombinantly into a cell constitutively expressing the Ca^{2+}-sensitive photoprotein aequorin targeted towards the mitochondrial matrix compartment (A). Cells had been seeded in 1536-well plates at 500 cells per well the day before. In the assay cells were stimulated by the ligand for the target receptor (B, left: target signal) before ATP was injected to activate endogenous P_2Y receptor also coupling to intracellular Ca^{2+} release (B, right; ATP signal). Inhibition of the target receptor caused the first signal to decrease while enhancing the amplitude of the ATP response. In contrast, unspecifically acting and toxic compounds reduced both signals. The scatterplot (C) summarizes the data from an HTS campaign comprising more that 1.5 million compounds. The ATP reference signal employed clearly differentiated two distinct pools of hit compounds. The ensemble of luminescence transients in (B) shows traces in the presence of increasing concentrations of a competitive target receptor antagonist. Note that the ATP signal is changed only in amplitude.

toolbox can be employed to render targets not naturally coupling to a cellular Ca^{2+} signal amenable to this readout technology. Co-transfection with promiscuous G proteins ($G\alpha_{16}$ or $G\alpha_{15}$), for example, has been used to force recombinant target receptors which couple normally G_i-adenylate cyclase-coupled receptors to phospholipase C and intracellular Ca^{2+} release from $InsP_3$-sensitive stores [43, 44]. In other cases, C-terminal fusion proteins of a GPCR and $G\alpha_{16}$ have been employed to force the receptor to elicit intracellular Ca^{2+} signals upon activation [45]. In addition, the Ca^{2+} permeability of cyclic nucleotide-gated channels has been exploited to convert intracellular cAMP or cGMP signals into Ca^{2+} signals readily resolvable by fluorescent or luminescent Ca^{2+} indicators.

Online recordings of cellular Ca^{2+} signals with either fluorescent or luminescent indicators provide kinetic data which reveal additional information on the potential molecular mode of action of added test compounds. The measured luminescence signal approximates the temporal derivative of the corresponding signal measured with fluorescent indicators. This behavior, together with the fast Ca^{2+} binding kinetics of the photoproteins, renders the bioluminescence readout a very sensitive indicator of the underlying cellular Ca^{2+} fluxes (Fig. 4.7). In contrast, slow changes in intracellular Ca^{2+} concentration might not be adequately resolved using photoproteins. The signals generated by the rapidly responding intracellular Ca^{2+} signaling machinery typically precede agonist–receptor equilibrium. Moreover, the intracellular Ca^{2+} release transients not only increase in amplitude with increasing receptor agonist concentration but also display accelerated kinetics (Fig. 4.7 B). Capturing this kinetic information also reveals information on the possible mode of action of test compounds. Competitive inhibitors, for example, should not only reduce the signal amplitude but also delay and slow the upstroke of the luminescence transient (Fig. 4.7B).

4.5
Conclusions

Improved readout techniques together with the rich molecular biology toolbox have rendered cell-based assays an important methodology for the targeted discovery of pharmacological lead compounds in HTS. The designed cell-based HTS assay ideally combines high specificity for and superior sensitivity towards the targeted receptor. In addition, measuring receptor function rather than binding allows one to monitor all possible drug–receptor interactions including allosteric modulation and reveals additional information on ligand efficacy, i.e. agonism or antagonism. Exploiting temporally resolved cellular signals or integration of multiple signal recording steps in one assay can reveal further information on the mode of action of test compounds. Generally, target-specific hits are rare events. Assays delivering a large number of unspecific hits acting through other mechanisms generally require additional high-throughput experimental approaches to identify positively the few meaningful target modulators within large hit pools. The often limited resolution of such "off assays" together with the risk of producing false negative results complicates this approach when testing comprehensive corporate compound collections.

References

1 Colquhoun, D., Sakmann, B., *Neuron 20* (**1998**) 381–387.
2 Maehle, A.-H., et al., *Nat. Rev. Drug Discov. 1* (**2002**) 637–641.
3 Lester, H. A., *Science 241* (**1988**) 1057–1063.
4 Luyten, W. H. M., Leysen, J. E., *Trends Biotechnol. 11* (**1993**) 247–254.
5 Hertzberg, R. P., *Curr. Opin. Biotechnol. 4* (**1993**) 80–84
6 Grynkiewicz, G., et al., *J. Biol. Chem. 260* (**1985**) 3440–3450.
7 Minta, A., et al., *J. Biol. Chem. 264* (**1989**) 8171–8178.
8 Kenakin, T., *Receptors Channels 10* (**2004**) 51–60
9 Henry, B., *IDrugs 7* (**2004**) 819–821.
10 Presland, J., *Curr. Opin. Drug Discov. Dev. 8* (**2005**) 567–576.
11 Rees, S., et al., *Receptors Channels 10* (**2004**)
12 Hill, S. J., et al., *Curr. Opin. Pharmacol. 1* (**2001**) 526–532.
13 Gorman, C. M., et al., *Mol. Cell Biol. 2* (**1982**) 1044–1051.
14 Markova, S. V., et al., *J. Biol. Chem. 279* (**2004**) 3212–3217.
15 Li, X., et al., *J. Biol. Chem. 273* (**1998**) 34970–34975.
16 Zhang, J., et al., *Nat. Rev. Mol. Cell Biol. 3* (**2002**) 906–918.
17 Chamberlain, C., Hahn, K. M., *Traffic 1* (**2000**) 755–762.
18 Zlokarnik, G., et al., *Science 279* (**1998**) 84–88.
19 Knapp, T., et al., *Cytometry 51A* (**2003**) 68–78.
20 Kunapuli, P., et al., *Anal. Biochem. 314* (**2003**) 16–29.
21 Oosterom, J., et al., *Assay Drug Dev. Technol. 3* (**2005**) 143–154.
22 Dolmetsch, R., et al., *Nature 392* (**1998**) 933–936.
23 Li, W.-H., et al., *Nature 392* (**1998**) 936–941.
24 Baker, J. G., et al., *Mol. Pharmacol. 64* (**2003**) 679–688.
25 Pallotta, B. S., *FASEB J. 5* (**1991**) 2035–2043.
26 Cestèle, S., Catterall, W. A., *Biochemie 82* (**2000**) 883–892.
27 Hille, B., *Ion Channels of Excitable Membranes*, 3rd edn. Sinauer, Sunderland, MA, **2001**.
28 Perez, D. M., Karnik, S. S., *Pharmacol. Rev. 57* (**2005**) 147–161.
29 Denyer J., et al., *Drug Discov. Today 3* (**1998**) 323–332.
30 Gill, S., et al., *Assay Drug Dev. Technol. 1* (**2003**) 709–717.
31 Loew, L. M. in: *Fluorescent and Luminescent Probes for Biological Activity*, Mason, W. T. (ed.). Academic Press, Cambridge, **1999**, pp. 210–221.
32 González, J. E., Tsien, R. Y., *Chem. Biol. 4* (**1997**) 269–277.
33 González, J. E., Maher, M. P., *Receptors Channels 8* (**2002**) 283–295.
34 http://www.auroradiscovery.com.
35 http://www.molecular.devices.com.
36 Krahn, T., et al., Patent WO 9745739, Bayer AG (**1997**).
37 Sala-Newby, G. B., et al., in: *Fluorescent and Luminescent Probes for Biological Activity*, Mason, W. T. (ed.). Academic Press, Cambridge, **1999**, pp. 251–272.
38 Button, D., Brownstein, M., *Cell Calcium 14* (**1993**) 663–671.
39 Chiasa, A., et al., *Biochem. J. 355* (**2001**) 1–12.
40 Illarionov, B. A., et al., *Gene 153* (**1995**) 273–274.
41 Inouye, S. and Tsuji, F. I., *FEBS Lett. 315* (**1993**) 343–34641.
42 http://www.axxam.com.
43 Stables, J., et al., *Anal. Biochem. 252* (**1997**) 115–126.
44 Kostenis, E., et al., *Trends Pharmacol. Sci. 26* (**2005**) 595–602.
45 Milligan, G., et al., *Curr. Pharm. Des. 10* (**2004**) 1989–2001.

5
Biochemical Assays for High-throughput Screening

William D. Mallender, Michael Bembenek, Lawrence R. Dick, Michael Kuranda, Ping Li, Saurabh Menon, Eneida Pardo, and Tom Parsons

5.1
General Considerations for Biochemical High-throughput Screening

Genomic research has identified a treasure trove of potential new drug targets and progress has been made in relating many of these targets to their role in disease biology. The biochemistry shared by target families allows one, in many cases, to leverage existing assay technology to screen for compounds directed against a drug target with unprecedented rapidity. Indeed, today drug screening is often completed in parallel with target validation studies allowing identified compounds to influence project progression or potentially aid in the advancement of the target biology.

Biochemical high-throughput screening (HTS) generally refers to the use of a purified or partially purified target protein to screen small molecule compound libraries to find candidates for drug development. In principle, it is the simplest experimental system for this purpose allowing for the uncomplicated evaluation of the interaction between a target and test molecule. Thus, membrane permeability, serum binding and compound metabolism, which are common problems in cell-based or *in vivo* studies, are not assay variables. Often, such complexities can mask or prevent the identification of weak inhibitors that might allow a drug discovery or medicinal chemistry program. With primary biochemical data in hand and lead compounds identified, one can use cell-based and other more complex secondary assay formats to predict compound behavior better in a therapeutic setting.

Biochemical assay systems, because of their inherit simplicity, are more robust, easier to automate and perform better statistically than typical cell-based or *in vivo* assays. However, the superior performance of a biochemical approach requires isolation of the target and reconstitution of its activity in a way that can be easily monitored and quantitated. This generally means following a signal related to enzyme-catalyzed product generation or substrate depletion. Additionally, the signal may be related to target–ligand binding. For the assay employed for a given target, a change in the signal as a result of compound addition can be taken as

High-Throughput Screening in Drug Discovery. Edited by Jörg Hüser
Copyright © 2006 WILEY-VCH Verlag GmbH & Co. KGaA, Weinheim
ISBN: 3-527-31283-8

indicative of a compound–target interaction. The quantity of target protein required for large-scale screening efforts often necessitates the use of recombinant protein expression techniques. Truncation, fusions or other modifications of the target may be required to gain solubility or catalytic efficiency. In addition, screening large compound libraries requires the use of artificial substrates to make rapid activity assessments. These compromises must be considered relative to the behavior of the target *in situ* and biological relevance of the HTS assay should be demonstrated experimentally whenever possible.

As with other screening approaches, the number of compounds that need to be tested using a biochemical approach leaves us with the challenge of achieving data quantity without sacrificing quality. Assays need to be simple, sensitive and robust to accommodate automated processing in microplates of 384- or 1536-well or higher density formats. To permit this rapid parallel testing of compounds, the enzyme reaction must be detected as an optical measurement, which can take the form of luminescence, fluorescence, absorbance or light emission via the scintillation effect for measuring radioactivity. These measurements allow for the use of array detectors or plate imaging systems. The signal modalities available for a target are specific for a given enzyme class and are frequently used repeatedly to assess multiple family members.

With automation, most modern HTS laboratories can complete the testing of a million compounds in less than 1 month. The prelude to HTS always takes significantly longer. Fortunately, even when working with previously uncharacterized targets, selection and development of an assay protocol have become significantly more efficient owing to the realization that the tools available for HTS can be employed to assess rapidly matrices of assay variables and substrate libraries to yield the most sensitive and relevant assay conditions. Experimental design software coupled with automation further allows for the comprehensive examination of key assay parameters such as salt, detergent, pH, substrates, metal or cofactor concentrations and their potential interdependencies. Enzyme characteristics such as K_M, V_{max} and turnover number can be rapidly determined using microplate technology to guide decisions including selection of the final assay format used for HTS. Even with these refinements, completion of HTS assay development combined with production and testing of reagents is a significant endeavor which usually takes 6–9 months to complete. The unique challenges and considerations are examined here for each major class of biochemical target in order for it to become HTS ready. Although many would class binding assays for targets such as receptors or protein complexes as biochemical, we focus exclusively on the prosecution of enzyme targets. These targets include peptidases, oxidoreductases, synthases, lipid-modifying enzymes, transferases and kinases. The diversity of enzyme mechanisms encompassed by these classes is well represented in currently marketed drugs [1]. It is our hope that the assay approaches described here will enable HTS efforts for related targets emerging from genomics.

5.2
Expression and Purification of Recombinant Enzymes

To permit enzyme characterization, assay development and implementation of HTS, researchers face the daunting task of generating proteins in large quantities that are enzymatically active, pure and stable. This involves using recombinant protein expression and purification technologies for addressing key issues including construct design, expression testing and optimization, purification and final formulation.

We have developed an efficient process to allow the expression and purification of a wide variety of novel enzymes, of sufficient quantity and purity and with the required functional enzymatic activity to source HTS. This process is described below and the success rate for each enzyme class is shown in Table 5.1.

Table 5.1 Successes in expression and purification projects to source HTS.

	Oxidoreductases	Protein Kinases	Transferases, Synthetases	Hydrolases, Lipases	Proteases
Total	11	33	20	20	21
Successful	5	20	10	6	8
Not successful	6	13	10	14	13
% successful	45	61	50	30	38

Success is defined by the following criteria: did we successfully express and purify an active enzyme, did we develop an assay suitable for the screen and did we source the enzyme reagents to complete this screen? For simplicity, these enzyme efforts have been summarized into five general classes (specifics of particular enzymes from each class will be discussed): (1) oxidoreductases (including reductases, dehydrogenases, oxidases, desaturases and aldoketoreductases), (2) protein kinases (including serine–threonine kinases and tyrosine kinases), (3) transferases and synthetases (including arginine methyl transferases, glycosyl transferases and fatty-acyl-CoA synthetases), (4) hydrolases and lipases (including phospholipases, glycosidases and sulfatases) and (5) peptidases (including Ser trypsin-like proteases, carboxypeptidases, matrix metalloproteases and aminopeptidases).

Each enzyme class (and frequently each enzyme) offers unique challenges in its expression, purification, demonstration of functional activity and development of an efficient assay format suitable for the screen. These activities are usually referred to collectively as assay configuration. The class of enzyme, time frame for each configuration attempt, expression, purification and sourcing of the successful screen and the cell production system are indicated in Fig. 5.1.

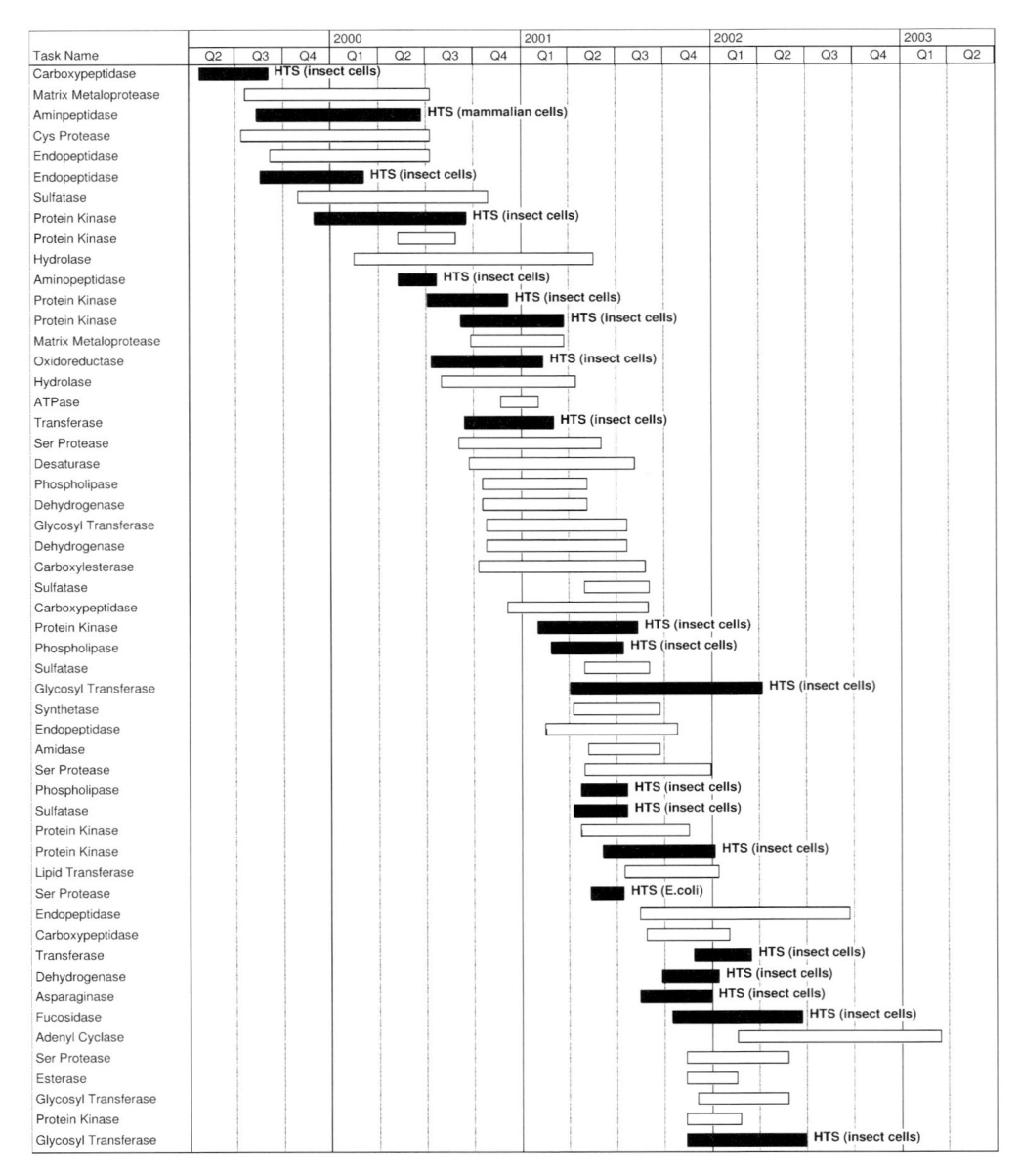

Figure 5.1 Expression and purification projects to source HTS. The class of enzyme and time frame for each "configuration attempt" (expression, purification and sourcing of the successful screen) and the cell production system are indicated. A solid bar indicates a successful screen and an open bar indicates that initiation of a screen was not successful (failure in either assay configuration or recovery of suitable amounts of enzyme to source the screen).

5.2 Expression and Purification of Recombinant Enzymes | 97

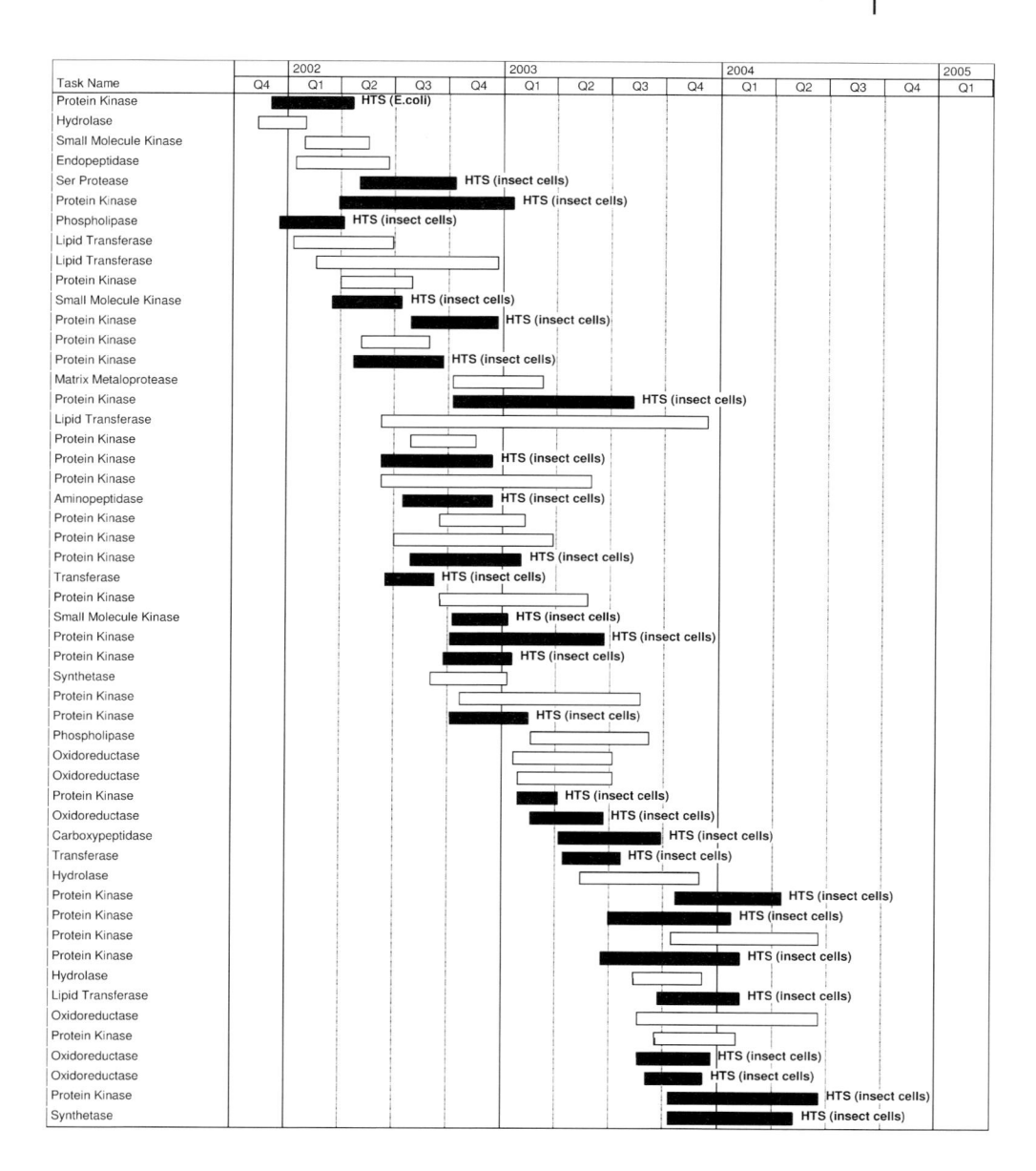

Task Name	2002 Q4	2002 Q1	Q2	Q3	Q4	2003 Q1	Q2	Q3	Q4	2004 Q1	Q2	Q3	Q4	2005 Q1
Protein Kinase		██	██ HTS (E.coli)											
Hydrolase	▭													
Small Molecule Kinase			▭											
Endopeptidase			▭											
Ser Protease					██ ██ HTS (insect cells)									
Protein Kinase					██ ██ ██ HTS (insect cells)									
Phospholipase			██ HTS (insect cells)											
Lipid Transferase		▭												
Lipid Transferase														
Protein Kinase				▭										
Small Molecule Kinase				██ HTS (insect cells)										
Protein Kinase					██ HTS (insect cells)									
Protein Kinase				▭										
Protein Kinase				██ HTS (insect cells)										
Matrix Metaloprotease					▭									
Protein Kinase					██ ██ ██ ██ HTS (insect cells)									
Lipid Transferase				▭▭▭▭▭▭▭▭										
Protein Kinase														
Protein Kinase				██ HTS (insect cells)										
Protein Kinase				▭										
Aminopeptidase				██ HTS (insect cells)										
Protein Kinase					▭									
Protein Kinase														
Protein Kinase				██ HTS (insect cells)										
Transferase				██ HTS (insect cells)										
Protein Kinase														
Small Molecule Kinase				██ HTS (insect cells)										
Protein Kinase				██ ██ HTS (insect cells)										
Protein Kinase				██ HTS (insect cells)										
Synthetase				▭										
Protein Kinase														
Protein Kinase				██ HTS (insect cells)										
Phospholipase														
Oxidoreductase					▭									
Oxidoreductase														
Protein Kinase					██ HTS (insect cells)									
Oxidoreductase					██ HTS (insect cells)									
Carboxypeptidase					██ HTS (insect cells)									
Transferase					██ HTS (insect cells)									
Hydrolase						▭								
Protein Kinase						██ HTS (insect cells)								
Protein Kinase						██ HTS (insect cells)								
Protein Kinase						▭								
Protein Kinase						██ HTS (insect cells)								
Hydrolase						▭								
Lipid Transferase						██ HTS (insect cells)								
Oxidoreductase						▭▭▭								
Protein Kinase						▭								
Oxidoreductase						██ HTS (insect cells)								
Oxidoreductase						██ HTS (insect cells)								
Protein Kinase						██ HTS (insect cells)								
Synthetase						██ HTS (insect cells)								

Figure 5.1 (continued)

5.2.1
Design of Expression Constructs

The generation of quality recombinant enzymes starts with addressing aspects including the enzyme class of the target protein, its protein structure, its cellular location, its requirement of interacting protein(s) and its dependence on cofactors or modifications.

Based on the enzyme class and the structure of the protein, normally more than one construct needs to be considered and tested for expression. There are several reasons for this. First, although it is often desirable to have the full-length protein, truncated versions around the catalytic domain should also be tested because they can give increased solubility and better purification yields relative to the full-length protein. Truncated versions can also have constitutive activities due to the removal of regulatory domains. Second, different fusion tags and their position should be tested. The primary reason for a fusion tag is to facilitate protein purification and identification. Different fusion tags and their positioning can have an impact on both solubility and activity. For short tags such as 6× His (Invitrogen) and Flag (FLAG®, Sigma-Aldrich), one might want to test the effect of an N-terminal fusion versus a C-terminal one. Other commonly used tags include GST (glutathione transferase, Invitrogen), MBP (maltose binding protein, New England Biolabs) and Trx (thioredoxin, Novagen). Third, enzymes such as type-2 transmembrane serine proteases require precise processing for activity [2]. In these cases, sequence variations at the junction of the fusion tag and the catalytic domain can affect the precision of the processing and consequently, the generation of the active enzyme.

The commonly used recombinant protein expression systems include *Escherichia coli*, yeast, baculovirus/insect cells and mammalian cells. *E. coli* has the advantage of a quick turnaround time and high expression levels, but often suffers from protein aggregation in the form of inclusion bodies. The lack of post-translational modifications can also lead to inactive proteins. Mammalian cells would be the ideal host for recombinant proteins as they potentially provide for natural post-translational processing of the target protein. The limitations of the mammalian expression are that it is time consuming and the level of protein expression is generally low. The baculovirus/insect cell system has the combined benefits of both *E. coli* and mammalian cells. This system allows robust, high-level protein expression and is straightforward to scale up. Also, many of the protein processing and modification processes are conserved between mammalian and insect cells. Therefore, there is a high likelihood of producing properly folded and adequately modified active enzymes.

Because different fusion tags and the choice of expression system can vary from protein to protein, one more factor to consider in construct design is the engineering options built into the vectors. It is desirable to choose vectors that allow for easy exchanging of fusion tags and the flexibility of switching between expression systems. The commercially available vector systems providing such transferability include the Gateway™ (Invitrogen) and the Creator™ (Clontech) systems.

5.2.2
Expression Assessment and Optimization

In most cases, primary expression assessment can be done by Western blot analysis of the total cell lysate and the soluble fraction using antibodies recognizing the fusion tag. This analysis will identify specific construct(s) with good levels of protein expression and solubility. Very often, further optimization of expression conditions is required to ensure reproducible yield and solubility. For the *E. coli* expression system, this typically involves testing factors which include the effect of bacteria strains (e.g., DE3 vs. DE3-pLysS, Novagen), induction mode (auto-induction vs. IPTG, Novagen), induction temperature and time of induction. For the baculovirus system, it is highly recommended to generate plaque purified virus. In addition, expression can be further optimized by testing factors such as virus load (multiplicity of infection) and duration of infection. One can also optimize soluble protein expression through co-expression with molecular chaperons [3, 4].

For some enzymes, addition of biosynthetic intermediates or inhibitors to the culture medium can significantly increase the level of enzymatically active proteins. For heme-containing enzymes, adding the heme precursor δ-aminolevulinic acid to the culture medium has proven to be helpful [5]. For cell cycle kinases, brief treatment of the cells with a phosphatase inhibitor, okadaic acid, can lead to the generation of active kinases [6].

5.2.3
Purification

As stated above, the generation of quality recombinant enzymes, especially novel enzyme targets with unknown natural substrates and products, requires addressing several uncertainties. In general, to express and purify an active enzyme suitable for HTS, several requirements must be met: (1) the cell expression system should allow the correct processing and post-translational modifications required for enzymatic activity, (2) the purified enzyme preparation should not be aggregated (for non-membrane preparations), (3) the enzyme preparation should be enzymatically clean (functional detectable activity is a result of the recombinantly expressed enzyme), (4) the enzyme preparation should be as pure as reasonably possible (for non-membrane preparations, usually > 95% purity of the recombinant enzyme allows assurance that the functional detectable activity is a result of the recombinantly expressed enzyme), (5) the scale of the cell production system and purification procedures should efficiently allow the purification of the quantity of enzyme required for the screening of the number of compounds in the HTS and (6) the purified enzyme (and enzymatic activity) should be stable both in storage before the screen as well as during the screen.

A literature search of the particular enzyme or other members within the enzyme class can yield important information for the purification, structure characterization and stabilization of the enzyme. For example, cellular location will help define the requirement for reducing reagents during the purification process. For those enzymes that are extracellular, secreted or lysosomal or in the endoplasmic reticulum reducing reagents are not required (disulfide bonds are probably present). This would include extracellular proteases [7] and glycosyl transferases [8] as typical examples. Conversely, for those enzymes found in the cytoplasm, nucleus or mitochondrial matrix, typically seen with kinases [9], reducing reagents are probably required. The enzyme class and primary sequence can indicate if the enzyme is membrane bound, either with multiple membrane spanning segments or with a single membrane tether. The number of membrane spanning segments in turn helps determine whether a "membrane preparation", often with multi-spanners [10], a "detergent-solubilized" full-length construct, typically with 1% NP40™ or 1% Tween 20™ [11] or a "cytoplasmic" or "intracellular" domain construct lacking the membrane-spanning tether [12] can be appropriate for production and purification to source the screen. The enzyme class will also help define the requirements for buffer components during the purification process. If phosphorylation of particular activation site(s) of kinases is required for enzymatic activity, phosphatase inhibitors such as sodium fluoride, sodium orthovanadate and β-glycerol phosphate should be added to the purification buffers [9]. If cofactors, such as flavin mononucleotide (FMN), flavin adenine dinucleotide (FAD), ubiquinone (Q), nicotinamide adenine dinucleotide (NAD/NADP), pyridoxyl phosphate (PLP) or metal ions such as iron, zinc or copper, are required for enzymatic activity, they could be added to the purification buffers to help stabilize the enzyme during purification.

As stated above, to yield active enzyme suitable for the screen, the purified enzyme preparation should not be aggregated, should be >95% pure and should be stable during storage. Several considerations to help ensure the purification of a quality enzyme preparation (for non-membrane bound enzymes) include: the choice of a buffer pH at least one pH unit away from the calculated pI of the enzyme construct, the addition of detergents (0.01–0.1% Tween 20™), the addition of 10% glycerol, the choice of appropriate protease inhibitors and, whenever possible, the addition of at least 150 mM monovalent salt (such as NaCl or KCl) in the buffers. Each of these components may affect the assay readout and quality of the enzymatic activity and therefore care must be taken to assess their addition to a specific enzyme target. Generic purification protocols include initial capture to an affinity resin specific for the engineered tag on the construct followed by a gel filtration chromatographic step to assess and fractionate the non-aggregated preparation [11] and, if needed, an ion-exchange chromatographic step.

Characterization of the enzyme preparation includes both physical properties of the protein and demonstration of enzymatic functional activity. Each enzyme class (and often each enzyme) offers unique challenges. Several generalizations regarding protein characterization can be made. Purity and verification of the enzyme construct should be demonstrated by assessing the appropriate apparent molecular weight by SDS polyacrylamide gel electrophoresis and Western blot analysis. If proteolytic processing is required to generate functional activity, verification of N-terminal processing by Edman degradation [13] and verification of C-terminal processing by tandem mass spectrometry [14] are commonly used. If covalent post-translational modification is required to generate functional activity (such as glycosylation, covalent incorporation of flavins or pyridoxyl phosphate or phosphorylation of particular activation sites of kinases), tandem mass spectrometry can be utilized to verify protein size and the nature of the modification [15, 16].

Most importantly, enzymatic activity of the various purification fractions should always be assessed: Does the recombinant enzyme protein (by SDS gel electrophoresis) follow the enzymatic activity readout? Does increased enzymatic specific activity follow increased purity? Does the expression and purification of an "irrelevant protein control" (using the same expression cell system and purification process) yield negligible activity?

The final enzyme preparation (and enzyme activity) should be stable during storage. It should be noted that the buffer composition required for stability will probably be different to the buffer composition for the actual assay reaction. The concentration of salt, reducing reagents and detergents may be decreased in the reaction buffer for optimum activity and detection in the assay format (see the following examples for specifics). Generally, the considerations of the composition of the final storage buffer follow the same considerations for the purification buffers, with the final protein concentrations ranging between 0.1 and 5 mg mL^{-1}. For all enzyme preparations, stability to freeze–thaw procedures should be assessed. A common matrix of freeze–thaw tests include: ± detergents at –80 °C, ± 10% glycerol at –80 °C, ± detergents and 10% glycerol at –80 °C and ± detergents and 50% glycerol at –20 °C.

The quantity of enzyme (and the corresponding volume of production cell culture) required to source the screen depends on (1) the expression and purification yield, (2) the turnover number of the enzyme, (3) the concentration of enzyme required to yield a functional readout in the assay, (4) the number of compounds and the number of wells to be screened, (5) the reaction volume per well and (6) an estimate of "overage" (i.e. the amount of extra volume required for liquid handling). The scale of the cell production and purification process should allow the yield of enough enzyme to source the number of compounds in the screen (typically 250,000–1,000,000 compounds in 250,000–1,000,000 wells, with reaction volumes between 5 and 70 µL and enzyme concentrations of 1–50 nM). Our experience indicates that 10–100 mg of enzyme (from 10 to 50 L of production cells) is required to source a typical screen. The cell production and purification process should be scaled sufficiently to allow the yield of enough enzyme in a minimum number of "batches" (typically 1–5 "batches"). All "batches" should be assessed for purity and activity, pooled appropriately and the pool reassessed for purity and activity. The final enzyme preparation should be aliquoted appropriately to allow efficient operation of the screen, with minimal manipulations, dilutions and transfers.

5.3
Peptidases

Enzymes that hydrolyze peptide bonds (peptidases) are numerous and perhaps the most extensively studied class of enzyme. This abundance of knowledge [17] provides for straightforward guidelines as to the expression, purification and HTS of known peptidases. In addition, the wealth of commercially available substrates and inhibitors for the known peptidases, combined with some educated guessing and a little bit of luck, can often quickly allow the characterization and assay development necessary to pursue novel peptidases as drug targets.

5.3.1
Application of Fluorogenic Substrates to Configure Peptidase Screens

The use of fluorogenic substrates to measure peptidase activity dates back almost 30 years [18] and has been well reviewed by Knight [19]. These substrates include those based on peptidyl derivatives of fluorescent amines, such as the 4-methyl-7-coumarylamides, and peptides containing a fluorescent donor that is quenched by an acceptor chromophore via resonance energy transfer. Examples of the latter include substrates with the Dabcyl–Edans [20] or the 2,4-dinitrophenyl-7-methoxycoumarin [21] acceptor–donor pairs. Identification of a fluorogenic substrate for a peptidase of interest is the most logical starting point because assays configured with these substrates are, in general, sufficiently sensitive and robust that they can be readily adapted to HTS.

The 4-methyl-7-coumarylamides generate a useful change in fluorescence when the amide bond connecting the peptide to the fluorophore is hydrolyzed to release free 7-amino-4-methylcoumarin (AMC). Both the methylcoumarinylamide in the substrates and the AMC product are highly fluorescent, but the excitation and emission of AMC are red shifted relative to the substrates, making it possible to detect low concentrations of AMC (10 nM) in the presence of high concentrations of a given 4-methyl-7-coumarylamide substrate (10 µM), provided that proper filters are installed on the fluorimeter. It is possible to assemble diverse panels of 4-methyl-7-coumarylamide substrates from commercial vendors and screening a peptidase of interest against that panel to identify a suitably active substrate is straightforward; one simply measures AMC production upon mixing the substrate with the peptidase and the activities against all of the substrates can be quantified by reference to a single AMC calibration curve for the instrument employed.

Figure 5.2 shows an example of the application of 4-methyl-7-coumarylamides to a novel peptidase that bioinformatic analysis indicated was a serine protease structurally related to trypsin (EC 3.4.21.–) and with similar specificity. Initially a large panel of 4-methyl-7-coumarylamides containing P1 Lys or Arg residues was tested and 10 of them showed readily detectable activity with the enzyme. However, the enzyme that had been expressed contained a presequence that needed to be removed in order for the enzyme to be active (i.e. we had expressed the zymogen form) and this was accomplished by treatment with enterokinase. The caveat then

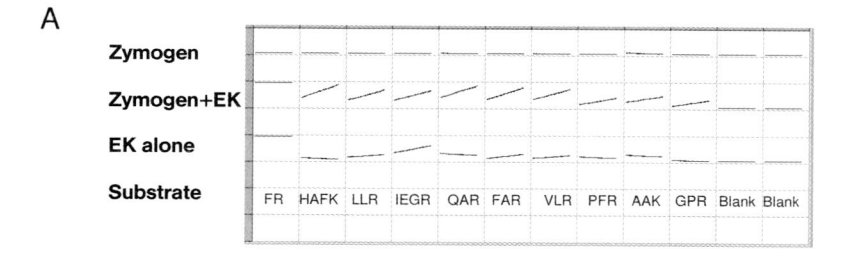

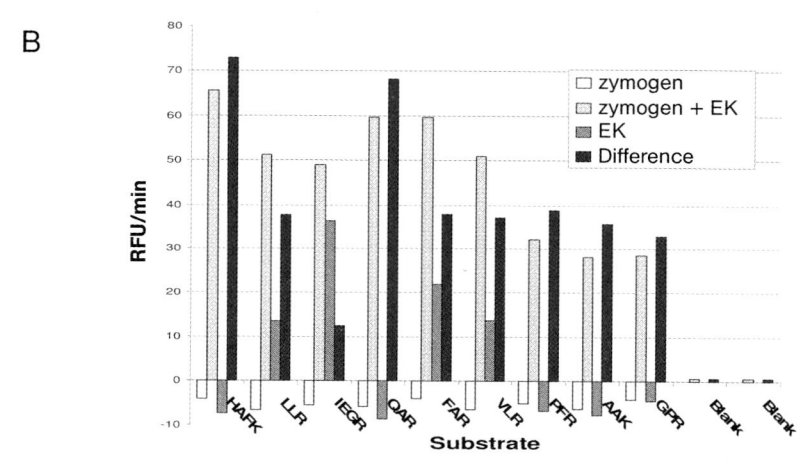

Figure 5.2 Screening a set of 4-methyl-7-coumarylamides to identify substrates of a novel trypsin-like protease. (a) Progress curves of AMC production recorded on a BMG Fluostar plate reader. The different peptides in each column are indicated by the single-letter amino acid code. The top row shows progress curves for the inactive zymogen form of the enzyme of interest at 120 ng per well. The bottom row shows progress curves for enterokinase at 0.6 ng per well. The middle row shows progress curves for the zymogen preincubated with 1:200 (w/w) enterokinase overnight before the assay. (B) Bar graph summarizing the data in (A). The bar height corresponds to the slope of the progress curves derived by linear regression of fluorescence on time.

with the initial data was the inability to distinguish whether or not a particular substrate was being cleaved by the enzyme of interest or the residual enterokinase in the sample. Figure 5.2A shows a series of progress curves (fluorescence plotted against time) collected in a 96-well plate for the 10 different substrates with the zymogen (row A), enterokinase (row E) or the enterokinase-treated zymogen (row C = the activated enzyme of interest). The slopes of the progress curves are plotted in Fig. 5.2B and it is clear that certain of the 4-methyl-7-coumarylamides (IEGR, FLR, LLR and VLR) are substrates for enterokinase, whereas the others are specific for the enterokinase-activated) enzyme of interest.

There are many peptidases whose activity against a peptide substrate requires interactions of the enzyme with amino acid residues in the substrate on both sides of the peptide bond that is being cleaved. For these enzymes and for carboxypeptidases that require free C-termini in their substrates, the 4-methyl-7-coumarylamides will not work. However, the use of quenched fluorescent peptide substrates can overcome these limitations.

Figure 5.3 shows two fluorogenic substrates for the carboxypeptidase ACE2 (an angiotensin-converting enzyme homolog) (EC 3.4.15.1) that are based on the 2,4-dinitrophenyl–7-methoxycoumarin acceptor–donor pair. Initial characterization of the cleavage of various biological peptides by ACE2 using mass spectrometry [22] identified its carboxypeptidase activity and indicated that a proline residue in P1 and that a bulky hydrophobic residue, leucine or phenylalanine, in P1' were tolerated. The larger of the two substrates in Fig. 5.3 was initially prepared as a caspase I (EC 3.4.22.36) substrate [23] and is commercially available. We hypothesized that it might be a substrate for ACE2 because it presents a free C-terminus, a proline in P1 and a bulky residue in P1'. The latter is a lysine residue modified on

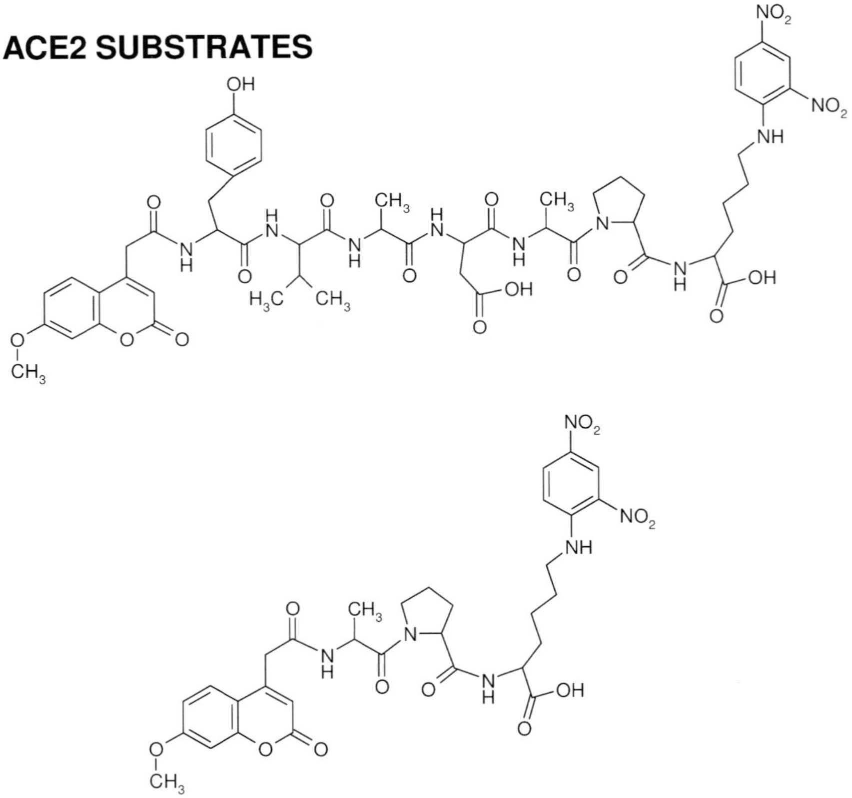

Figure 5.3 Two quenched fluorescent peptide substrates each containing the 2,4-dinitrophenyl–7-methoxycoumarin acceptor–donor pair at their C- and N-termini, respectively. Both were found to have utility for assaying activity of the carboxypeptidase ACE2. The smaller of the two provided for greater sensitivity and specificity as described in the text.

its side-chain with the 2,4-dinitrophenyl quencher. This substrate was tested and was found to be hydrolyzed by ACE2 [12]. The sight of peptide bond hydrolysis between the Pro and Lys(Dnp) was confirmed by mass spectrometry and the fluorescence signal was calibrated by subjecting a fixed amount of substrate to complete hydrolysis by the enzyme.

The smaller of the two ACE2 substrates in Fig. 5.3 was designed with the goal of increasing the assay sensitivity and thus minimizing the amount of enzyme required for HTS. As is apparent, the smaller substrate is just the larger one minus the four N-terminal amino acid residues and the two are identical in the region immediately adjacent to the peptide bond that ACE2 cleaves. The simple thought behind making this substrate was to decrease the distance between the donor and the acceptor so that the quenching would be more efficient resulting in a greater fluorescence increase upon hydrolysis. Complete hydrolysis of 40 µM of the smaller substrate by ACE2 resulted in a 300-fold increase in fluorescence over background. Complete hydrolysis of the larger substrate by ACE2 under identical conditions yielded a fluorescence increase of only 21-fold over background, although this substrate was still of considerable utility in early work on the characterization of the enzyme [12]. A fortuitous finding was that the turnover number of the smaller substrate was ~10 times greater than that of the larger substrate and, together with the increase in sensitivity, this translated into significant savings by lowering the enzyme consumption in HTS. In addition, the smaller substrate was selective enough for ACE2 that it found utility in measuring the enzyme in crude biological samples [24].

The application of quenched fluorescent peptide substrates is more complex than that of the 4-methyl-7-coumarylamides because the change in fluorescence that will be observed upon hydrolysis of a peptide bond between a given donor and acceptor depends on the structure of the peptide. The primary reason for this is that the quenching (by fluorescence energy transfer) depends on both the distance between and the relative orientations of the donor and acceptor [25]. Hence different peptides containing the same donor–acceptor pair may differ considerably in sensitivity (i.e. the magnitude of the fluorescent signal generated upon hydrolysis of the substrate), as shown in the example above. As was mentioned, a single AMC calibration curve is sufficient to quantify activity measurements for all 4-methyl-7-coumarylamide substrates. Conversely, each quenched fluorescent peptide substrate and the corresponding product containing the fluorescent donor are unique, so that conversion of the measured fluorescence changes into concentration of product requires that a different instrument calibration be applied to each individual substrate. This situation complicates the application of these types of substrates to discern substrate structure–activity relationships. Nevertheless, when screening a panel of quenched fluorescent peptide substrates to identify one that will be suitable for HTS, the primary goal is to identify those that will provide for the greatest assay sensitivity (i.e. that give the largest fluorescence change in the least amount of time with the least amount of enzyme). For a given donor, the optimal instrument settings to detect donor fluorescence are the same, so the screening can be done in a single plate.

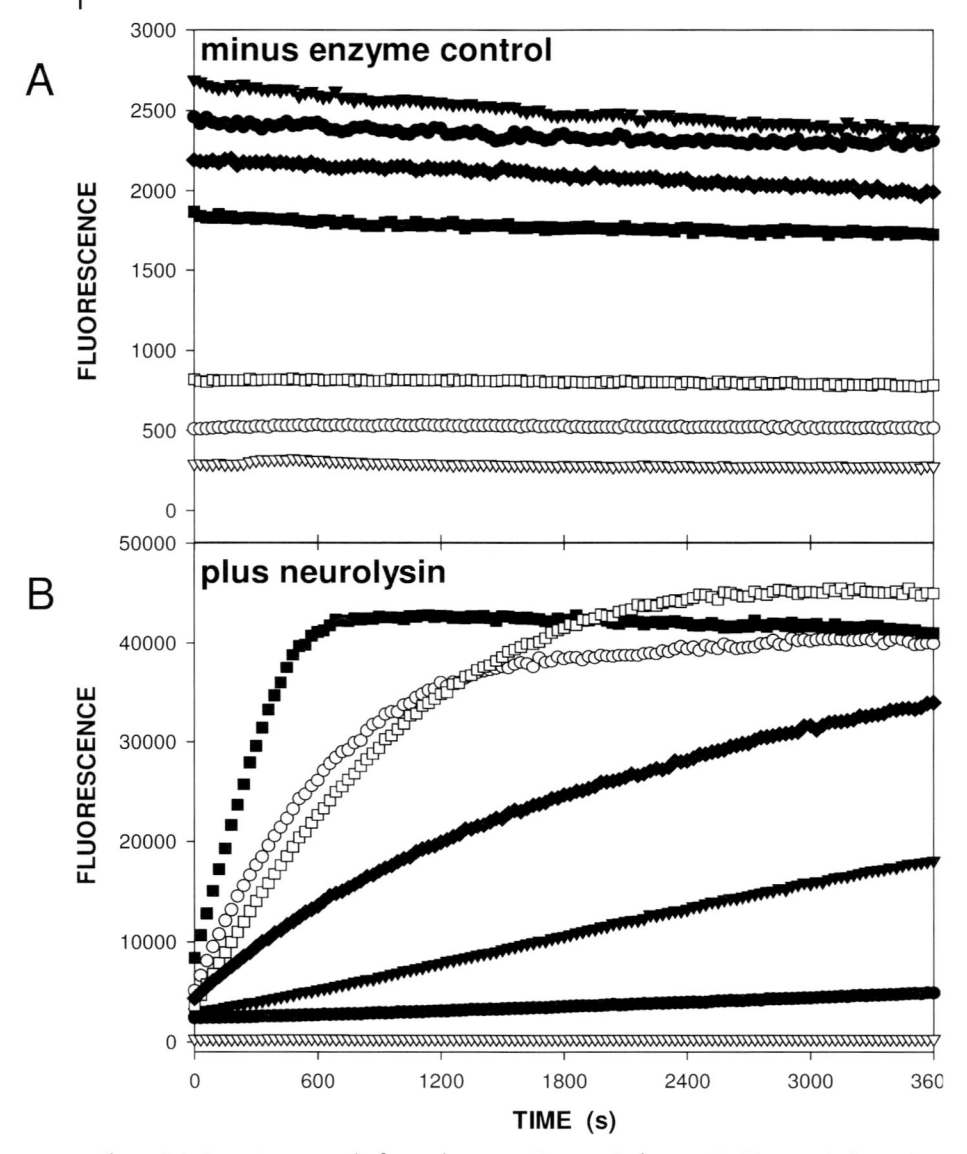

Figure 5.4 Screening a panel of peptides containing the Dabcyl–Edans donor–acceptor pair to identify substrates of the metalloprotease neurolysin. The substrates were purchased from Bachem: ■, M-2060; ○, M-1900; □, M-2120; ◆, M-2155; ▼, M-1940; ●, M-1865; ▽, M-2050. (A) The fluorescence (background) of the substrates was monitored for 1 h in the absence of enzyme. (B) Progress curves for the hydrolysis of the substrates by neurolysin.

Figure 5.4 shows an example of screening a panel of peptide substrates (7–11 amino acid residues in length) each containing the Dabcyl–Edans acceptor–donor pair. This panel was assembled from a commercial vendor [26]. The enzyme of interest in this example was a well-characterized metalloproteinase, neurolysin (EC 3.4.24.16). Figure 5.4A shows progress curves for the seven substrates at 50 µM concentration incubated in the absence of enzyme. The background fluorescence of the different substrates exhibits substantial variation, primarily reflecting the different efficiencies of quenching of the Edans fluorescence by the Dabcyl acceptor. Nevertheless, this background is stable over time. Figure 5.4B shows progress curves for the same substrates incubated with a relatively high concentration (100 nM) of neurolysin. A broad spectrum of activities for the different substrates was observed from virtually zero for M-2050 to very robust for M-2060, which was completely hydrolyzed in about 10 min. This initial screening for substrates which included an additional seven (not shown) provided several options and the choice of the "best" substrate for HTS was based as much on cost of production as it was on sensitivity and kinetic parameters.

5.3.2
The Value of Continuous Assays

The greatest advantage of applying fluorogenic peptidase substrates in HTS efforts is that they are adaptable to continuous assays and thus facilitate the collection of reaction progress curves. This statement seems counterintuitive because practical considerations (time and cost) preclude the ability to collect progress curve data for primary screening. In general, the data from a primary screen will be an end-point measurement. However, continuous assays and the progress curves that they afford are invaluable tools for what comes before and after HTS. Leading up to HTS, the configuration of an HTS assay will typically include optimization of the enzyme reaction buffer components and assessment of the enzyme«s tolerance to the test compound vehicle, most often DMSO. Continuous assays simultaneously provide a measurement of activity and enzyme stability and thus allow one to optimize the assay with respect to both parameters. This will insure that the endpoint measurement employed in the final HTS protocol is the best possible estimate of the initial reaction velocity (see discussion in [27]). Assay artifacts from test compounds that have intrinsic fluorescence or that absorb strongly at the excitation or emission wavelengths employed inevitably result in false positives. These problem compounds can be quickly identified by counter screening in a continuous assay. Reversible versus irreversible inhibition is also readily distinguished by analysis of reaction progress curves. Wherever possible, continuous assays should be developed for peptidases or any enzyme of interest in a drug discovery process.

5.4
Oxidoreductases

The oxidoreductase (EC 1.–.–.–) enzyme superfamily contains a vast number of enzyme subfamilies that carry out a diverse set of reactions, all involving the

1   $AH_2 + B \rightleftarrows A + BH_2$

2   $AH_2 \rightleftarrows A^{--} + 2H^+$
    $A^{--} + B \rightleftarrows A + B^{--}$
    $B^{--} + 2H^+ \rightleftarrows BH_2$

3   $AH_2 \rightleftarrows AH^- + H^+$
    $AH^- + B \rightleftarrows A + BH^-$
    $BH^- + H^+ \rightleftarrows BH_2$

Figure 5.5 Schematic representation of reaction carried out by oxidoreductase enzymes. Reaction (1) shows the complete, generally reversible, reaction where a reduced donor (A) transfers reducing equivalents (electrons and hydrogen atoms) to an oxidized acceptor (B). Reaction (2) shows the redox reaction where direct electron transfer occurs between A and B. Reaction (3) shows the redox reaction where hydrogen atoms are transferred in series between A and B.

oxidation or reduction of molecules via the transfer of electrons, hydrogen or hydride atoms (see [28] for general mechanism discussion) (Fig. 5.5). The members of this superfamily are classified by what molecules they use as electron donors and acceptors and what cofactors, such as metals, flavins, quinones and nicotinamide adenine dinucleotide (phosphate) [NAD(P)], they employ in their reaction scheme [29, 30]. While a variety of methods exist for tracking the activity of oxidoreductase enzymes, the bulk of methods that are amenable to HTS involve spectrophotometric-, fluorometric- or luminescence-based technologies. With the large number of distinct enzymes contained in this superfamily (>500), this section will focus on methods for three general classes: NAD(P)-dependent enzymes, non-NAD(P) cofactor-dependent enzymes and oxygen-utilizing enzymes.

5.4.1
NAD(P)-dependent Oxidoreductases

These enzymes use the cofactor NAD or NADP to shuttle electrons and hydrogen atoms to and from another substrate molecule serving as either an oxidizer (NAD^+/$NADP^+$) or as a reducer (NADH or NADPH). An example of this class of oxidoreductase is NADPH-dependent *trans*-2-enoyl coenzyme A reductase (EC 1.3.1.8) [31]. This enzyme, part of the chain elongation pathway for fatty acids, catalyzes the addition of the double bond at the 2-position of long-chain (>4 carbon atom acyl chains) fatty acyl coenzyme A molecules using NADPH as the reducing agent. The activity of this enzyme can be monitored directly as NADPH has significant absorbance at 340 nm ($\varepsilon \sim 6220$ cm^{-1} M^{-1}) and fluorescence near 460 nm when excited at 340 nm. As a side note, NADH can be monitored similarly owing to its absorbance maxima at 340 nm ($\varepsilon \sim 6290$ cm^{-1} M^{-1}) and fluorescence emission maximum near 460 nm. Monitoring NADPH consumption using one of these methods during assay development is convenient as both end-point and continuous measurements can be made. Unfortunately, NADPH stability and sensitivity can be strongly dependent on the buffer and reaction conditions. Furthermore, the near-UV wavelengths needed for spectrofluorimetric detection of NADPH and its relatively low quantum yield can make HTS assays difficult to interpret as many

5.4 Oxidoreductases

A

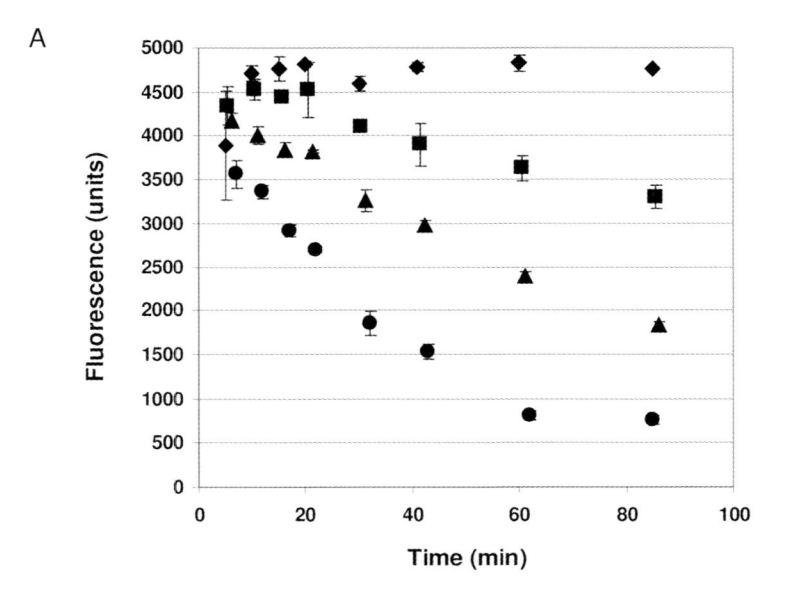

[Graph: Fluorescence (units) vs Time (min), showing data points with symbols ■, ▲, ●, ♦]

B

Primary Enzyme Reaction

Trans-2-decenoyl-CoA → Decanoyl-CoA
+   +
NADPH   *Trans-2-enoyl-CoA Reductase*   NADP⁺

Coupling (STOP) Enzyme Reaction

Resazurin → Resorufin
+   *Diaphorase*   +
NADPH   NADP⁺

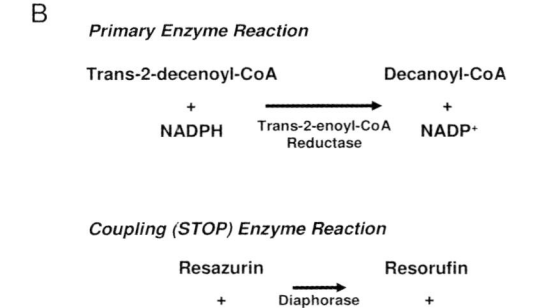

Figure 5.6 NADPH-based coupling assay for *trans*-2-enoyl coenzyme A reductase activity. (A) A typical reaction assayed in triplicate of a 70 μL reaction containing 50 mM Na phosphate (pH 7.0), 5 mM EDTA, 0.01% Tween-20, 15 μM NADPH, 15 μM *trans*-2-decenoyl coenzyme A with (■) 80, (▲) 160 or (●) 320 ng of reductase enzyme or (♦) an irrelevant control. At the indicated time point on the graph, a 30 μL solution of 0.04 units mL^{-1} diaphorase and 40 μM resazurin is added to convert remaining NADPH rapidly to NADP⁺ and generate a resorufin signal. One unit is defined as the amount of enzyme necessary to oxidize 1 μmol of NADPH per minute. Resorufin product is measured by the increase in fluorescence emission at 590 nm after excitation at 544 nm. This signal reports the amount of NADPH remaining after the primary reductase enzyme reaction at the indicated time points. (B) Schematic depicting the primary enzyme reaction (*trans*-2-enoyl coenzyme A reductase) and the coupling (or stopping) enzyme reaction (diaphorase). The reaction is considered "stopped" owing to the complete consumption of NADPH by diaphorase.

compound libraries are filled with colored or UV-absorbing molecules. To avoid these difficulties, the amount of NADPH or NADH remaining in the reaction can be quantified using enzyme coupling systems utilizing the commercially available enzymes diaphorase (EC 1.6.99.2) [32–34] or FMN reductase (EC 1.5.1.29)–luciferase (EC 1.14.14.3) [35, 36]. Diaphorase can use either NADH or NADPH as a reductant along with several dye molecules as electron acceptors, including resazurin (RZN), 2,6-Dichlorophenolindophenol (DCPIP) and various tetrazolium molecules (MTS and MTT). These dyes display either a significant shift in their absorbance spectra (DCPIP and tetrazolium dyes) or in their fluorescence spectrum [RZN is converted to the highly fluorescent resorufin (RFN); excitation ~560 nm, emission ~590 nm]. The bacterial luciferase system utilizes FMN and FMN reductase to accept electrons from NADH in the presence of a long-chain aldehyde and oxygen to emit light. Using a large excess of diaphorase and limiting concentration of RZN, a high-throughput end-point assay was developed for *trans*-2-enoyl coenzyme A reductase that was sensitive, stable, linear with respect to enzyme concentration and capable of detecting enzyme inhibition (Fig. 5.6).

5.4.2
Non-NAD(P) Cofactor-dependent Oxidoreductases

Several members of the oxidoreductase family rely on cofactors other than NAD(P) for the ability to transfer reducing equivalents to a separate oxidized acceptor molecule. These cofactors include metal ions (iron, zinc, copper), flavin mononucleotide (FMN), flavin adenine dinucleotide (FAD), quinone (Q) and pyrroloquinoline quinone (PQQ) [37]. The final acceptor molecule is often another electron transport protein or cofactor and sometimes O_2. Although some of these cofactors permit spectrophotometric detection of the enzyme in solution (FMN or FAD), their absorbance coefficients are too small for the development of sensitive enzymatic activity assays where turnover of substrate can be monitored over time. Fortunately, many enzymes using these cofactors can be monitored using surrogate chemical electron acceptors such as phenazine ethosulfate (PES), phenazine methosulfate (PMS) [38] or ferricenium hexafluorophosphate (FC) [37]. Whereas reduction of FC can be monitored directly by changes in absorbance at 300 nm, reduced PEM and PMS must pass on their extra reducing equivalents to a terminal electron-accepting molecule, usually some kind of spectrophotometric (DCPIP, MTS, MTT or cytochrome *c*) or fluorescent (RZN) dye. Using a chemically coupled PES–RZN system, a spectrofluorimetric assay was developed for choline dehydrogenase (EC 1.1.99.1) enzyme activity [39, 40] (Fig. 5.7). The assay displayed excellent substrate and enzyme dependence and linearity over time. Furthermore, it was shown that a choline analog known to suppress activity from the enzyme, dimethylethylamine, displayed inhibitor activity in the assay [41]. For reasons that are not known, not all non-NAD(P) cofactor-dependent oxidoreductases can be successfully chemically coupled to a spectrofluorimetric readout, so it is imperative that each coupling system be implicitly and rigorously tested.

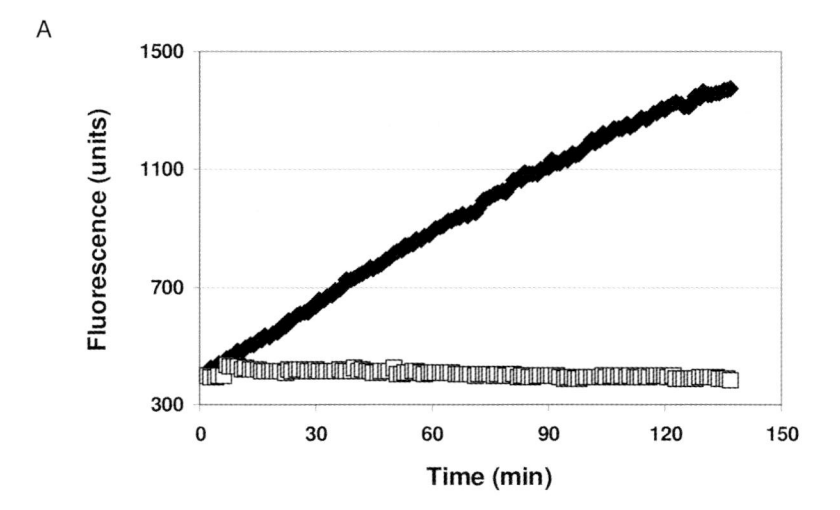

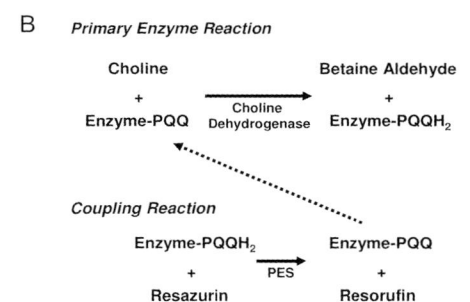

Figure 5.7 Phenazine ethosulfate (PES)-based coupling assay for choline dehydrogenase activity. (A) A typical reaction assayed in triplicate of a 70 µL reaction containing 50 mM Na phosphate (pH 7.5), 10 mM NaCl, 0.01% Tween-20, 300 µM PES, 100 µM resazurin with 1.25 µg of partially purified choline dehydrogenase, (◆) with and (□) without 15 mM choline chloride. Resorufin product is measured by the increase in fluorescence emission at 590 nm after excitation at 544 nm. Choline dehydrogenase was partially purified from a cellular extract by hydrophobic interaction chromatography using phenyl Sepharose. (B) Schematic depicting the primary enzyme reaction (choline dehydrogenase) along with the chemical coupling reaction (PES and resazurin).

5.4.3
Oxidases or Oxygen-utilizing Oxidoreductases

Oxidases are enzymes that use O_2 as the terminal electron/hydrogen acceptor to generate hydrated and reactive species such as H_2O_2 and OH^-. These enzymes can involve all the cofactors mentioned here, but O_2 is the ultimate acceptor molecule. For these enzymes, high-throughput assay methods have focused on the ability to monitor the production of H_2O_2 spectrofluorimetrically using enzymatic coupling systems. The primary enzyme for these coupling systems is horseradish peroxidase

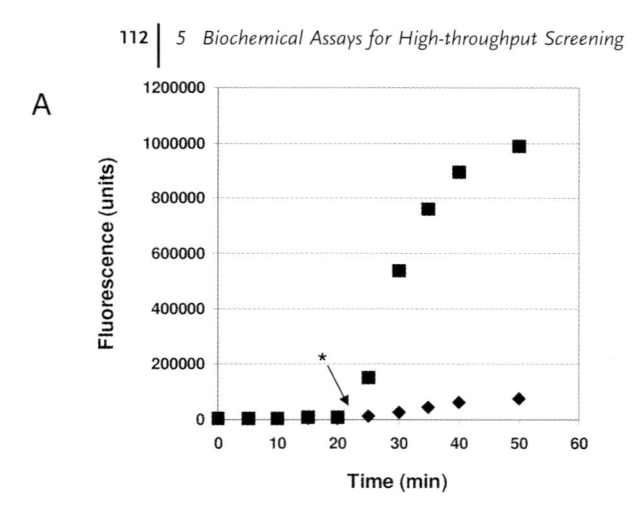

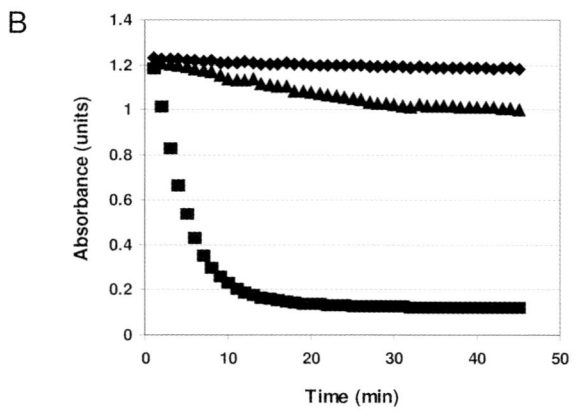

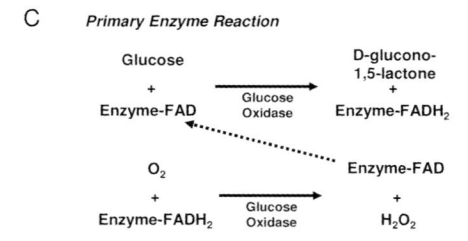

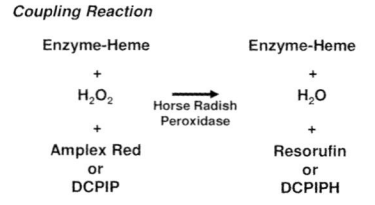

Figure 5.8 Horseradish peroxidase (HRP)-based coupling systems for detection of glucose oxidase activity. (A) A typical reaction assayed in triplicate of a 70 µL reaction containing 20 mM HEPES (pH 7.5), 10 µM Amplex Red, 7 units mL^{-1} glucose oxidase and 0.01 units mL^{-1} HRP, (■) with and (◆) without 100 µM glucose. HRP is added when indicated (*). One unit is defined as the amount of enzyme necessary to oxidize 1 µmol of glucose per minute. Resorufin product is measured by the increase in fluorescence emission at 590 nm after excitation at 544 nm. (B) A typical reaction assayed in triplicate of a 70 µL reaction containing 20 mM potassium phosphate (pH 7.1), 200 µM DCPIP, 0.01 units mL^{-1} HRP with (■) 5 units mL^{-1} glucose oxidase, (△) no glucose oxidase and (◆) buffer control. One unit is defined as the amount of enzyme necessary to oxidize 1 µmol of glucose or H$_2$O$_2$ per minute for glucose oxidase and HRP, respectively. Resorufin product is measured by the increase in fluorescence emission at 590 nm after excitation at 544 nm. DCPIPH product is measured by the decrease in absorbance at 600 nm. (C) Schematic depicting the primary enzyme reaction (glucose oxidase) along with the enzymatic coupling reaction (HRP and Amplex Red or DCPIP).

(HRP) (EC 1.11.1.7), which uses a reduced donor molecule to convert H_2O_2 to H_2O along with generation of the oxidized donor product [42]. The reduced donor molecule that is employed for most coupling reactions is a spectrofluorimetrically active dye molecule so the output is absorbance, fluorescence or luminescence. Examples of such dyes are Amplex Red® (Molecular Probes), Quanta Blue® (Pierce Chemical), DCPIP, tetramethylbenzidine (TMB), 4-aminoantipyrine (AAP), 3-(*p*-hydroxyphenyl)propionic acid (HPPA) and 2,2'-azobis-3-ethylbenzothiazoline-6-sulfonic acid (ABTS). Under certain conditions, some of these dye molecules can react directly with the oxidase-produced H_2O_2 or OH^-, but the activity of HRP can increase the efficiency and sensitivity of the assay [43, 44]. Several examples of enzyme assays involving this technology include glucose oxidase (EC 1.1.3.4), D-amino acid oxidase (EC 1.4.3.2), L-amino acid oxidase (EC 1.4.3.3) and pyruvate oxidase (EC 1.2.3.3). In the case of these enzymes, specifically glucose oxidase and pyruvate oxidase, a spectrophotometric assay using DCPIP and a fluorimetric assay using Amplex Red can be used where HRP is acting on the H_2O_2 generated from the primary oxidase activity [45, 46]. Although some activity can be detected from the oxidase alone using Amplex Red, the activity of HRP serves to increase greatly both the sensitivity and signal window of the assay (Fig. 5.8).

5.4.4
General Considerations

Although the examples presented here suggest that this broad enzyme family is readily assayable, there are a number of issues that must be considered. First, many of these assay systems can be sensitive to the buffer system utilized and the environment surrounding the reaction. Some of the dyes can show spectral drift over time due to very low oxidation or reduction rates. Reducing agents in the buffer required for enzyme activity and stability can cause nonspecific reactions in some detection systems. Some coupling assay enzymes may have high substrate-independent background rates. Second, if the ultimate goal of an assay development project for an oxidoreductase is to screen a compound library for inhibitors, compounds which have chemical redox capabilities will cause spurious results in the assay. It is important to counter-screen inhibitory compounds in control assays to ensure that the compound itself is not responsible for the detected signal changes in the original assay. Third, for many members of the oxidoreductase family, substrate identification and selection for assay are an important step in the assay development process. Many oxidoreductases can utilize multiple substrates with vastly different kinetic properties which may influence detection methods and signal windows. For example, aldoketo reductase subfamily members (EC 1.1.–.– and EC 1.2.–.– members) utilize NADPH to reduce an incredibly broad spectrum of aldose-, aldehyde- and carbonyl-containing substrates [47]. A single enzyme may be able to utilize several dozen distinct substrates. It is necessary to develop, screen and characterize a panel of substrates to determine which will give the assay properties required for a successful high-throughput screen.

The methods and examples described in this section demonstrate that oxidoreductases play two major roles in the pharmaceutical industry. First, they have been and continue to be identified as relevant drug targets in various diseases. Second, oxidoreductases are also an important biochemical tool that scientists can harness for the development of enzyme-based coupling systems for assaying novel disease targets. Developing an assay for an enzyme that has no direct output can be challenging, but oxidoreductases may serve the role as coupling reagent if they can utilize a substrate or product of the primary reaction.

5.5
Transferases, Synthetases and Lipid-modifying Enzymes

In general, screening of transferases, synthetases or enzymes involved in lipid metabolic reactions is limited by using the assay technologies that are amenable to HTS. Several technical hurdles exist in using plate-based formats for these targets. Most are related to the detection of the reaction products. Here, we will focus upon the utilization of selective capture chemistries of radiolabeled substrates or products of these enzymatic reactions. Capture chemistry can be subdivided into three basic formats: affinity, ionic and hydrophobic interactions. In this section, the utility of each of these interactions will be described for the assay development of transferases, synthetases and lipid-metabolizing enzymes.

Affinity capture relies on the ability of molecules containing site-specific tags to be bound by a complementary molecule. For example, one of the most widely used binding pairs employ a biotin tagged molecule and streptavidin as the capturing agent and will be used as an illustration discussed below. Other types include affinity binding partners such as surface-affixed nickel ions (or copper ions) and a 6-histidine tagged protein, protein A capture of an antigen specific IgG or other complementary antibody–antigen-based formats and, finally, lectin-based capture of complex carbohydrates.

5.5.1
Streptavidin–Biotin Capture

An example of the use of the affinity binding between biotin and streptavidin was employed in one of our first efforts to develop an assay for a novel member of the human polypeptide–N-acetylgalactosaminyltransferase family (EC 2.4.1.41, ppGal-Nac-T12). This enzyme catalyzes the O-glycosylation of exogenous mucin-related proteins producing a peptide–Thr–α-1-N-acetylgalactosamine linkage [48]. The activity can be monitored using the tight binding of the streptavidin–biotin pair (illustrated in Fig. 9A). The activity of the enzyme transfers a radiolabeled group from a donor substrate to an acceptor peptide. This peptide possesses a biotin tag permitting capture by streptavidin-labeled beads. These beads and the high-affinity interaction of biotin and streptavidin are the keys to scintillation proximity assay (SPA) technology (GE Healthcare) [49]. SPA beads contain a scintillant that emits

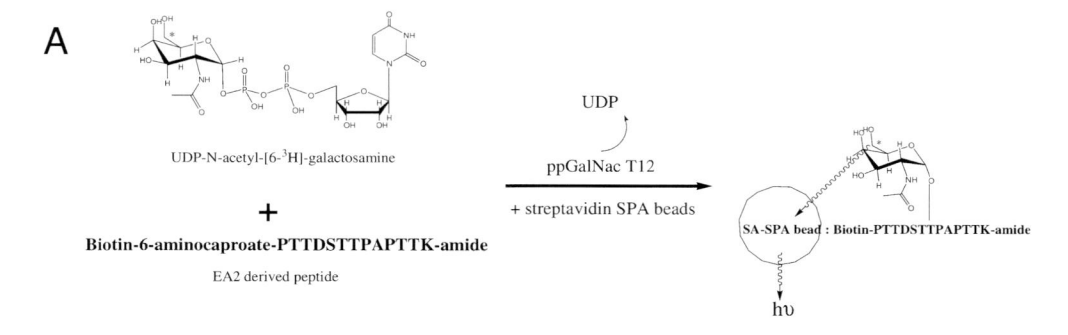

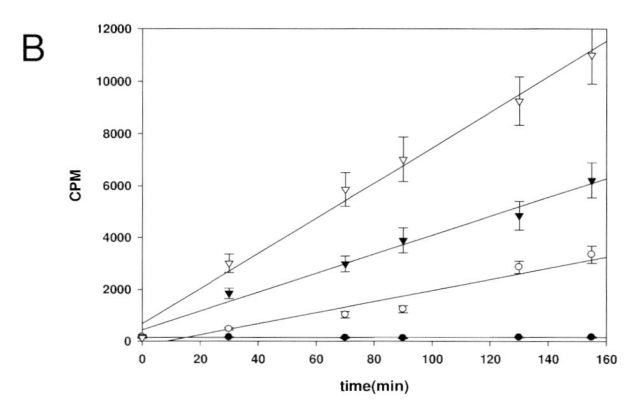

Figure 5.9 Reaction schematics and progress curves of recombinant human pp-GalNAc-T12 and CARM1. (A) The reaction scheme and capture of the product of human pp-GalNAc-T12 by using streptavidin SPA beads. (B) A typical reaction assayed in triplicate of a 50 μL reaction containing 20 mM HEPES–NaOH (pH 7.5), 10 mM MnCl$_2$, 0.1% Triton X-100, 0.05% BSA, 0.5 mM 2-ME, 0.5 μM UDP–[^{3}H] GalNac (at 300 nCi per reaction), 0.5 μM biotinylated EA2 peptide in 0.5 mL at (■) 20, (▲) 40 and (◆) 80 ng of pp-GalNAc-T12 or (○) of an irrelevant control enzyme. As a function of time 30 μL were removed from the main reaction and quenched with an equal volume of 60 mM EDTA (pH 10) containing 300 μg of streptavidin-coated SPA beads in triplicate in a 384-well white microtiter plate. Individual wells were radioassayed using a TriLux instrument. (C) The reaction scheme and capture of the product of CARM1 by using streptavidin SPA beads. (D) CARM1 activity profiles determined in 70 μL volume reactions containing 25 mM HEPES (pH 7.5), 200 mM NaCl, 4 mM EDTA, 0.6 μM biotinylated peptide, 3 μM [^{3}H]-S-adenosyl-l-methionine (50 nCi per reaction), 200 μg of SA-SPA beads and (◇) 15, (▼) 30 or (▽) 60 nM of CARM1 or (■) 60 nM of a control protein purified in the same way as CARM1. The reactions were stopped by adding a solution of 6 M CsCl to 120 μL final volume. Beads were allowed to float to the surface and radioassayed with a PerkinElmer TopCount.

light when in close proximity to a radioactive isotope and permits direct detection of the enzyme-catalyzed reaction product. In this instance, that isotope has been captured to the bead surface via the biotin–streptavidin complex. A primary advantage of the SPA format is that it is homogeneous and no separation of product from substrate is required. Thus, recombinantly derived human pp-Gal-

C

S-[Methyl-³H]-Adenosyl-L-Methionine

+

Biotin-X$_n$PRX$_n$-amide

Poly-A binding protein derived peptide

→ CARM1, SA-SPA bead →

S-Adenosyl-L-Homocysteine

+

SA-SPA bead | Biotin-X$_n$PRX$_n$-amide — CH$_3$ *

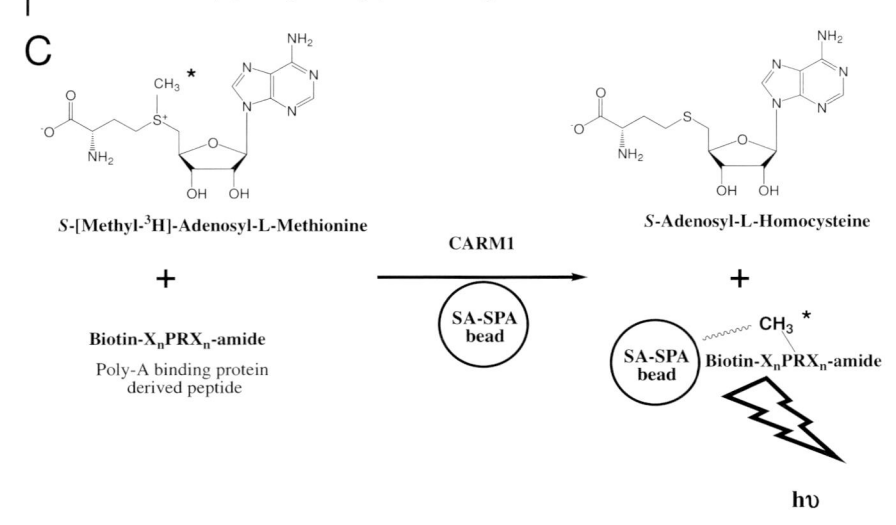

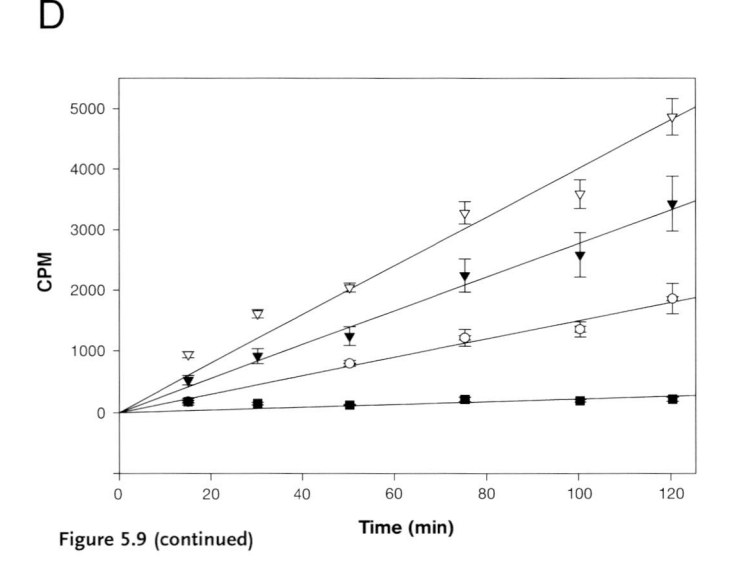

hυ

D

Figure 5.9 (continued)

NAc-T12 transfers [³H]GalNAc from UDP–[³H]GalNac to a biotin labeled mucin-derived peptide from EA2 and product formation could be easily assessed by using an SPA bead format (Fig. 5.9B). The assay displays significant signal-to-noise ratio and produces reaction rates that are constant and dependent upon enzyme concentration over the 150 min reaction time course.

Another example of the use of affinity capture, the activity of the protein arginine methyltransferase, CARM1, was monitored. Protein arginine methyltransferase (EC 2.1.1.125) catalyzes the transfer of the methyl group from S-adenosyl-l-methionine to the guanidino group of l-arginine within protein substrates. CARM1 was first shown to catalyze the transfer of methyl groups on to histone H3 [50]. In developing a high-throughput assay for the recombinant human CARM1, the transfer of a [^{3}H]methyl group to a biotin-labeled peptide derived from the polyadenylate-binding protein nuclear 1 (PABPN1) was measured using streptavidin SPA bead capture (Fig. 5.9C and D). This assay format is also homogeneous and produces reaction rates that are constant and dependent upon enzyme concentration over at least 120 min. Moreover, the signal-to-background ratio in this homogeneous format is exceptionally robust owing to the lower energy radioisotope employed (tritium).

5.5.2
Ionic Capture

Ionic capture is generally less discriminating than utilizing specific capture mechanisms but does have effectiveness when selectivity of capture and the appropriate signal-to-background ratio can be achieved. For instance, the use of yttrium silicate SPA beads, which tend to have a higher affinity for phosphate-containing molecules relative to carboxylic acids or other anions, can provide selectivity when either a radioactive phosphate-containing substrate or product of an enzymatic assay can be ionically captured [51–53]. Hence in terms of selectivity, this capture format affords some ability to control the capture of target molecules. Utilizing this method, the activity of human long-chain fatty-acyl-CoA synthetase 5 (LACS 5) (EC 6.2.1.3) was monitored successfully [11]. LACS 5 catalyzes the transfer of long-chain fatty acyl chains to coenzyme A with the formation of high-energy fatty-acyl–coenzyme A thioester bonds (Fig. 5.10A). By manipulating the capture conditions with detergent and a low pH, the product of the reaction, [^{3}H]palmitoyl-CoA, could be selectively trapped on YSi SPA beads and segregated from the substrate, [^{3}H]palmitate. Moreover, the reaction was constant and displayed enzyme concentration dependence over the course of 4 h (Fig. 5.10B).

5.5.3
Hydrophobic Capture

Hydrophobic interactions can be used as a selection method provided that the capturing surface can produce the appropriate signal of either the enzyme reactant or product. For this format it is essential to define the minimal required assay conditions in which the maximum hydrophobic interactions occur. Therefore, this particular format has been most successfully used for enzymes involved in lipid metabolism. For example, we can monitor the activity various isozymes of phospholipase C (PLC) (EC 3.1.4.3) by employing a hydrophobic scintillation proximity plate-based surface for the capture of the various substrates, [*myo*-inositol-2–^{3}H(N)]-phosphatidylinositides [54, 55]. As shown in Fig. 5.11A, this format is a loss of signal assay and monitors the release of aqueous soluble product, [^{3}H]inositol, from [^{3}H]phosphatidylinositol attached hydrophobically to the surface of the microtiter well. Because PLCs are processive enzymes and react between the lipid and aqueous

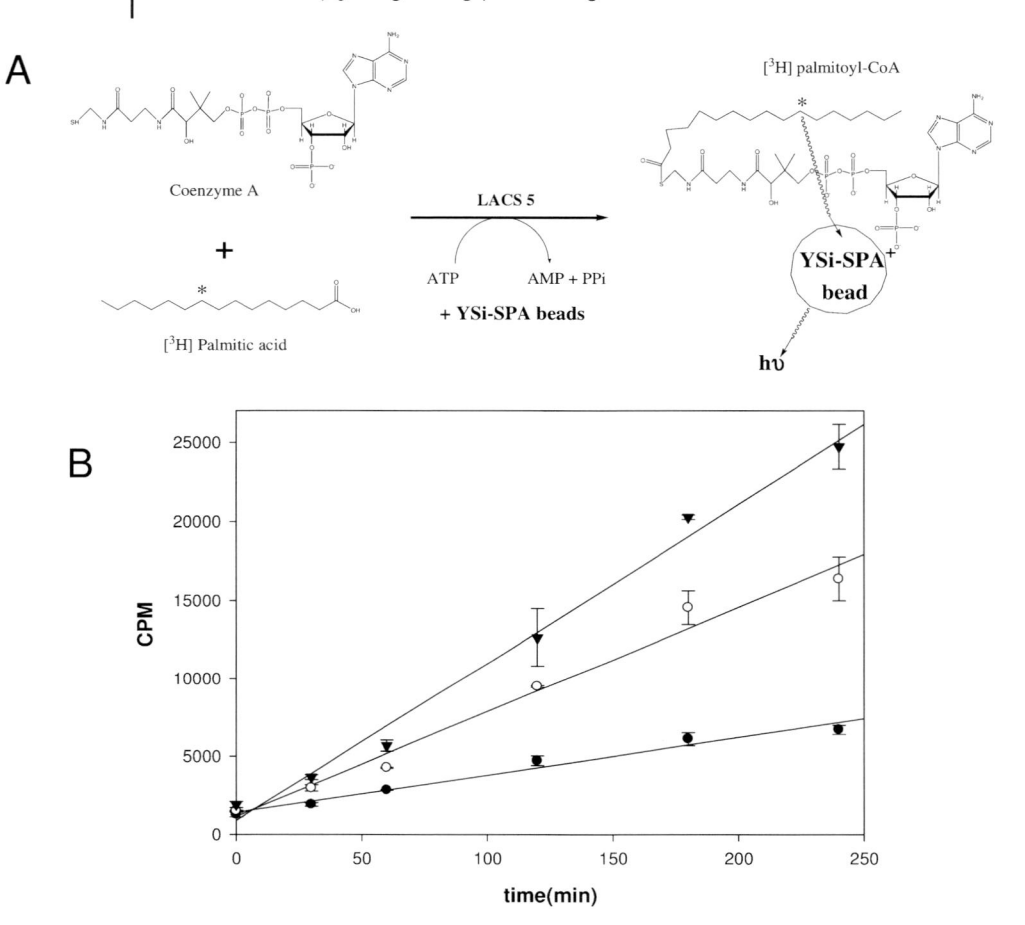

Figure 5.10 Enzyme titration and kinetics of the LACS 5 reaction. (A) The overall reaction pathway and capture of the product of the LACS 5 reaction using the phosphate ionic capture by the YSi-SPA method. (B) A representative reaction of duplicate determinations which were carried out at room temperature in 0.05 mL of 25 mM HEPES–Tris (pH 7.5), 0.125 M NaCl, 10 mM MgCl$_2$, 0.01 % BSA, 2 mM 2-ME, 5 µM [^{3}H]palmitate (at 300 nCi per well), 25 µM coenzyme A and 0.25 mM ATP and the reactions initiated by the addition of 1 µL of either (●) 50, (○) 100 or (▼) 200 ng of LACS 5. The reaction was stopped at the times indicated with 40 µL of 10 mg mL^{-1} YSi-SPA beads in 25 mM Na acetate (pH 5.6), 0.2 % Triton X-100, 0.05 % NaN$_3$. The beads were allowed to settle overnight and the CPM were monitored with a PerkinElmer TopCount.

surface interface, the reaction kinetics can be nonlinear. Nevertheless, inhibitors of the reaction can be identified through HTS campaigns, as illustrated by the observed titration effects of calcium concentration as shown in Fig. 5.11B.

The assays discussed above all utilize radiometric measurements of enzymatic reactions based on scintillation proximity effects. Microtiter plate-based assays of this type are most easily formatted on the basis of the current capture technologies

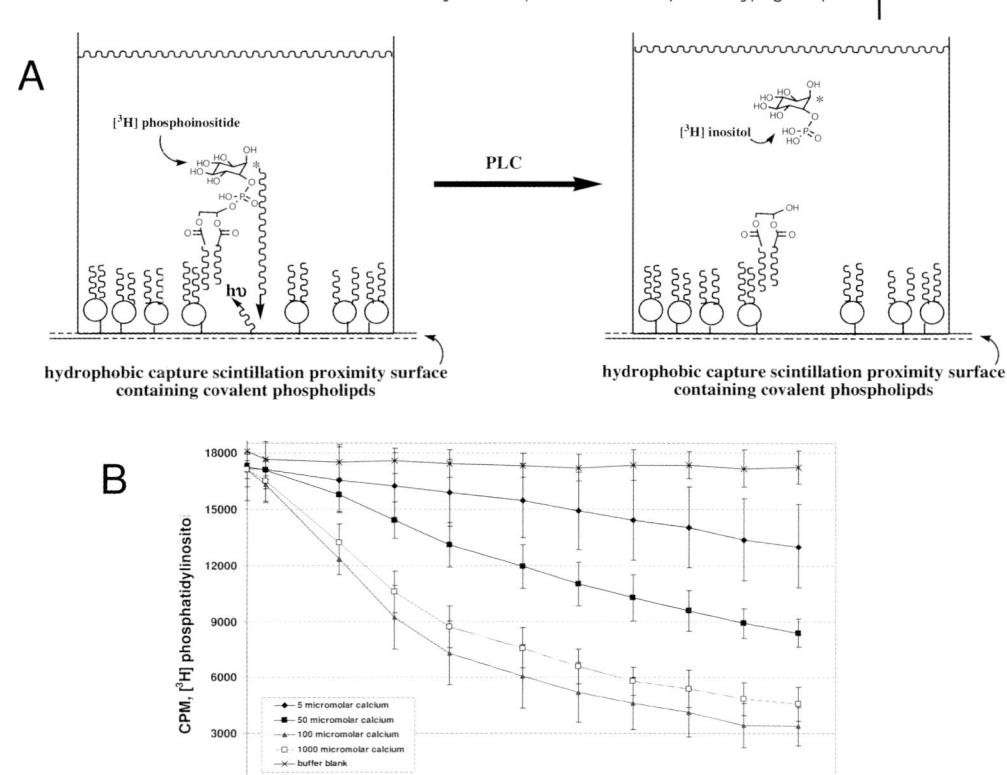

Figure 5.11 Schematic of the PLC reaction and kinetics of a typical PLC assay using a phospholipid Flashplate at various calcium concentrations. (A) A depiction of the hydrophobic capture surface on a scintillation-based format of [^{3}H]phosphatidylinositol and the PLCδ reaction. (B) A representative reaction of triplicate determinations which was carried out at room temperature in a 96-well phospholipid Flashplate at a final volume of 0.2 mL of 50 mM HEPES–Tris (pH 7.3), 0.15 M NaCl, 2 mM DTT, 0.1 mM EGTA, at various [CaCl$_2$]s, 0.01% acetylated BSA and the reaction initiated by addition of PLCδ. Titration of calcium was according to methods described previously [54] and are shown at (◆) 5, (■) 50, (▲) 100 or (□) 1000 μM Ca$^{2+}_{free}$ or (×) buffer alone. CPM were monitored with a PerkinElmer TopCount over the times indicated.

which inevitably rely upon surface chemistry for capture and separation of either the substrate or product of an enzymatic reaction from the bulk solution. Thus, separation allows for the effective utilization of the homogeneous format. This then results in the conversion of low-throughput or bench-top assay into a higher throughput format which is amenable to HTS.

5.6
Kinases

Technically classified as transferases, kinases comprise one of the largest enzyme classes in the human genome (>500 identified) [56] and represent a high percentage of targets currently entering into HTS. Kinases catalyze the transfer of the terminal phosphate group from an ATP molecule to an appropriate acceptor molecule. Different subfamilies exist where the acceptor molecule can be a wide variety of proteins or small molecules. Owing to the exquisite specificity of most kinases for their acceptor substrate, a key component in developing an HTS assay is the identification of an HTS-amenable substrate. For kinases that phosphorylate peptides and proteins, a popular approach is to screen through a peptide panel that contains peptides derived from human phosphorylation sites and randomly generated peptides to find the best surrogate substrate for screening [57]. For this purpose, the peptides are usually tagged with biotin, which is subsequently captured by either streptavidin-coated flashplates or on a streptavidin-coated membrane after phosphorylation using radiolabeled ATP. The specificity and strength of the biotin–streptavidin interaction provide a reliable, reproducible and high-affinity capture system for quantitating radiolabeled product. Once a surrogate substrate is identified, further assay optimization is done utilizing design of experiments software before it enters screening.

During the past few years, several new assay technologies have emerged for screening kinases. These offer the advantages of increased sensitivity with decreased interference from compounds and also improved signal-to-background ratios and a smaller number of assay steps. For the purpose of this discussion, we will focus on three formats (one radiometric and two fluorescent based) that we utilized for running primary screens. These assay formats are SPA, homogeneous time-resolved fluorescence (HTRF) and the pyruvate kinase (PK)–lactate dehydrogenase (LDH) coupled assay

5.6.1
Streptavidin–Biotin Capture

As described previously, SPA technology is ideal for monitoring reactions where small, radiolabeled donor groups are transferred to biotinylated acceptor molecules, such as glycol- and methyl-transferases. This technology has also been used to develop kinase-specific assays. In this specific case, the radiolabeled donor group is the γ-phosphate (^{32}P, ^{33}P) from ATP, which is transferred by the kinase to a hydroxyl group (Ser, Thr or Tyr residue) on a biotinylated peptide. This generates the phosphorylated and radiolabeled substrate, which is then captured by streptavidin-coated SPA beads. Again, this makes it a homogeneous technique since no separation of product and substrate is required. SPA kinase assays therefore show high precision and reproducibility and are easy to automate for assaying a large number of samples in a short time. SPA was used to develop a signal increase assay (product capture) for the recombinant human homolog related to casein kinase (EC 2.7.1.–) [58]. Since the

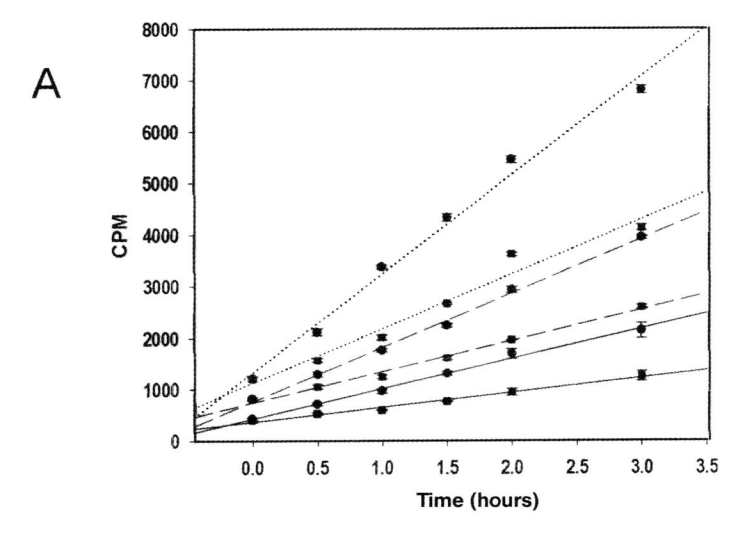

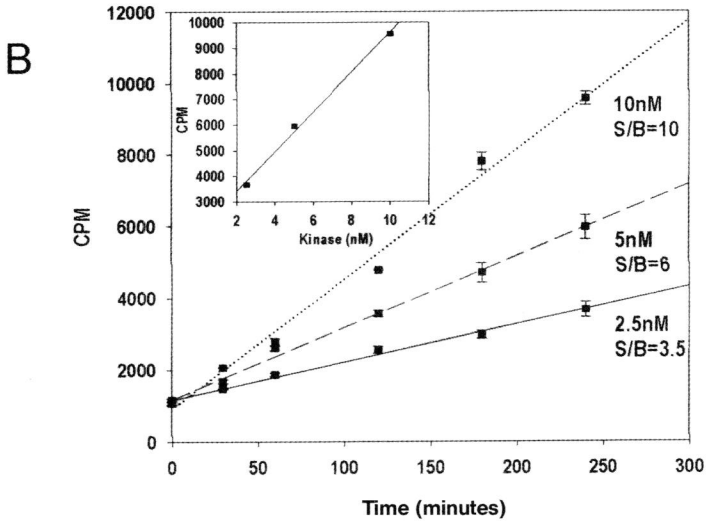

Figure 5.12 Biotin SPA capture and enzyme optimization for maximum signal window. (A) A 70 μL reaction containing 10 mM HEPES (pH 8.0), 3 mM MgCl$_2$, 5 mM DTT, 0.005% Triton X-100, 0.5 mg mL^{-1} BSA (35 μL of enzyme mix and 35 μL of substrate mix, 0.1 μCi [γ-^{33}P]ATP), 1 μM peptide and 2 μM ATP at 5 and 10 nM enzyme at room temperature. The reaction was stopped at 0, 0.5, 1, 1.5, 2 and 3 h with (—) 0.1, (---) 0.2 and (⋯) 0.3 mg per well of SPA beads, spun at 3200 rpm for 10 min, settled overnight and read on a PerkinElmer Topcount. (B) A 70 μL reaction was run in 10 mM HEPES (pH 8.0), 3 mM MgCl$_2$, 5 mM DTT, 0.005% Triton X-100, 0.5 mg mL^{-1} BSA (35 μL of enzyme mix and 35 μL of substrate mix, 0.1 μCi [γ-^{33}P]ATP), 1 μM peptide and 2 μM ATP at (—) 2.5, (---) 5 and (⋯) 10 nM enzyme at room temperature. The reaction was stopped at 0, 0.5, 1, 1.5, 2 and 3 h with 30 μL of 10 mg mL^{-1} SPA beads (0.3 mg per well), spun at 3200 rpm for 10 min, settled overnight and read on a PerkinElmer Topcount.

binding capacity on the bead surface is finite, it was important to balance the quantity of bead with the concentration of substrate to provide an assay which shows good linearity and enzyme dependence. A typical example is illustrated in Fig. 5.12.

5.6.2
Homogeneous Time-resolved Fluorescence (HTRF)

In an effort to move to nonradioactive technology platforms, TR-FRET (that takes advantage of fluorescent properties of the lanthanides) has been introduced as a tool for screening a number of tyrosine kinases. LANCE reagents (PerkinElmer) [59, 60] are based on the proximity of the Eu-labeled phosphotyrosine antibody and the streptavidin–allophycocyanin (SA–APC) brought together in the presence of the phosphorylated product of the reaction. The success of this assay not only is determined by the phosphorylation of the substrate peptide but also by the interaction of the Eu-labeled antiphosphotyrosine antibody with the phosphorylated peptide. The complex is completed when an SA–APC binds to the biotin on the substrate peptide. When the Eu label and APC protein are in close proximity, fluorescence energy transfer can occur, which is monitored spectrofluorimetrically. The long fluorescence lifetime of Eu-based fluorescence improves the signal specificity of the assay. This assay format was utilized for monitoring the activity of a novel Cdc-42-associated tyrosine kinase (EC 2.7.1.112) [61]. The activity was linear over 120 min and the assay exhibited a low baseline signal and an overall high signal-to-background ratio (Fig. 5.13).

5.6.3
Pyruvate Kinase–Lactate Dehydrogenase Assay System

Kinase reactions are generally not monitored directly by simple fluorescence. However, hydrolysis of ATP can be coupled to the formation or the disappearance of NADH. Following each cycle of ATP hydrolysis, the regeneration system consisting of phosphoenolpyruvate (PEP) and pyruvate kinase (PK) converts one molecule of PEP to pyruvate when the ADP is converted back to ATP. The pyruvate is subsequently converted to lactate by lactate dehydrogenase, resulting in the oxidation of one NADH molecule. The rate of NADH loss can be measured by absorbance or fluorescence decrease, as previously described in Section 5.4, and is proportional to the rate of steady-state ATP hydrolysis. The constant regeneration of ATP allows monitoring of the ATP hydrolysis rate over the entire course of the assay. One of the main advantages of this assay format is that it can be used to monitor the kinase reaction continuously. Another advantage is that it can be employed for developing assays for kinases that use acceptor molecules other than proteins and peptides where biotin capture methods are not feasible. This format was utilized for screening inhibitors of orphan serine–threonine kinases similar to the human doublecotin-like kinase (EC 2.7.1.–) [62]. The assay was linear for almost 1 h under the optimized buffer conditions that were compatible with the standard kinase coupling system (Fig. 5.14A).

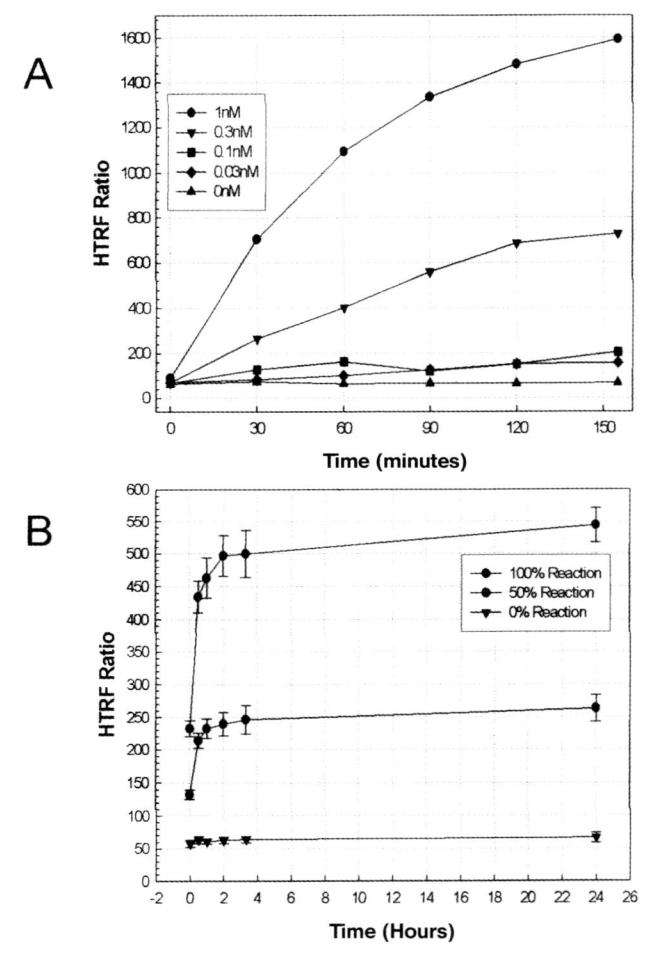

Figure 5.13 Enzyme titration and optimization of signal development time for optimum signal in HTRF format. (A) A 30 μL reaction was run in 50 mM HEPES (pH 7.5), 3 mM MnCl$_2$, 0.005% Tween 20, 0.02% BSA, 0.4 nM kinase, 0.03 μM biotinylated peptide, 5 μM ATP. The reaction was stopped with 15 μL of anti-phosphotyrosine stop solution containing 50 mM HEPES (pH 7.5), 10 mM EDTA, 0.02% BSA, 1 nM Eu-tagged antiphosphotyrosine, 2.5 μg mL^{-1} streptavidin–APC and read on an LJL-Analyst with donor excitation 330–380 nm, donor emission 620 nm, delay time 100 μs, at 50 flashes per well. The acceptor emission was set at 665 nm and an integration time of 1000 μs. The time course and linearity of the reaction were monitored at (◆) 0.03, (■) 0.1, (▼) 0.3 and (●) 1 nM enzyme. (B) The reaction was run under the same conditions as in (A) at 0.4 nM enzyme. The optimum time for the signal development after addition of stop was determined by reading the reaction at 0.5, 1, 2, 3 and 24 h after addition of stop.

As previously described, there are several variations on this assay theme which yield a fluorescent or luminescent readout. The oxidation of NADH by diaphorase can be coupled to the reduction of resazurin to resorufin, which leads to a gain of

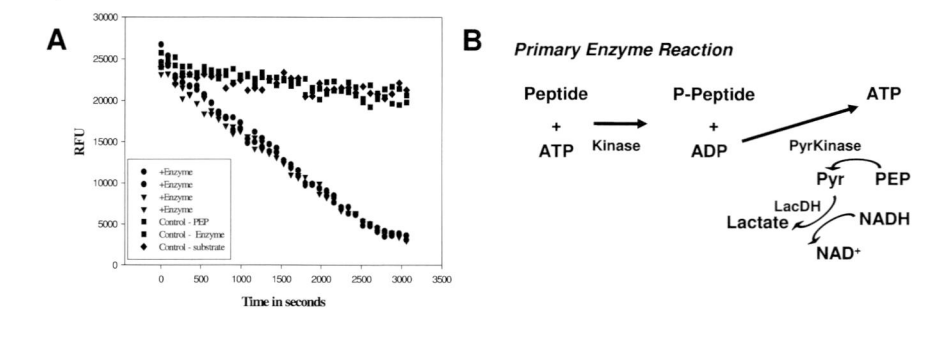

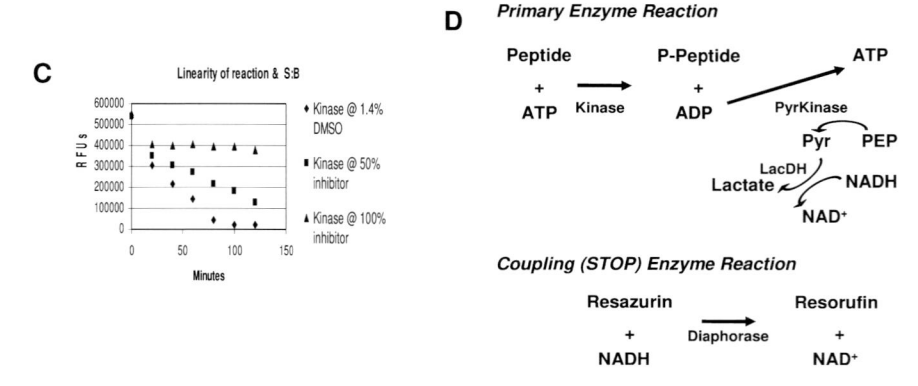

Figure 5.14 Reaction scheme and time course of the PK–LDH system. (A) and (B) a 50 μL reaction containing 50 mM HEPES (pH 7.5), 100 mM NaCl, 10 mM MgCl$_2$, 1 mM MnCl$_2$, 2.5 mM DTT, 0.025% BSA, 0.05% NP-40, 25 μM NADH, 100 μM ATP, 2 mM PEP, 30 μM biotinylated peptide substrate, 2 U mL^{-1} PK–LDH and 30 nM enzyme. The reaction was monitored on a BMG plate reader using a 107 gain setting at 340 nm excitation and 460 nm emission. (C) and (D) reaction scheme and time course of the PK–LDH system using coupled system: The reaction contained 20 mM HEPES (pH 7.5), 10 mM MgCl$_2$, 1 mM MnCl$_2$, 0.01% Tween 20, 0.01% BSA, 1 U mL^{-1} PK–LDH, 100 μM ATP, 1 mM PEP, 35 μM biotinylated peptide, 12 μM NADH, 0.5 U mL^{-1} diaphorase, 6 μM resazurin and 25 nM enzyme in a final volume of 50 μL. The reaction was monitored on a Wallac Victor II plate reader at 560 nm excitation and 590 nm emission. The graph shows the loss of signal in an end-point format at (▲) 100, (■) 50 and (◆) 0% inhibition.

fluorescent signal. This coupling system allows rapid clearance of NADH and prevents its accumulation, thus facilitating the primary reaction in the forward direction. This assay was used for screening inhibitors of a novel kinase involved in T cell activation (EC 2.7.1.–) [63]. The reaction was linear over 90 min when followed in the kinetic and end-point mode (Fig. 5.14B). Kinase activity can also be monitored by coupling this system with a luciferase reaction and measuring the loss of luminescence. An advantage of these methods is that kinase inhibitors increase luminescence or fluorescence (by reversing the enzyme-dependent loss of NADH) and therefore can be easily distinguished from quenchers, which reduce luminescence or fluorescence. Coupling enzyme inhibitors can be identified through counter-screening hits from the primary screen in a control assay.

For the purposes of this discussion, we have limited ourselves to SPA, HTRF and the PK–LDH coupled system and its variations for assaying protein kinases. These homogeneous formats are ideal for HTS since they are amenable to miniaturization in 384- and 1536-well formats and involve a minimal number of steps: addition of substrate and necessary cofactors to the microplate is followed by initiation of the reaction by addition of enzyme. At the end of the reaction, stop and detection reagents are added and the signal is read. Since these assays combine simplicity with high reproducibility, reliability and robustness, they are the technologies of choice for HTS of kinases.

5.7
Pitfalls and Reasons for Assay Development Failures

We have described here HTS assay methodologies for several major enzyme classes that have historical precedent in drug discovery. Our expectation is that these methods will serve as useful guides to enable novel class members as new disease annotations emerge and interest in the target for drug development is realized. Implementation of these methods at the HTS scale, however, can be challenging. The inability to express and purify an active target protein can end an effort at an early stage. Complicating the activity assessment is the possibility that the properties of the target itself can be atypical of previously investigated family members exhibiting low turnover or unprecedented substrate specificity. Even if an active enzyme is achieved, both the protein and the reagents used for its assay must be produced at a scale suitable for the HTS effort envisioned. One of the most appealing facets of a well-designed HTS campaign is that, in principle, with appropriate planning and quality control, the HTS portion of a drug discovery project should never fail. Assay development may fail for the reasons stipulated above, but when failed during the course of a rigorous process as outlined in this chapter, this failure often results in the conservation of valuable resources. Post-HTS activities may need to be terminated owing to issues with chemistry, pharmacology, efficacy *in vivo* and toxicology in more complex screening cascades. The HTS itself, however, is run only after rigorous testing and trouble-shooting activities have been completed, effectively removing all potential causes of failure. In our experience, achieving a high-quality biochemical HTS assay using the approaches outlined here is the rule rather than the exception.

Acknowledgments

We acknowledge the assistance of Joan Abrams, Sue Fish, Sadhana Jain, Yune Kunes, Zhi Li, Andrea Messina, Mihail Rokas, Neil Rollins, Norman Rousseau, Rebecca Roy, Jin Tang and Kara Queeney in vector construction, protein production and purification. We wish to thank Craig Karr for contributions to assay development of the glycosyltransferase and Paul Hales and Chris Tsu for contributions to peptidase assay development. We also wish to thank James Gavin, Cynthia Barrett and Suresh Jain for sharing their data and experiences using the HTRF and coupled assay formats.

References

1 J.G. Robertson, *Biochemistry* **2005**, *44*, 5561–5571.
2 G. Salvesen, J. J. Enghild, *Biochemistry* **1990**, *29*, 5304–5308.
3 K.E. Amrein, B. Takacs, M. Stieger, J. Molnos, N. A. Flint, P. Burn, *Proc. Natl. Acad. Sci. USA* **1995**, *92*, 1048–1052.
4 P.T. Wan, M. J. Garnett, S. M. Roe, S. Lee, D. Niculescu-Duvaz, V. M. Good, C. M. Jones, C. J. Marshall, C. J. Springer, D. Barford, R. Marais, *Cell* **2004**, *116*, 855–867.
5 T. Tanaka, S. Sato, H. Kumura, K. Shimazaki, *Biosci. Biotechnol. Biochem.* **2003**, *67*, 2254–2261.
6 S. Gaudet, D. Branton, R. A. Lue, *Proc. Natl. Acad. Sci. USA* **2000**, *97*, 5167–5172.
7 P. Towler, B. Staker, S. G. Prasad, S. Menon, J. Tang J, T. Parsons, D. Ryan, M. Fisher, D. Williams, N. Dales, M. Patane, M. W. Pantoliano, *J. Biol. Chem.* **2004**, *279*, 17996–18007.
8 N. Shiraishi, A. Natsume, A. Togayachi, T. Endo, T.Akashima, Y. Yamada, N. Imai, S. Nakagawa, S. Koizumi, S. Sekine, H. Narimatsu, K. Sasaki, *J. Biol. Chem.* **2001**, *276*, 3498–3507.
9 M.E. Bembenek, S. Schmidt, P. Li, J. Morawiak, A. Prack, S. Jain, R. Roy, T. Parsons, L. Chee, *Assay Drug Dev. Technol.* **2003**, *1*, 555–563.
10 A.J. Petrolonis, Q. Yang, P. J. Tummino, S. M. Fish, A. E. Prack, S. Jain, T. F. Parsons, P.Li, N. A. Dales, L. Ge, S. P. Langston, A. G. Schuller, W. F. An, L. A. Tartaglia, H. Chen, S. B. Hong, *J. Biol. Chem.* **2004**, *279*, 13976–13983.
11 M. E. Bembenek, R. Roy, P. Li, L. Chee, S. Jain, T. Parsons, *Assay Drug Dev. Technol.* **2004**, *2*, 300–307.
12 C. Vickers, P. Hales, V. Kaushik, L. Dick, J. Gavin, J. Tang, K. Godbout, T. Parsons, E. Baronas, F. Hsieh, S. Acton, M. Patane, A. Nichols, P. Tummino, *J. Biol. Chem.* **2002**, *277*, 14838–14843.
13 D.F. Reim, D. W. Speicher, *Curr. Protocols Protein Sci.* **2002**, *2*, 11.10.1–11.10.38.
14 T.A. Addona, K. R. Clauser, *Curr. Protocols Protein Sci.* **2002**, *2*, 16.11.1–16.11.19.
15 B. Sullivan, T. A. Addona, S. A. Carr, *Anal. Chem.* **2004**, *76*, 3112–3118.
16 A.P. Snyder, *Interpreting Protein Mass Spectra A Comprehensive Resource*, Oxford University Press, New York, **2000**.
17 A.J. Barrett, N. D. Rawlings, J. F. Woessner (eds.), *Handbook of Proteolytic Enzymes*, Academic Press, London, **1998**.
18 M. Zimmerman, E. Yurewicz, G. Patel, *Anal. Biochem.* **1976**, *70*, 258–262.
19 C.G. Knight, *Methods Enzymol.* **1995**, *24*, 18–34.
20 E.D. Matayoshi, G. T. Wang, G. A. Kraft, J. Erickson, *Science* **1990**, *247*, 954–958.
21 C.G. Knight, F. Willenbrock, G. Murphy, *FEBS Lett.* **1992**, *296*, 263–266.
22 M. Donaghue, F. Hsieh, E. Baronas, K. Godbout, M. Gosselin, N. Stagliano, M. Donovan, B. Woolf, K. Robison, R. Jeyaseelan, R. E. Breitbart, S. Acton, *Circ. Res.* **2000**, *87*, E1–E9.
23 M Enari, R. V. Talanian, W. W. Wong, S. Nagata, *Nature* **1996**, *380*, 723–726.
24 M. Donaghue, H. Wakimoto, C. T. MacGuire, S. Acton, P. Hales, N. Stagliano, V. Fairchild-Huntress, J. Xu, J. N. Lorenz, V. Kadambi, C. I. Berul, R. E. Breitbart, *J. Mol. Cell. Cardiol.* **2003**, *35*, 1043–1053.

25 J.R. Lackowicz, *Principles of Fluorescence Spectroscopy*, Plenum Press, New York, **1983**.
26 Bachem, *2003 Product Catalog*, Bachem Biosciences, King of Prussia, PA **2003**.
27 R.A. Copeland, *Enzymes: a Practical Introduction to Structure, Mechanism, Data Analysis*, 2nd edn., Wiley-VCH, Weinheim, **2000**.
28 M. Dixon, E. C. Webb, *Enzymes*, 3rd edn, Longman, Harlow, **1979**.
29 A. Bairoch, *Nucleic Acids Res.* **2000**, *28*, 304–305.
30 E. Gasteiger, A. Gattiker, C. Hoogland, I. Ivanyi, R. D. Appel, A. Bairoch, *Nucleic Acids Res.* **2003**, *31*, 3784–3788; http://www.expasy.org/enzyme.
31 A.K. Das, M. D. Uhler, A. K. Hajra, *J. Biol. Chem.* **2000**, *275*, 24333–24340.
32 F. Mashiga, K. Imai, T. Osuga, *Clin. Chim. Acta* **1976**, *70*, 79–86.
33 G.H. Czerlinski, B. Anderson, J. Tow, D. S. Reid, *J. Biochem. Biophys. Methods* **1988**, *15*, 241–247.
34 P.C. Preusch, D. M. Smalley, *Free Rad. Res. Commun.* **1990**, *8*, 401–415.
35 S.E. Brolin, E. Borglund, L. Tegner, G. Wettermark, *Anal. Biochem.* **1971**, *42*, 124–135.
36 B. Lei, S. C. Tu, *Biochemistry* **1998**, *37*, 14623–14629.
37 P. Roberts, J. Basran, E. K. Wilson, R. Hille, N. S. Scrutton, *Biochemistry* **1999**, *38*, 14927–14940.
38 W.S. McIntire, *J. Biol. Chem.* **1987**, *262*, 11012–11019.
39 M.C. Barrett, A. P. Dawson, *Biochem. J.* **1975**, *151*, 677–683.
40 H. Tsuge, Y. Nakano, H. Onishi, Y. Futamura, K. Ohashi, *Biochim. Biophys. Acta* **1980**, *614*, 274–284.
41 D.R. Haubrich, N. H. Gerber, *Biochem. Pharmacol.* **1981**, *30*, 2993–3000.
42 W.G. Gutheil, M. E. Stefanova, R. A. Nichols, *Anal. Biochem.* **2000**, *287*, 196–202
43 H. Maeda, S. Matsu-Ura, Y. Yamauchi, H. Ohmori, *Chem. Pharm. Bull.* **2001**, *49*, 622–625.
44 D.M. Obansky, B. R. Rabin, D. M. Simms, S. Y. Tseng, D. M. Severino, H. Eggelte, M. Fisher, S. Harbon, R. W. Stout, M. J. DiPaolo, *Clin. Chem.* **1991**, *37*, 1513–1518.
45 M. Zhou, Z. Diwu, N. Panchuk-Voloshina, R. P. Haugland, *Anal. Biochem.* **1997**, *253*, 162–168.
46 K. Tittmann, D. Proske, M. Spinka, S. Ghisla, R. Rudolph, G. Kern, *J. Biol. Chem.* **1998**, *273*, 12929–12934.
47 T. O'Connor, L. S. Ireland, D. J. Harrison, J. D. Hayes, *Biochem. J.* **1999**, *343*, 487–504.
48 J.M. Guo, Y. Zhang, L. Cheng, H. Iwasaki, H. Wang, T. Kubota, K. Tachibana, H. Narimatsu, *FEBS Lett.* **2002**, *524*, 211–218.
49 J. Bertogglia-Matte, *US Patent* 4 568 649, **1986**.
50 D. Chen, H. Ma, H. Hong, S. S. Koh, S-M., Huang, B. T. Shurter, D. W. Aswad, M. R. Stallcup, *Science* **1999**, *284*, 2174–2177.
51 G. Vanhoof, J.Berry, M. Harvey, M. Price-Jones, K. Hughes, *A high throughput screening assay for thiamin pyrophosphate kinase activity using SPA*, Amersham Biosciences website, http://www1.amershambiosciences.com, **2003**.
52 P.E. Brandish, L. A. Hill, W. Zheng, E. M. Scolnick, *Anal. Biochem.* **2003**, *313*, 311–318.
53 R. Macarron, L. Mensah, C. Cid, C. Carranza, N. Benson, A. J. Pope, E. Diez, *Anal Biochem.* **2000**, *284*, 183–190.
54 T.R. Mullinax, G. Henrich, P. Kasila, D. G. Ahern, E. A. Wenske, C. Hou, D. Angentieri, M. E. Bembenek, *J. Biomol. Screen.* **1999**, *4*, 151–155.
55 M.E. Bembenek, S. Jain, A. Prack, P. Li, L. Chee, W. Cao, R. Roy, S. Fish, M. Rokas, T. Parsons, R. Meyers, *Assay Drug Dev. Technol.* **2003**, *1*, 435–443.
56 P. Cohen, *Nat. Rev. Drug Discov.* **2002**, *1*, 309–315.
57 B.D. Manning, L. C. Cantley, *SciSTKE* **2002**, *162*, PE49.
58 M.E. Olsten, D. W. Litchfield, *Biochem. Cell Biol.* **2004**, *82*, 681–693.
59 D.E. Biazzo-Ashnault, Y. W. Park, R. T. Cummings, V. Ding, D. E. Moller, B. B. Zhang, S. A. Qureshi, *Anal. Biochem.* **2001**, *291*, 155–158.

60 B. Bader, E. Butt, A. Palmetshofer, U. Walter, T. Jarchau, P. Drueckes, *J. Biomol. Screen.* **2001**, *6*, 255–254.

61 W. Yang, A. R. Cerione, *J. Biol. Chem.* **1997**, *272*, 24819–24824.

62 M.A. Silverman, O. Benard, A. Jaaro, A. Rattner, Y. Citri, R. Seger, *J. Biol. Chem.* **1999**, *274*, 2631–2636.

63 P. Kueng, Z. Nikolova, V. Djonov, A. Hemphill, V. Rohrbach, D. Boehlen, G. Zuercher, A. C. Andres, A. Ziemiecki, *J. Cell. Biol.* **1997**, *139*, 1851–1859.

6
Image-based High-content Screening – A View from Basic Sciences

Peter Lipp and Lars Kaestner

6.1
Introduction

High-throughput screening (HTS) is mainly employed to obtain end-point data and often solely from a single quality such as a single fluorescence, luminescence, etc. In contrast, high-content screening (HCS) methods are meant to extract a variety of data from a single experimental or screening run. A property common to most of the HCS systems available today is that they provide temporally resolved information about the parameters they record, such as fluorescence or luminescence over time. Among the most important features of HCS systems are the following:

1. *Analysis on the level of single cells*

   Owing to the spatially resolved data acquisition, these HCS methods are able to perform analysis of the chosen set of data qualities on the level of individual cells. Especially in the case of HCS imaging, this requires providing a cell population in which single cells can easily be distinguished. With electrophysiology, this property of HCS is inherent.

2. *Sub-population analysis*

   This is potentially a very important property of HCS imaging systems since it helps in eliminating shortcomings of both transiently transfected and stable cell lines. This is particularly important since even the latter samples are well known to display varying degrees of overexpression of the particular proteins. Systems that allow for identification of single cells are inherently able to distinguish the entire population of cells into meaningful sub-populations and the analysis might therefore much more reflect the physiological behavior of living cells in an organism. As an example, not all cells in a well are synchronized in their cell cycle, etc. In addition, such a kind of analysis helps in identifying false-positive and false-negative populations and thus increases sensitivity (see below).

High-Throughput Screening in Drug Discovery. Edited by Jörg Hüser
Copyright © 2006 WILEY-VCH Verlag GmbH & Co. KGaA, Weinheim
ISBN: 3-527-31283-8

3. *Analysis on the subcellular level*
   Despite the fact that those imaging HCS systems discussed here (see below) utilize different technologies for optical sectioning, resulting in different degrees of axial and lateral resolution (as discussed in greater detail below), all systems are able to provide sufficient optical resolution to perform analysis not only on the cellular but very importantly also on the subcellular level. Especially this property opened (and still opens) an entirely new area of cellular responses that can be analyzed in the framework of HCS systems, such as the localization of proteins, translocation of proteins, integrity of subcellular organelles (e.g. mitochondria) or physiological parameters of such structures (e.g. mitochondrial membrane potential).
4. *Acquisition and/or analysis of multiple parameters*
   Although HTS systems are able to record more than one parameter, often these systems are set up to detect a single data quality. HCS systems and here in particular those using imaging techniques can combine multi-wavelength recordings with the acquisition of many more parameters, basically all data qualities inherent in recording images and temporally resolved data. Interestingly, many analysis options can be changed retrospectively, e.g. the analysis of novel aspects of the images that do not necessarily have to be determined *a priori*. This leaves a huge parameter space to be explored, a property that is rather challenging, though. Prospectively, one can even imagine combining various cell types in a single HCS imaging analysis, given that the cell types can readily be distinguished. This will further increase the capabilities of HCS and allow us to explore more comprehensively the potential of compounds to be tested in a single HCS screening run.
5. *Data with temporal resolution*
   Temporal data are one of the data qualities often absent in HTS systems but inherent in all imaging HCS solutions. Here, it is not only important to obtain knowledge about the detailed kinetics of the particular cellular response but there are indeed numerous responses that appear transient and not coordinated between all cells, hence end-point detection will underestimate the effect of certain compounds and will thus help in eliminating false-positive or in fact negative results (see below).
6. *Data qualities can be related to one another*
   When acquiring only a single end-point data quality, there is obviously nothing to relate to another quality. This is certainly improved in HTS systems/assays, allowing the acquisition of multiple end-point data qualities, but we think it is already obvious from the explanation above that HCS systems employing imaging techniques offer a much greater number of parameters, the information content of which can be explored by combining and relating qualities to one another (see also below). This is certainly an extremely important field in HCS that in our opinion still remains to be fully appreciated and scrutinized. Taking into account the recent establishment of HCS technologies and our growing understanding of cellular and

subcellular cell biology, we strongly believe in the potential of these techniques yet to be explored.

7. *Decreased rate of false positives or false negatives*

   In HTS, much effort goes into the detection of false-positive or false-negative results since a single compound usually has only "one chance" to display its potential in a single screening run. In contrast, in HCS systems employing imaging and electrophysiological methods, each cell is the "test-tube" and can therefore be treated individually as a sort of "independent" experiment. In this context, the identification and analysis of cell sub-populations appear very important since they will foster the effort of identifying false positives or false negatives. In addition, the inclusion of other data qualities in the analysis (data relations) will enable us to further verify and support positive or negative results.

8. *Increased physiological relevance*

   For studying cellular reactions in response to the stimulation of signaling cascades, it has often been considered to be sufficient to analyse a single data quality as a sort of digital end-point readout. In the past, that has been proved to be very helpful and has certainly fostered the entire field of HTS, but in the last 15 years it became increasingly obvious that cellular responses are anything but straightforward single modal reactions. Cellular responses to stimulations are often multi-modal, employing complex signaling networks in which a multitude of intracellular signaling cassettes speak to one another and influence each other in a positive or negative way. Thus, in addition to recording multi-modal data (see above), the analysis and interpretation of HCS results reach more reliable levels despite the fact that such complexity has not eased the analysis algorithms (see below). Nevertheless, acquisition of multi-modal data sets allows a better characterization of the behavior of the cellular system and thus increases the physiological relevance of the results obtained in HCS. It should be mentioned, however, that mathematical algorithms to interpret those data are still in their adolescence but are a prerequisite to explore the full potential of such types of HCS in the future. At present, many of these analysis steps still require manual (i.e. human) input. Here, the intense exchange of scientific results between the HCS and the scientific community is imperative and certainly fostered by those scientific laboratories that employ HCS methods on their own to study physiological and pathophysiological phenomena in living cells.

9. *Increased sensitivity*

   HCS systems will inherently offer higher sensitivity when planned carefully, although the data acquisition itself is not necessarily more sensitive. It is in particular the combination of recording multi-modal data qualities that allows us, by combination of such qualities, to increase the sensitivity of the screen. In addition, all the properties of HCS discussed so far, in combination, foster further increases in sensitivity. Here, for example, the reliable omission of false positives or false negatives increases the effective sensitivity of the screen.

10. *Extended possibilities to perform post-screen data mining*
    We have intentionally put this particular point at the end of the list of HCS features, since the area of data post-processing is certainly one of the most interesting but clearly also one of the most challenging aspects of HCS. From basically all points (1–9) mentioned above, it becomes obvious that probably most of the power lies hidden in the post-screen (i.e. after experimental run) analysis. For this, all of the HCS systems we will discuss below incorporate a sophisticated database system. This is utilized not only for straightforward data storage and archiving but more and more also for post-processing of the data, such as relational analysis and data mining. Data obtained from high-content screens contain a wealth of information, probably today much more than can be analyzed and extracted. Hence sophisticated algorithms are an important, if not the most important, feature of HCS systems. In essence, the success of HCS systems depends almost entirely on the availability of appropriate analysis tools either provided by the companies producing the HCS hardware or by the end-user with appropriate programming tools built into the HCS software suite.

The scientific background of the increasing importance and attraction of HCS methods can be found in the enormous scientific advances in understanding cellular and subcellular processes, especially during signal transduction, that took place in the last 15 years.

6.2
HCS Systems Employing Confocal Optical Technologies

The common property of all HCS systems that we will discuss here is their ability to reject out-of-focus light emitted from the specimen. Figure 6.1 briefly describes this confocal principle. Excitation light (green) is focused on to the specimen (Fig. 6.1A) and emission light from the plane of focus (plane of focus, orange dashed line; emission light, red) is separated from the excitation light by means of a dichroic mirror (dark gray) before being refocused on to a photon detector. Emission light from the plane of focus can pass through the pinhole without further obstruction (Fig. 6.1A). In the case that emission light originates from fluorophores out of focus, it is also refocused in the direction of the detector yet its image plane lies well out of the plane of the pinhole and therefore most of the emission light is blanked by this mask (Fig. 6.1B). In general, the more the light originates from out of focus, the more it is rejected by the pinhole. Effectively, this spatial filter (i.e. pinhole) determines the optical performance of the system, i.e. its ability to reject out-of-focus light. From this, it follows that the confocal system's ability to reject out-of-focus light is fostered by decreasing sizes of the pinhole. This process is eventually restricted by physical limitations that depend on the objective's numerical aperture (NA) and on the particular wavelength used. In general, when imaging is performed in the visible range of wavelengths, the optical resolution, defined as point spread

6.2 HCS Systems Employing Confocal Optical Technologies | 133

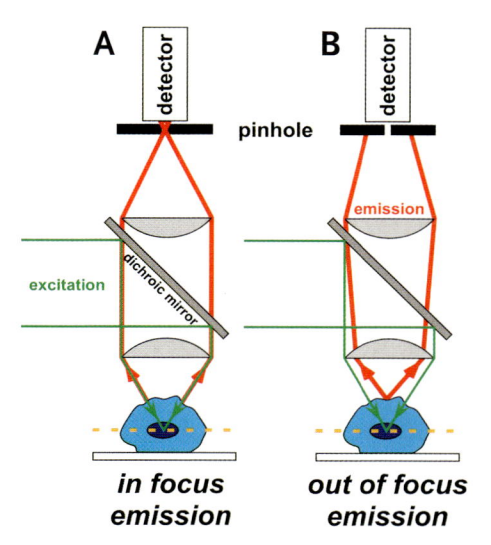

Figure 6.1 The confocal principle. Excitation light (green) is focused on to a specimen with a microscope objective and the emitted fluorescence (red) is returned, separated from the excitation light by means of a dichroic mirror and focused on to a detector [e.g. photomultiplier tube (PMT)]. If fluorescence is collected from within the plane of focus (A) then the emitted fluorescence will pass through the pinhole, which is a spatial filter. If fluorescence originates from out-of-focus planes (B), then the emitted light will not be focused though the pinhole and most of the light will be blocked by the pinhole.

function (PSF), is at best around 250 nm × 250 nm × 750 nm in the x, y and z directions, respectively. It can be seen that the axial z-resolution is decreased around threefold in comparison with the lateral x/y resolution. Since this is only applicable in combination with high-NA oil-immersion objectives, such theoretical values are not offered in currently available HCS systems. In fact, these low values are not necessary for most of the HCS tasks performed so far, but might be desirable in the future with further improved assay designs.

The range of HCS systems discussed below make use of the confocal effect to various extents. In general, it appears fair to state that available HCS imaging systems offer anything from solely rejecting some of the stray background light up to presenting high-resolution subcellular imaging solutions. All of these modes have their own right since confocality generally comes at the cost of light efficiency: the higher the confocality, the lower is the amount of light that can be collected from a resolvable volume. Hence it is always advisable to spend some thoughts as to what degree of confocality is indeed needed for the particular screen, at least for systems that offer the property of changing the pinhole size (or slit size; see below) and thus confocality.

Today there are four major classes of confocal HCS systems available that can be distinguished based on their particular type of scanning system or image-generation system employed. We will discuss the basic principles of each of these groups and give a representative example of a commercially available HCS system. We will discuss inherent advantages and limitations of the particular technology used. This should by no means be a direct comparison of commercially available technologies; instead, we will try to present an objective case for all four classes. The disadvantages and advantages provided are from the point of view of a basic scientist working in the field of cellular signal transduction and applying various imaging and in particular different confocal technologies.

6.3
Single-point Scanning Technology

Single-point confocal laser scanners are amongst the most established technologies in confocal laser scanning. Here a single laser beam is scanned across the specimen in x and y directions, usually by means of galvanometrically controlled high-reflective mirrors. The confocal image is formed in the attached computer that puts all the single pixels together. This confocal technology generally offers the highest optical performance in terms of resolution, since it images the emitted light through a pinhole adjustable in size (see above). An example of this technology can be found in Molecular Devices' ImageXPress^{Ultra}. Here, the scanning is performed by a galvo mirror in combination with stage that allows $x/y/z$ positioning. During imaging, the mirror is responsible for scanning in the x direction (Fig. 6.2), while the y-scanning is performed by the entire stage. Since the emitted light is de-scanned by the same x-scanning mirror, the light can be detected by PMTs. Additional filters and dichroic mirrors in the emission light path allow for the simultaneous detection of various ranges of wavelengths. The flexibility of the HCS system is increased by the ability to combine a whole range of excitation light sources (405-, 488-, 532- and 635-nm solid-state lasers) with a multitude of filter cubes (a maximum of three) and dichroic mirrors (beamsplitters; a maximum of five). Solid-state lasers offer a superior beam shape and decreased laser noise in comparison with standard gas lasers such Ar lasers or Ar/Kr mixed gas lasers. In addition, such lasers are environmentally more friendly in that they do not produce as much heat output as gas lasers; this puts a much decreased requirement on the environmental control necessary for the housing of the system. Such lasers are available today with almost all power output ranges necessary for imaging of living cells.

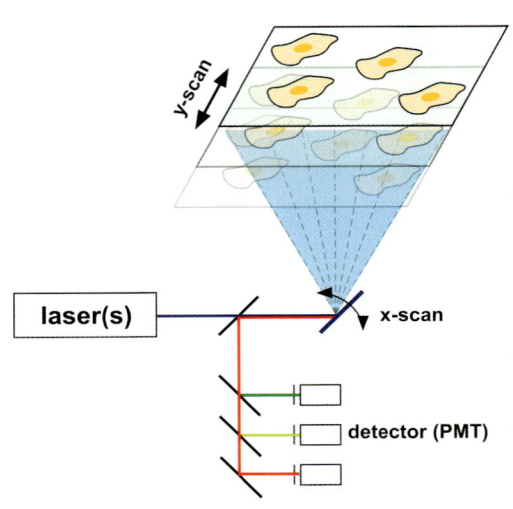

Figure 6.2 Principle optical setup of the ImageXpress^{Ultra} by Molecular Devices. A laser source generates the coherent excitation beam that is scanned in the x direction across the specimen. The necessary y scanning and z positioning are provided by the motorized scanning stage with the multi-well plate. Fluorescence detection is spatially filtered by a true pinhole. Emitted light can be multiplexed between up to three PMTs.

One of the additional key features that is necessary for 24-h HCS operations is the ability to incubate the cells and to keep them in a "physiological" environment, especially with respect to a carbonate-buffered culture medium. Hence for this HCS system and also for all the others that will be discussed here, upgrades are available that allow temperature, humidity and CO_2-content-control.

In comparison with standard video imaging, confocal imaging requires tight control of the positioning of the specimen in the x, y and z directions. Here, the positioning in the z direction and especially its reproducibility and stability over longer periods have been an essential demand for automatic HCS systems. The ImageXpress^{Ultra} described here (functionally similar to many other systems described below) offers the possibility of using a special laser for accurate z positioning. The major aspect in this is not necessarily the positioning accuracy offered by the mechanics; instead, when scanning a multi-well plate, repositioning back to the image plane used in the previous scans is the real challenge. The problem arises from the fact that the bottom of multi-well plates is inherently not designed to be absolute flat, the result of which is that the cells' z position unavoidably varies from well to well. In video microscopy that might be acceptable because such small variations can be minimized in the production process of the plates and the collection plane for light is much thicker, but, in contrast, in confocal microscopy mispositioning of the specimen in the micrometer range might produce false recordings and result in misinterpretation of the data. Usually laser-based positioning systems such that realized in the ImageXpress^{Ultra} offer very fast positioning times. They initially search for reflections of the air–plastic or plastic–culture medium phase transitions to reset themselves to this reference point. Repositioning is then achieved by shifting the specimen by a predefined distance in the z direction. This positioning is so accurate that even long-term recordings from an entire plate are possible without "losing" the initial focal plane.

Another important aspect of this HCS system is its ability to perform tasks unattended. For this, the ImageXpress^{Ultra}'s script wizard lets the user automate the acquisition of images. Here even complex tasks can be pre-programmed without the knowledge of a programming language.

In its "HCS mode", the software that comes with the system, MetaXpress, performs analysis off-line. For this a range of predefined analysis modules are available, including but not restricted to cell cycle stage, live/dead, neurite outgrowth and translocation. All of these analysis modules can be run across entire sets of data after the screen, e.g. also in an attempt to perform post-screen data mining. Interestingly, Molecular Devices also offers some of these analysis modules for higher throughputs as specialized HT modules. The code incorporated here has been particularly adapted to perform most of the analysis tasks online during the screen itself. These modules include cell proliferation and nuclear translocation.

The data and the analysis results are stored in a database that can also be shared across the network and thus allow for integration into more general data mining and data handling/archiving solution, in parts offered by Molecular Devices.

6.4
Line Scanning Technology

Another established confocal technology attempts to overcome one of the major problems with the single-point confocal scanner, namely the constraint on scanning speed, by scanning not a single point but an entire line across the specimen. Figure 6.3 gives a brief overview of a typical setup for a confocal line scanner. Usually the line to be scanned is generated by beam-shaping optics and the fast scanning is performed by a single galvanometrically controlled mirror. This mirror typically serves three major functions: (i) scanning the laser line across the specimen, (ii) de-scanning the emitted light from the specimen to generate a stationary emission line (the line is then usually spatially filtered by a slit aperture) and (iii) re-scanning the emitted line on to the camera(s). Such an optical arrangement ensures perfect synchrony between the scanning and the detection process. An HCS system employing such a line scanning confocal technology is the IN Cell Analyzer 3000 from GE Healthcare (formerly Amersham Biosciences). Similar to other line scanners, the IN Cell 3000 Analyzer also sacrifices optical resolution for gaining speed. This HCS system offers various degrees of confocality, i.e. effective optical resolution which can be user selected. Hand-in-hand with the decrease in optical resolution comes an increase in effective scanning speed. The single-point scanner introduced above needs to address every single pixel in succession and the computer takes the information collected by the detector (typically a PMT) and puts them together like in a jigsaw puzzle. The line scanner works more similarly to an analog camera in that it stores the entire information on the chip of a high-

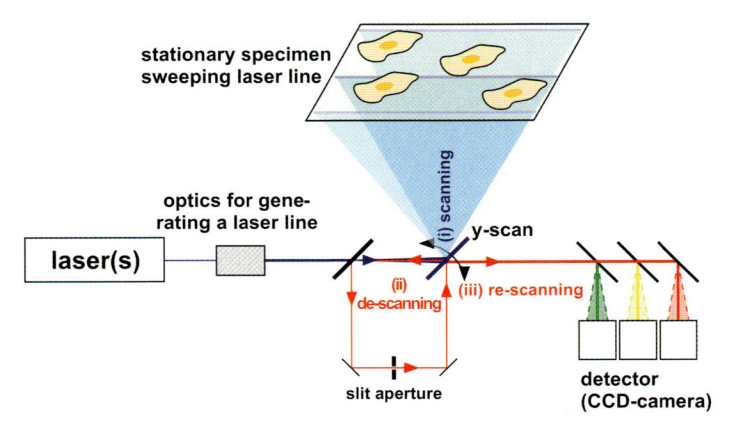

Figure 6.3 Optical principles and components of a line scanning confocal HCS system with the example of the IN Cell Analyzer 3000 from GE Healthcare. The laser source(s) generates a coherent excitation light beam that is transformed into a laser line by specialized beam-shaping optics (blue line). This line is scanned across a stationary specimen by means of a galvanometrically controlled scanning mirror (y scan). The emitted light (red line) is de-scanned by the scanning mirror, separated from the excitation light by a dichroic mirror and spatially filtered by slit aperture. Finally, the emitted fluorescence is re-scanned on to the multiplexed cameras on the reflective back side of the y-scanning mirror.

sensitivity charge-coupled device (CCD) camera. After illumination of the entire CCD chip, the information will be read out "at once" and stored on the controlling computer unit. In principle, the line scanner can obtain the final image by repetitive scanning of the entire field of view and accumulating the photons in each pixel of the CCD chip.

As mentioned above, the optical resolution of line scanners is reduced in comparison with the single-point scanners. The situation is in fact slightly more complex: since the emission is, strictly speaking, spatially filtered in only one direction, the optical resolution along the line will be rather low. Nevertheless, some of these limitations will be overcome by the implementation of a specially designed high-numerical aporture (NA) objective with a long working distance.

By means of this line scanning mode, the IN CellAnalyzer 3000 is able to scan over a fairly wide area of 750 µm × 750 µm in a single scan. This further fosters the speed of scanning an entire multi-well plate. The HCS system is designed for 96- or 384-well plates. These properties taken together allow for a rather high scanning speed, hence even faster cellular responses can be resolved temporally.

Similarly to the ImageXpress^{Ultra} system introduced above, also the line scanning HCS system allows full environmental control of the living cells to be studied, in that temperature, humidity and CO_2-content can be controlled and set by the controlling software and the user. The IN Cell Analyzer 3000 offers the ability to use three excitation wavelengths spanning the entire width of excitation, from UV (364 nm) for DNA stains such as DAPI, to the standard 488-nm argon line for fluorescein-based chemical stains or GFP/YFP (green fluorescent protein; yellow fluorscent protein). On the red side, the 647-nm krypton line offers excitation, e.g., for the long wavelength DNA stain DRAQ5, which can be very useful when utilizing, e.g., CFP/YFP-FRET (Förster resonance energy transfer between cyan FP and yellow FP). Three simultaneous emission measurements are possible because the system incorporates three highly sensitive CCD cameras. This can be particularly useful in situations where the temporal synchrony between emission images of varying wavelength is imperative.

The IN Cell Analyzer 3000 comes ready to allow for the application of various substances during the screen. This means that it is not only possible to apply a single substance but the included dispenser allows rather complex stimulation and application protocols to be performed, which can be important for particular screens. An important test is the reversal of stimulation regimes to verify that the effects recorded do not depend on the order of application but rather on the compound of interest.

The application suite that comes with the IN Cell Analyzer 3000 allows control of the instrument, analysis and data handling/mining at three different levels. Depending on the particular assays that should be performed, pre-packaged software comes with the instrument that allows setting up of the screen itself, acquiring the data and performing standard image analyses, such as protein translocation assays and cell proliferation transcription factor activation/deactivation assays. If the user needs more specific screens or adaptation of the screen setup offered "out of the box", then an IN Cell Developer Toolbox can be employed to fulfil these needs.

This HCS system also interfaces to software/hardware packages that are specialized for the storage of large data volumes and provide the possibility of performing screen-wide analysis and data mining. For this, the software discoveryHub™ interfaces between the various data sources (results from analysis algorithms, raw image data, data from databases, etc.) and systems performing higher level data analysis or visualization, including. but not limited to specialized data mining applications.

6.5
Multi-beam Technology

So far we have discussed HCS systems that are based on single-beam scanning, whether as a point or as a line. A completely different approach was introduced to the market based on a scanning mechanisms already invented almost a century ago by Paul Nipkow. The Nipkow disc was initially designed to transmit TV images but was rediscovered by Mojmir Petran's group in 1968 in a tandem scanning reflected light microscope. The basic principle of a Nipkow disc scanner is illustrated in Fig. 6.4, where it is also compared with a single-point scanner.

In comparison with single-point scanners, that have to scan every single pixel on the image one-by-one in succession, the Nipkow disc scanner utilizes an array of parallel excitation beams to excite several thousand points in the sample simultaneously. This array is generated out of a wide illumination, which can be either an illumination lamp or a laser (see below). By disc rotation, the array is functionally

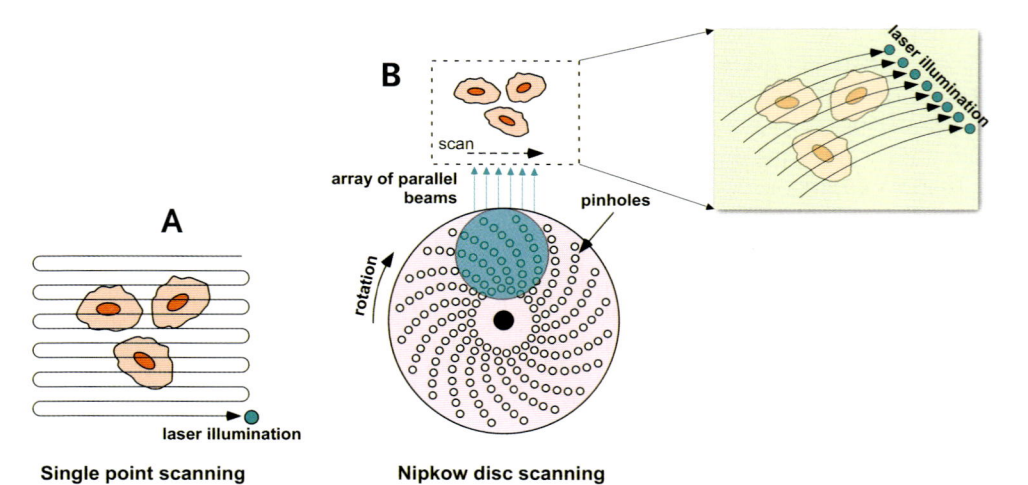

Figure 6.4 Comparison of the scanning process for a single-point scanner (A) and multi-beam scanner employing a Nipkow disc (B). Whereas the single beam scanner needs to address all pixels in a specimen in succession, the Nipkow disc generates a two-dimensional array of thousands of parallel laser beams that functionally scan across the specimen simultaneously (see inset in B).

scanned across the specimen in such a way that the entire sample is addressed (see inset in Fig. 6.4B). Usually a single rotation of the disc can generate several complete images and by rapidly rotating the disc image generation can easily achieve up to hundreds of images per second. Hence it becomes apparent that in comparison to single-point scanners, multi point scanners such as those employing Nipkow discs are able to produce live confocal images that are visible to the human eye in real time without the help of computers for constructing the image. Although this property is of limited use in high-throughput, industrial HCS systems, it is a very handy feature for the scientist performing high-content analysis with this type of confocal scanner.

The light that is emitted from the specimen is spatially filtered by each of the holes in the Nipkow disc serving the function of apertures as explained above for the single-point scanner. Similarly to the aforementioned detection of the emitted light by the human eye, a highly sensitive CCD camera is placed behind the disc to integrate and record the resulting confocal image.

The inherent advantage of the Nipkow disc over most other confocal technologies in terms of biocompatibility (bleaching, production of photo-products, etc.) is the very low bleaching rate achieved by the particular means of illumination. Whereas single-point scanners have to record the entire fluorescence of an excited volume element (voxel) in a single time point, multi-beam scanning achieves the same amount of returned fluorescence in small packages and the CCD cameras integrates the arriving photons from the specimen (similarly to analog film). To understand why the high-frequency excitation is beneficial in comparison to the "one-shot" excitation necessary for the single-point scanner assuming the same amount of resulting fluorescence in both cases, we need to look at the details of the excitation/fluorescence cycle. The excitation process involves the transition of the fluorophores from the ground state (S_0) with a low energy content into the excited state (S_1) in which the molecule contains more energy, i.e. the energy absorbed during the excitation process. After spending a very short amount of time in this excited state (usually in the nanoseconds range), the energy is returned as a photon of longer wavelength (emitted light) and the molecule returns to the S_0 state. In order to obtain good-quality images, the user needs either to increase the excitation time per pixel (called the dwell time) or to increase the excitation energy put into the excited volume. Let us first look at the latter situation. More light delivered to the sample generally results in more light coming out of the specimen. This relationship is only linear or even proportional over a certain range of excitation energies. One of the reasons for this is that the emitted fluorescence is a balance between relaxation processes of the fluorescent entity resulting in the emission of light and those processes resulting in the photochemical destruction of the molecule (i.e. bleaching). Photobleaching itself is an unavoidable side-effect of fluorescence; no fluorescence comes without the "cost" of bleaching, but the quantitative relationship between these two processes is nonlinear. With increasing excitation energies, the resulting fluorescence increases linearly and also the bleaching rate is proportional. Above a certain threshold that is characteristic of each fluorophore and the environment that the fluorophore encounters, the rate of bleaching grows nonlinearly or over-proportionally. This means that for a given increase one has to put in much more excitation energy in comparison with the linear region of this relationship. HCS, especially if run in a high-throughput screening environment or when imaging fast processes (or in fact both!), the user attempts to keep the dwell time as short as possible. In that situation, in order to obtain good quality images the only remaining option is to increase the excitation energy. This will gain sufficient photons per pixel that, for a single-point scanning microscope, has to be obtained in a single run.

In contrast, multi beam confocal scanners such as the Nipkow disc-based systems attempt to stay in the linear range of excitation/emission ratios by exciting the specimen repetitively with low excitation energies and integrating, i.e. collecting the returning fluorescence for a given period. Therefore, instead of obtaining the entire fluorescence in a single period, the functional dwell time is divided into many shorter intervals but at a much more reduced excitation energy. This phenomenon can be illustrated when considering the following simplified situation. Let us assume that we need to obtain an image of $1\,000 \times 1\,000$ pixels within one second. The single-point scanner will scan the specimen with such a speed that a single pixel will "receive" excitation light for roughly 1 µs; this is the dwell time. For this, of course, we assume no time for retracing or reverting the scanning direction at the end of each line of the scan. To collect an image of good quality in living cells, we need to increase the excitation energy significantly and therefore have to accept a certain bleaching of the sample, which is particularly harmful when collecting a long time series. Increased bleaching will not only decrease the image quality but will also contribute to the increased likelihood of irreversible cell damage. In contrast, with a multi-beam scanner the excitation process is pulsatile and the same amount of collected fluorescence will be recorded in many small portions. This is in favor of low bleaching since the process of excitation takes place in the time range of attoseconds whereas the emission (e.g. as fluorescence) displays a lifetime in the nanoseconds range. The bleaching of a fluorophore is favored when a photon hits the molecule in its S_1 state (i.e. excited state). Thus, by dispersing the excitation energy over many focal excitation points, a single excitation volume will receive less energy. We therefore have to increase the overall dwell time of the pixel within the image, but since at this low excitation energy bleaching processes are reduced, the result of such a high-frequency excitation process is reduced overall bleaching while maintaining good image quality and extended cell viability. With current Nipkow disc systems, the discs are rotating at frequencies that will generate images at rates of up to many hundred images per second. Thus, instead of collecting all of the fluorescence of an individual pixel within 1 µs (single-point scanner), the fluorescence can be collected in hundreds of small packages even with an effectively longer dwell or recording time per pixel (multi-point scanner).

In the living cell, these technologies allow data acquisition at higher frame rates over significantly increased recording periods. On the other hand, such scanning techniques can also be suitable for studying very fast processes requiring frame rates well above the video rate (30 frames per second) with sufficiently high excitation input.

A HCS system that uses Nipkow disc technology is the Pathway Bioimager from BD Biosciences. This HCS hardware is build around a confocal microscope design for the research environment by attoBioscience. The BD Pathway Bioimager allows real-time confocal recording of cellular and subcellular events in 96- and 384-well plates for higher throughput screening. In contrast to most other confocal designs using a Nipkow disc scanner the Pathway Bioimager employs a high-energy halide arc lamp delivering light over a wide range of wavelengths (360–720 nm) (Fig. 6.5). This range covers excitation wavelengths necessary for almost all of the fluorophores available today, whether organic compounds or protein based fluorophores. This certainly represents an important advantage over basically all other HCS confocal systems that require high-power laser sources for excitation. Although currently available lasers, both gas lasers and solid-state lasers, offer most excitation wavelengths, the white light source provides the possibility of fine tuning the excitation in situations where specimens are analyzed with multiple stainings and

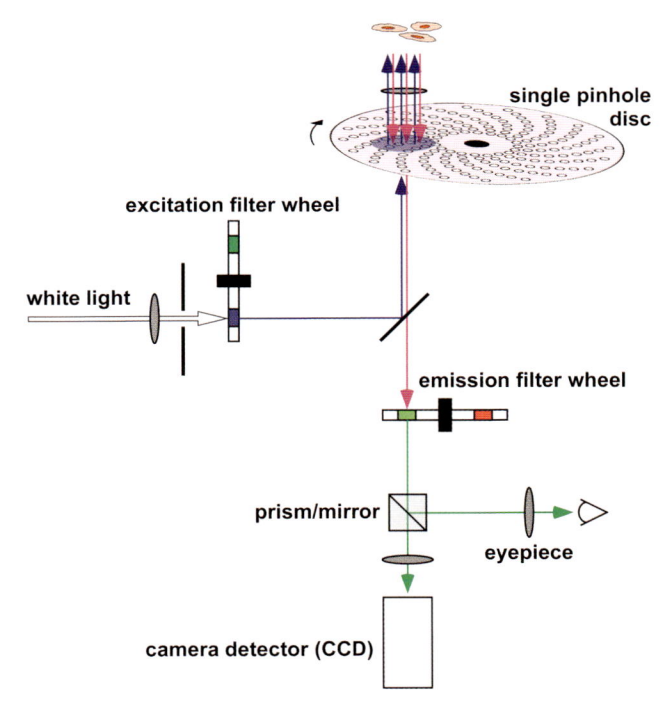

Figure 6.5 Optical principles and schematic setup of the BD Biosciences Pathway Bioimager. The Pathway Bioimager based on the attoBiosciences CARV confocal module employing a single Nipkow disc scanner. A powerful white light source provides the excitation light that is selected via an excitation filter wheel (blue line), directed and expanded on to a single Nipkow disc that generates a multitude of parallel excitation light beams that scan across the specimen upon rotation of the disc. Emitted fluorescence (pink line) is spatially filtered by the pinholes in the Nipkow disc, analyzed by an emission filter wheel. The generated confocal images can be visualized on a side port either for live viewing by the user or by a highly sensitive fast CCD camera.

the separation of the fluorophores based on sole emission might prove difficult. In addition, novel probes employing more unusual spectral properties might only require light adaptation with a white light source whereby development of a laser source appears difficult and costly. From this, it follows that the combination of a multitude of excitation wavelengths (currently 16 are supplied) and up to eight different emission wavelengths should allow the spectral separation of virtually all fluorophores available. Further flexibility can be achieved by simply exchanging excitation or emission filters and using appropriate dichroic mirrors.

In addition to these advantages, the Pathway Bioimager offers a dispensing system that provides the ability to monitor the specimen continuously before, during and after application of substances, a feature called "image-as-you-add". This is particularly important when imaging rapid responses in the seconds or sub-seconds time domain or when the cellular responses are expected to display a high degree of temporal dispersion. The specimens do not have to be taken out of the detection chamber for the application of compounds, instead they can remain under confocal inspection.

Similarly to the HCS confocal systems described so far, the BD Bioscience Pathway Bioimager also offers the full range of environmental control for the living samples under investigation.

On the software side, the Pathway Bioimager provides the usual set of sophisticated data analysis features that appear to be standard in HCS confocal systems of that kind: automatic cell identification, segmentation and identification of subcellular compartments, etc. In addition, the software offers a special pattern recognition algorithm that allows the automatic classification of cell responses within a single well and thus permits the analysis of subsets of cells. Furthermore, the high spatial resolution provided by the Nipkow disc scanner at the heart of the HCS system enables the user also to perform true three-dimensional (3D) or four-dimensional (3D over time) recordings and analysis. However, it should be mentioned that at least at present this is a feature aimed more at the research user than at HCS applications with high throughput.

Finally, the AttoVision and Cytoprint software packages enable the user to develop specialized analysis routines for novel assays which can be easily incorporated into daily routines.

Despite the large degree of flexibility provided by Nipkow scanners such as that incorporated in the BD Biosciences Pathway Bioimager, the single disc scanning system has a potential problem in that the excitation yield is drastically decreased by the necessary arrangement of the pinholes on the Nipkow disc itself. In order to provide good spatial separation, the pinholes ought to have a certain geometric arrangement, especially their distance from one another cannot be decreased below a critical value. From this, it follows that the excitation efficiency in terms of excitation energy put into the system in comparison with the energy that effectively reaches the specimen is particularly limited. This is system inherent since by far the greatest portion of the excitation energy is absorbed or reflected by the opaque surface of the Nipkow disc. This blanking can obscure more than 80% of the light energy put into the system for excitation.

6.5 Multi-beam Technology

A solution to increase the excitation efficiency significantly is to add a second Nipkow disc not equipped with pinholes but instead with micro-lenses (Fig. 6.6). In such a system, the micro-lenses considerably increase the excitation efficiency in that they collect more than 75% of the input excitation energy and focus the excitation light directly into the pinholes (see inset in Fig. 6.6) . On the way back from the specimen, the emitted light passes through the pinholes, which now serve the function of a spatial filter, similarly to the AttoBiosciences CARV design from above. They blank out emission light from out-of-focus planes in the specimen.

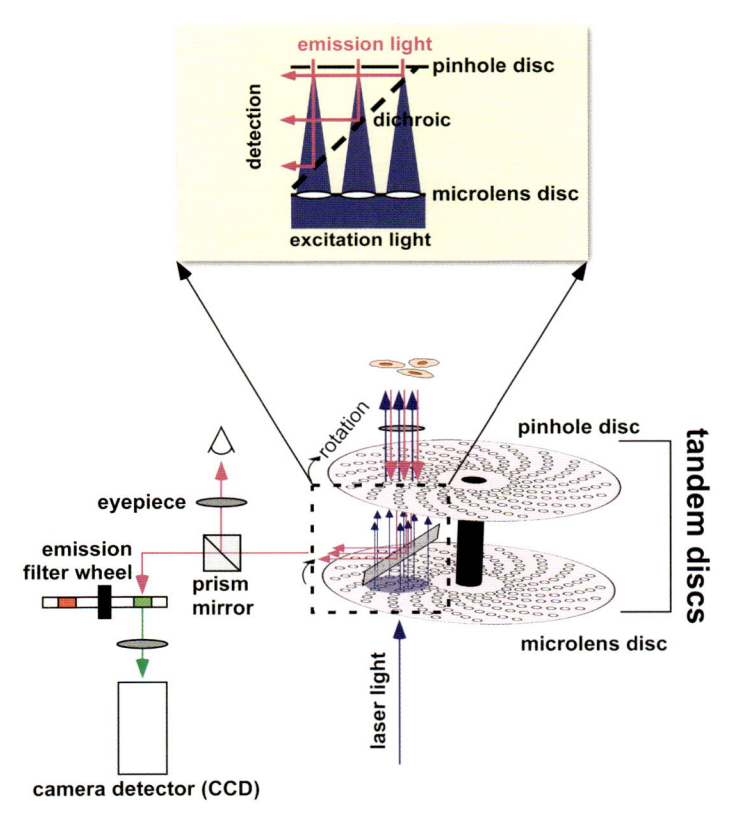

Figure 6.6 Optical principle of a tandem Nipkow disc scanner-based HCS system: the Opera from Evotec. Laser light (blue line) for excitation is beam shaped and directed on to the micro-lens disc of the tandem Nipkow disc scanner. The lenses effectively collect the excitation light and focus it down into the pinholes of the pinhole-disc (see also inset for optical details). The constraints associated with single Nipkow disc scanners are overcome by the second micro-lens disc and excitation efficiency is increased to more than 80%. The parallel beam of laser excitation scans across the specimen when the discs rotate in synchrony. Emitted fluorescence from the sample is separated by a dichroic mirror sandwiched between the two Nipkow discs and directed to the detection unit. Here the user can inspect a confocal image or, alternatively, the emitted light can be multiplexed by a series of dichroic mirrors and filter wheels in front of up to four highly sensitive fast CCD cameras.

The emission light is separated from the excitation light by means of dichroic mirrors and detection is achieved by highly sensitive high-speed CCD cameras.

Such a tandem Nipkow disc scanner was introduced by the Japanese company Yokogawa as their CSU10 confocal laser scanning head. It is probably fair to say that the introduction of this design and the incorporation of such a confocal scanner into PerkinElmer's confocal microscope systems such as the Life Cell Imager (LCI) thereafter really marked a revolution in life cell imaging. This introduction coincided with the success story of green fluorescent proteins and their other colorful siblings. The Nipkow tandem scanner and the ready availability of research assays rapidly developing around fluorescent protein technology provided novel and exciting new insights into cell signaling. These technologies delivered a multitude of tools to allow the researcher to use living cells as "test-tubes" rather then *in vitro* assays. The robust handling and reliable operation of the tandem Nipkow disc scanning head were convincing reasons to hope for an HCS system that would be build around such a system. Recently, Evotec Technolgies introduced their tandem Nipkow disc scanning confocal HCS system, named Opera.

The basic optical setup of the Opera is illustrated in Fig. 6.6. Although the system also offers a non-confocal detection mode for recording fluorescence that requires UV excitation, we would like to concentrate here on the confocal applications and technologies in this machine. With the maximum of four lasers as excitation sources, this HCS confocal system spans a broad range of wavelengths for excitation. The collection of lasers provide the most commonly required excitation lines at 405, 488, 532 and 635 nm and thus offer inherent excitation flexibility, although not as high a degree of flexibility as is displayed by the BD Biosciences Pathway Bioimager equipped with a white light source.

In comparison with the Pathway Bioimager, the Opera contains up to four cameras for multiplexing the emission detection and fosters data acquisition speeds and information content drawn from a single screen run even further. Another particularly interesting design detail of the Opera is the availability of an automatic objective changer that additionally incorporates the use of high-NA water immersion objectives. A specific design allows for the automatic application of the immersion medium and its replenishment during the screen.

In addition to the other HCS confocal systems introduced here, the Opera also easily integrates with other plate handling hardware and HTS systems. The Opera, very similar to the Pathway Bioimager, allows the use and setup of more complex application protocols for compounds even during an ongoing screen. For this a specific solution dispenser in included. Flexibility in integration into existing HTS or cell handling systems is fostered by the fact that the Opera can easily be adapted to handle many different kinds of multi-well plates from different manufacturers.

Evotec has paid particular attention to the design of an easy-to-use and easy-to-adapt and set up software system for HCS assays. For most of the "programming" of the HCS system the user can use a drag-and-drop methodology to build up the screen assay, the experimental run, data acquisition and data analysis including online analysis. Interestingly, the Opera software, such as Acapella, uses an XML (extended markup language) file format that permits integration and interfacing to a multitude of available data handling, archiving and data mining systems, such as the OME software (open microscopy environment). Systems like the OME software allow one to handle very large amounts of data (exceeding terabytes).

As with all the other systems, the Opera form Evotec also permits a large (and ever growing) number of predefined HCS assays including, but not restricted to, cell viability, calcium handling, membrane potential, gene activity and protein regulation assays, ligand binding, protein translocation and a number of cell morphology assays.

6.6
Structured Illumination

In the strict sense, structured illumination is not a confocal technique. However, it allows optical sectioning that comes close to the axial resolution of a pinhole-based confocal arrangement. Therefore, the term confocality as a measure of axial resolution can also be applied for images derived from structured illumination. The working principle of structured illumination is sketched in a simplified scheme in Fig. 6.7.

An optical grid is placed in the illumination beam in such a way, that the image of the grid is exactly in the focal plane of the microscope. The grid image needs to be moved laterally in the focal plane. This can be realized by lateral movement of the grid by a piezo (as in the initial setup by Toni Wilson) or by introducing a beam shift due to a swing of a glass plate in the excitation beam path as shown in Fig. 6.7A (and realized in Zeiss' ApoTome). Exposures are taken in at least three fixed grid image positions. Out of these three exposures, an image of an optical section is calculated.

The simplified principle of calculation (based on two grid positions) is visualized in Fig. 6.7B: the upper row shows sections of the x–z plane, with the grid image (dark ovals), resembling the excitation intensity distribution; the red and green beads are in the focal plane whereas the blue bead is out of the focal plane. Below there are schemes of the x–y plane to which the camera would be exposed. In the left column, the green bead is shaded by the grid image, whereas the red bead is perfectly illuminated and therefore produces a sharp fluorescence image on the camera. The blue bead is out of focus and therefore gives a blurred image on the camera. In the right column, the grid position has changed and therefore the red bead is no longer visible, while the green bead gives a sharp image. For the blue bead the situation is virtually unchanged and is therefore mapped exactly as for the first grid position. If now the two x–y exposures as shown in the middle row are subtracted from each other, structures that have been out of focus vanish, while the absolute value of the structures in focus gives the required optical section (bottom of Fig. 6.7B).

In contrast to confocal laser scanning, the structured illumination poses fewer requirements to the illumination source (no laser necessary) and is therefore less expensive. In terms of acquisition speed, structured illumination recordings can be faster than classical confocal recordings (see above). However, the acquisition speed is limited by the read-out of the camera used and since there are at least three exposures necessary for the calculation of one optically sectioned image it is inherently slower than other camera-based confocal systems such as Nipkow disc systems or slit scanners (see above). This extended data acquisition can partially be overcome by using the principle of the "running average", i.e. exposures 1–3 give the first image, exposures 2–4 the second and so on. In any case, if the process one is looking at is fast, e.g. calcium transients may rise in the range of milliseconds, there is not only a smear in the x–y plane but also a decrease in confocality since the algorithm relies on the assumption that there are no sample changes within the three exposures needed to calculate the image. When comparing structured illumination with the confocal imaging technique, one has to take into account that the structural illumination image is already a processed image and especially when taking z-stacks for 3D reconstruction, the deconvolution of confocal images leads to an increase in resolution.

Structured illumination was introduced into HCS in the form of the ArrayScan VTI HCS Reader by Cellomics. This instrument is based, in terms of structured illumination, on Zeiss' ApoTome (Fig. 6.7A). It is a fully automated platform including robot plate handling and software-controlled auto focus. The ArrayScan VTI HCS Reader comes with Cellomics' High Content Informatics (HCi^{TM}) suite, providing a software platform for managing, analyzing, visualizing and decision-making. Hardware and software are open for end-user customization.

The concept of the HCS multi-parametric platform based on the iMIC by TILL Photonics also uses structured illumination as the method to achieve optical sectioning. Although this is just a prototype, the concept is discussed here since

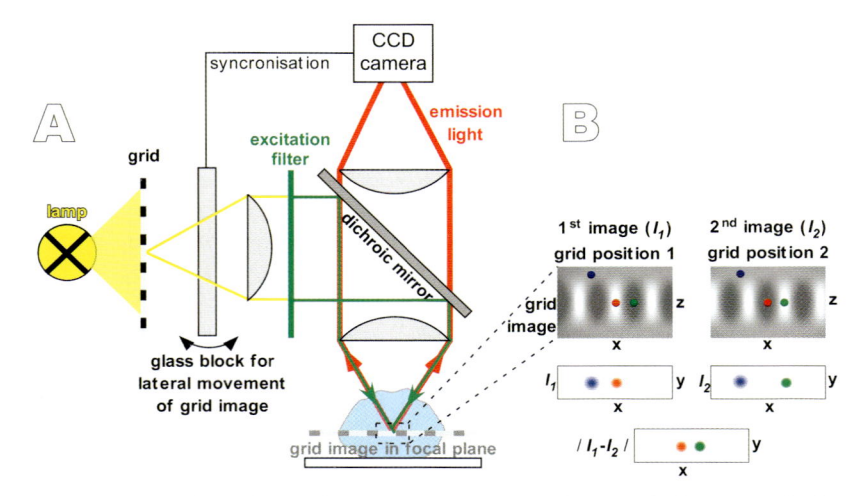

Figure 6.7 Simplified scheme of optical sectioning by structured illumination. (A) The modified optical arrangement as initially introduced by Neil, Juskaitis and Wilson in 1997. For a detailed description, see the text. (B) The principle of image calculation out of – for simplification – two exposures (courtesy of André Zeug, Center for Molecular Physiology of the Brain, Göttingen, Germany).

in terms of flexibility it goes beyond the instrumentation already on the market. The so-called Polytrope module provides this flexibility by allowing one to switch the illumination source within a millisecond into different illumination modes: in addition to the structured and wide-field illumination, there are also options for total internal reflection fluorescence (TIRF) and all other techniques requiring a rapidly positioned laser beam, e.g. fluorescence recovery after photo-bleach (FRAP) or laser micro-dissection.

6.7
Summary and Perspectives

We have introduced the currently available main types of HCS confocal imaging systems with emphasis on the underlying imaging technology rather than on the particular pros and cons for HCS or even HTS. All systems have been developed out of researchers' laboratories and have proven in that environment that they can operate reliably and handling of the equipment is reasonably robust. These are important features for the application of such systems in an industrial environment. Although all systems can be adapted to work in an industrial high-throughput workflow environment or in fact come supplied with this feature, some of them appear of the right design (i.e. size) for use also in more research laboratories, such as the Molecular Devices ImageXpress^{Ultra}. This is not to say that the other systems are inappropriate for routine laboratory work, but simply the price constrains the research locations that can afford an HCS confocal imaging system of that type.

Moreover, most of the systems' design is modular, thus permitting the inclusion of up-to-date novel confocal technologies such as the number of multi-beam scanners that have dominated the market recently. Amongst those most promising designs are systems employing the sweeping of a one- or two-dimensional field of parallel laser beams across the specimen. In this context, we should mention, e.g., the Vt$_{infinity}$ from VisiTech Int. Their novel design of a two-dimensional array of laser beams (around 2500 parallel beams) permits easy synchronization of image generation with image acquisition by means of modern high-speed CCD cameras. Another very interesting development is swept field confocal scanners such as that from Prairie Technologies (now marketed by Nikon. USA) that uses a line arrangement of laser beams to scan across the specimen, in the case of the Prairie design either in a single direction (high-speed mode, lower resolution) or in the x and y directions for higher resolution. Another manufacturer offering such confocal scanners is the German company LaVision Biotec. Their TriMScope is based on scanning a multi-beam laser line of high-energy pulses (pulse lengths are in the femtosecond time domain). The confocality in this system is achieved by utilizing the multi-photon excitation process that already constrains the excitation to a diffraction-limited volume, i.e. "confocal excitation". In contrast, Zeiss have approached that problem from a different angle. In their LSM 5Life confocal laser scanning microscope, they perform excitation scanning with a laser line (similar to the GE Healthcare IN Cell Analyzer 3000) but detection is performed by a specialized high-speed line-CCD element. This allows very fast acquisition rates with a minimized loss of resolution.

It will be interesting to see which of these technologies will make it into HCS confocal systems in the future.

In addition to all the technical achievements presented so far, there is of course also the other aspect, namely that HCS has a number of as yet problematic limitations that need to be recognized and addressed either by the system manufacturers or by the user. HTS systems and the assays used, also for ultra-HTS (UHTS), have been optimized for single-run screens, i.e. the cell-to-cell or well-to-well variabilities are extremely small (from the point of view of a basic scientist, amazingly small). In HCS, when analyzing single cells with assays that are simply less robust than such end-point assays used in HTS or UHTS, cell variability can become a problem. As soon as more complex biological systems are analyzed and the responses depend on entire signaling cascades, cells will display their "individuality". Parameters that are important in this context include stage in cell cycle, expression of endogenous proteins and expression of exogenous proteins. It should be kept in mind that even cells in culture comprising a single cell type display vastly different genotype depending on the gene of interest. Even after clonal picking we have found that within three passages HeLa cells lost their initial genotypical homogeneity and were back to their heterogeneous genotype determined just before clonal picking. The result of this will be a much increased variability of imaging results. The way to tackle such problems might be found in intelligent algorithms that are able, e.g., to identify automatically classes of responses or cells. These algorithms might classify data based on the cellular response alone or alternatively based on a secondary property of the cell that is also recorded in the assay. Certainly the best strategy might be to combine as many of the recorded parameters as possible for this classification and to attempt to find rules within this parameter relationship matrix which might be helpful for the analysis of other data sets in the same screen, in future screens or in fact also in screens that have been performed already. With sufficiently intelligent assays, post-processing even weeks or months after the actual screens or data mining (i.e. combining different screens) might provide deeper insights into the extraction of additional information and the correct interpretation of such data sets.

In the context of cell type variability, another aspect is highly desirable in HCS systems, namely the use of primary cultured cells in stead of cell lines. Although primary cultures of a variety of cell types are routine in modern laboratories, many cell types of interest such as neurones, cardiac cells, smooth muscle cells and blood cells are extremely difficult to handle, which is already true for the laboratory environment. Nevertheless, HCS assays with increasing physiological relevance will have to perform the transition from utilizing simple cell lines to primary cell cultures. The problem with primary cells such as neurones or muscle cells is their inertial behavior against classical transfection methods necessary to induce the generation of protein-based fluorescence, such as fluorescence protein–protein fusion. Here, virus-based transfection methods might be the right solution. They offer almost 100% transfection efficiency, high expression rates and easy handling in a routine environment. The downside of viral transfection methods is that the expression of fluorescence protein or other exogenous proteins usually takes 1–2 days, depending on the type of virus and culture conditions to keep primary cells in culture without imposing major physiological and/or morphological changes. This is, of course, not true for fluorescence that can be chemically loaded into the cells.

Currently available HCS confocal assays very often involve protocols that do not really require high-resolution imaging equipment. Examples for this are monitoring mitochondrial membrane potential with TMRE fluorescence, calcium fluxes, nuclear translocations and many other simple translocations. All of the signals involved in these assays simply rely on clean recording conditions but do not really rely on optimal confocal recording conditions. Very often the signal quality of HCS systems can be increased significantly by technically easy approaches to minimize or reduce background fluorescence, such as particular dispense regimes and choice of compounds. If one wants to put it in a provocative statement: many assays do not really depend on confocal recording technologies; optimized video imaging can be more than adequate. To continue with these thoughts: now that confocal HCS technology is available, true confocal-based assays, and we mean those that essentially depend on the application of high-resolution confocal scanning, need to be developed. Assays that probably come the closest to this situation are those that, e.g., attempt to characterize granularity of vesicles or granules in living cells.

Up to now, our notion is that many assays cannot be performed because adequate computing power and intelligent algorithms are still being developed. These software solutions are the key for HCS confocal systems; they will offer the robustness and ease of the screening itself. They are also the key to transforming HCS systems into higher throughput techniques. New systems need to address the following points, amongst others: (i) data handling; (ii) data storage and archiving, accessibility at the corporate level; (iii) tools for data mining that can handle terabytes of data for analysis; and (iv) easy methods to develop novel assays or to adapt existing assays.

In conclusion, confocal technology has indeed made it into the HCS field. Robust instruments have been developed around basically all technologies formerly present solely in researchers' laboratories. Nevertheless, the true success story has still to unfold. In our opinion, this is due in part to the lack of primary cell culture systems that are necessary to reveal the full potential of highly multiplexed HCS confocal systems: recording physiological signaling in physiological test-tubes, the individual cells. For this, sophisticated cell handling and cell manipulation techniques need to be refined from currently available techniques.

Although many reports forecast exceptionally high growth rates for HCS screens in large (150% increase) and small–medium pharmaceutical companies (140% increase), no survey as yet has specifically addressed the proportion of HCS confocal system in that. Nevertheless, there is general interest in developing assays on cellular signal transduction, a perfect field for confocal HCS techniques. From this, it follows that very likely HCS technologies requiring high-resolution confocal methods will develop into an important part of high-content screening and might even develop into an integral part of high-throughput screening workflow.

**Part IV
Data Analysis**

7
Methods for Statistical Analysis, Quality Assurance and Management of Primary High-throughput Screening Data

Hanspeter Gubler

7.1
Introduction

7.1.1
Overview

After a general introduction into the varied problems that can arise during the analysis of primary high-throughput screening (HTS) data, a more in-depth look into the data quality-relevant aspects of assay development (Section 7.2) and into different types of normalizations (Section 7.4) is presented. Robust and outlier-resistant statistical methods play a very important practical role in this field and are reviewed and illustrated in Section 7.5. It is well known that good visualizations and graphical representations of the measured data and the derived quality indicators readily allow quick identification of experimental problems and artifacts. Several types of useful visualizations are illustrated with a realistic simulated primary HTS data set in Section 7.6. Different mathematical and statistical methods for the correction of the systematic background response variations and for hit identification are presented and discussed in Sections 7.7 and 7.8. Some of the correction methods and their effects on the primary activity data are again illustrated through visual representations. Most of the methods lend themselves to the setting up of a fully automated HTS data analysis and quality assurance workflow which can aid scientists to focus quickly on problem results.

7.1.2
Problems during the Analysis of Primary HTS Data

In this chapter, the most important aspects of primary HTS data management, from the point of view of visualization and quality control (QC), are addressed and the necessary statistical methods for data reduction and correction of systematic errors are described. Adequately adapted methods are crucial for extracting optimal

High-Throughput Screening in Drug Discovery. Edited by Jörg Hüser
Copyright © 2006 WILEY-VCH Verlag GmbH & Co. KGaA, Weinheim
ISBN: 3-527-31283-8

information and reaching high levels of confidence in the hit selection process from the large body of, often noisy, experimental HTS data. Today's industrial HTS laboratories are producing massive amounts of primary screening data, as typical screening library sets of 0.5–2 million compounds and samples from other sources are measured in an average of some 60–200 HTS assays per year [1, 2]. Many of the newer assay technologies result in multiparameter readouts [3, 4], that is, multiple numerical values per assay sample. As an illustration of the amounts of newly created data that need to be managed, properly processed and analyzed by the systems and scientists in those laboratories on a yearly basis, we make the following crude estimate: primary HTS experiments with, for example, 1 million compounds in 100 screens with an average 1.5-fold readout multiplicity result in ~150 million data points distributed over ~270 000 datasets (or data files) if half of the assays are being run in either 384- or 1536-well plates, respectively. In such an environment, more that 1000 data sets need to be processed, analyzed and assessed every single working day. Modern ultra-high-throughput screening (uHTS) equipment can easily screen several hundred thousand wells in 24 h [5, 6]. It is therefore clear that streamlined, highly controlled and optimized processes for data management, quality control, data analysis and visualization are absolute necessities in such laboratories. In the following sections of this chapter we focus on the important properties of HTS data analysis systems and methods for today's standard HTS technologies which are based on microplate experiments. The basic statistical principles can also be applied to other types of experiments, however.

Automated software for quality control, detection, correction or at least flagging the presence of systematic errors in the data sets (that is having an automated HTS quality assurance process) is necessary to process the assay results and focus the attention of the screening scientists to the problematic cases at an early stage. Systematic errors or response variations across the plate area or response shifts as a function of time, especially between different experimental batches, are often observed in plate-based HTS runs. Such systematic errors or repetitive patterns can result from reagent degradation over time, temperature or evaporation gradients, variations in incubation time, inhomogeneous cell dispensing or protein coating, liquid handling equipment malfunction, reader edge effects, spillover, compound cross-contamination and other sources. Various types of visualizations and numerical measures of data quality derived from the raw, normalized and corrected data together with representations of the detected spatial or time-dependent error patterns should allow for quick review and approval of the results on the one hand and possibly a diagnosis of the process-related origin of these systematic effects on the other. Some of the observed error patterns can usually be directly linked to distinct process steps in the experiments, for example, particular systematic patterns arising from non-optimal functioning of liquid handling equipment, whereas others are not so easily explained without further, more in-depth investigation. For cost and time reasons, data that suffer from "repairable" systematic errors are usually not discarded or repeated, but corrected with suitable statistical methods to allow for proper hit identification.

The possibilities for data correction and for optimization of the experiments due to insight into the presence and possible sources of systematic errors should finally lead to data and to decisions of higher quality. It is important to extract the most meaningful information from the HTS data even in the presence of such systematic errors. Statistical methods that minimize both the false-positive and false-negative hit identification rates across the whole assay in a global fashion need to be reliably applied to speed up the HTS process. Since it is impossible for a human operator to adjust and optimize the hit selection threshold on a plate-by-plate basis, the process of quality assessment and reliable correction of systematic errors needs to be largely automated.

Whereas the detection of the most active hits from a primary screen is usually not a problem, the identification of hits with lesser activity is best done through the application of more objective automated correction and selection procedures. Simply applying a global threshold to the normalized data for all plates can easily introduce a bias due to the presence of systematic trends and edge effects. It is the aim of automated quality assurance procedures to eliminate as many false positives as possible while at the same time recover as many false negatives as possible. In addition to the immediate goal of obtaining data of optimal quality for the identification of hits in a single assay, generating results with the least amount of bias and systematic errors is also important because the data will be stored in databases for subsequent retrieval and analysis. Increasingly, HTS data are being used repeatedly for comprehensive mining of chemical and biological information and for large-scale cross-assay analyses (bio-profiling), thus emphasizing the need for well-defined processing standards and measures of data quality.

Although the large amounts of data generated in industrial HTS laboratories make it necessary to set up a highly integrated and streamlined data processing and visualization infrastructure, the data analysis methods described in the following sections also have fundamental merits in any other biological screening laboratory dealing with plate-based experiments.

This chapter is largely structured in the way that it provides a "guided walk" through the data analysis and diagnostics steps by applying increasingly more complex, but at the same time more powerful, algorithms for data reduction, quality assessment and correction of systematic errors.

Whereas up to a few years ago systems for HTS data analysis, large-scale data quality control, diagnostics and, occasionally, correction of systematic errors were mostly created by the in-house HTS informatics and data analysis groups in several of the larger pharmaceutical companies [7–11], there are now companies in the market that provide specialized software tools and integrated systems targeting these problems [12, 13, 16, 18]. Other tools for HTS data quality monitoring originate from academic sources [19–21]. More generic visualization and statistics tools [22, 23] can also be applied in the HTS data quality control and monitoring area, but they do not possess any readily available methods for data correction. Furthermore, there are the general assay data management solutions which include some graphics components for visual data quality assessment [24, 25]. The available commercial software systems that include HTS data correction possibilities [14, 16] contain different proprietary algorithms that have not been disclosed. Basic descriptions of the algorithms used in the software programs that are available from the academic groups have been published [19–22]. Also, the fundamental aspects of the methods implemented in the HTS data analysis system developed by the research statistics group of Merck pharmaceuticals have been described [7, 8].

In some of the later sections, the most important aspects of the published methods will be described. Furthermore, a series of methods and algorithms for quality control, quality assurance and plate-array-based data correction that draw on a series of basic methods which are being successfully employed in the well-established areas of image analysis, image processing, geo-statistics and the general area of analysis of observational data in the earth sciences will be explained and illustrated (see Section 7.7). Several data reduction and signal processing methods from the latter area are well suited to also address the needs of plate-based HTS data analysis because of the parallels in the fundamental aspects of the data types. The microtiter plate grid coordinates play the role of the spatial coordinates, whereas the aspects of time dependence of the observed patterns enter through the measurement sequence dependence of the HTS plate measurements. While the analysis of time series data in the earth sciences allows the investigation of many important natural phenomena and their underlying physical mechanisms, the interpretation of such derived data in HTS remains at the phenomenological level, but nonetheless allows insight into the behavior of the assay and associated experimental equipment.

The data correction methods which are described in Section 7.7 can relatively easily be implemented and integrated in a software system for plate-based HTS data analysis because they are mostly based on established numerical algorithms that can often be adapted and modified to create outlier-resistant variants.

Remark on variable index notations The following sections deal more or less formally with mathematical and statistical aspects of plate data, that is, readout values x_{ij} in a two-dimensional grid arrangement of n_{rows} rows with index i and n_{cols} columns with index j. For a sequence of n_{plates} plates with index k this third index is added: x_{ijk} is then the value at grid point (i, j) on the kth plate (Fig. 7.1). The triple index notation is used if the sequence-neighborhood context is important and needs to be taken explicitly into account, otherwise this index is dropped with the understanding that the results apply to all data sets or plates with index k in the same fashion. If the grid position is not important, then the elements of the data set (or data vector) $\boldsymbol{x}$ can be indexed linearly as x_i. The set of all values of $\boldsymbol{x}$ is also denoted by $\{x_i\}$. For the treatment of some specific problems, continuous functions of two (or three) dimensions that have defined values for all real values of the spatial coordinates x, y and the measurement time point t are being introduced: $f(x, y, t)$. As above, the notation is simplified to $f(x, y)$ if the time or sequence information is irrelevant. Given the equivalence pairs (column coordinate $j \leftrightarrow x$ direction; row coordinate $i \leftrightarrow y$ direction; plate sequence index $k \leftrightarrow$ time t) f has the value f_{ijk} at the discrete grid points (i, j, k) or simply f_{ij}, if the sequence index k is not

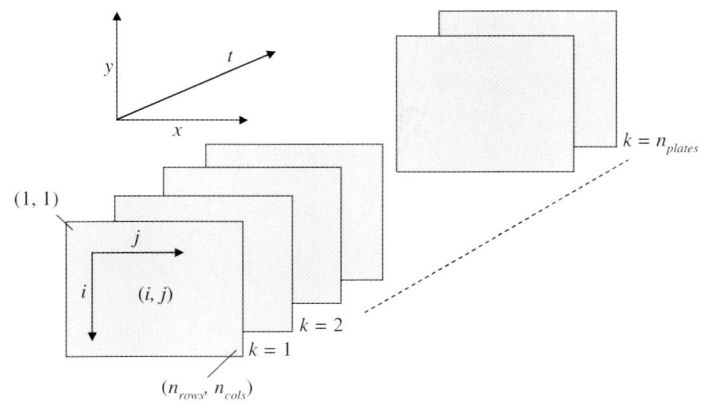

Figure 7.1 Representation of the plate and plate series coordinate and index system used in this chapter. Discrete plate grid coordinates (i, j) follow the usual plate row and column numbering convention. k indicates the index in the measurement sequence. The coordinate system of functions $f(x, y)$ that depend on the equivalent two-dimensional spatial coordinates x and y is also indicated. The measurement time t is often an auxiliary data dimension and functions f may explicitly depend on it, as for example in $f(x, y, t)$. The transition between the continuous spatial coordinate system and the grid indices is trivial.

explicitly needed. The continuous spatial coordinate x and the numerical values x_{ij} (for example, absolute plate readouts or normalized % values) are separate quantities. The meaning should be obvious from tehe respective usage context.

7.2
Statistical Considerations in Assay Development

While the main emphasis of this chapter lies in presenting and explaining various methods for analysis, correction and visualization of the "production" data of an HTS campaign, the basic features of what makes up the "quality" of such data need, of course, to be addressed during the assay development phase. Typical quantities being optimized during that stage are the signal-to-noise ratio (SNR) or signal-to-background ratio (SBR) (that is, measures related to the dynamic range of the assay readout or the "assay sensitivity"):

$$\text{SNR} = (H-L)/sd_L \tag{1}$$
$$\text{SBR} = \varrho_{HL} = H/L \tag{2}$$

where H and L are high control and low control (background) signal averages, respectively, and sd_L the low (background) signal standard deviation calculated from all the wells of the corresponding sample type on a test plate. In biochemical assays, these values could correspond to the measured reader output when the signal of interest is not inhibited (high control H, neutral control, zero effect level: for example, wells containing assay vehicle without any compound) and when it is fully inhibited (low control L, positive control: for example, a solution containing a potent reference compound) or simply corresponds to a basal background reading. In an agonist assay these labels can be exchanged, H now signifying the readout level of a reference agonist and L corresponding to the "zero" effect or the basal value. ϱ_{HL} denotes the SBR value and is used in certain data correction equations that will be encountered later in this chapter.

Another quantity of interest is the coefficient of variation, CV (%), defined as

$$CV\,(\%) = 100 sd/\bar{x} \tag{3}$$

where sd is the standard deviation and $\bar{x}$ the mean of the set of data values $\{x_i\}$. A standard combined measure of the quality of an assay, taking into account the variability of the readout values as compared with the separation of the H and L levels, is the well-established *assay window* coefficient Z' [26]:

$$Z' = 1 - (3sd_H + 3sd_L)/|H-L| \tag{4}$$

where sd_H and sd_L are the standard deviations of the high and low control values, respectively. Z' can take on any value in the interval $(-\infty, 1)$. Z' values >0.5 are generally accepted as being the hallmark of an excellent or good assay. Values of Z' between 0.5 and 0 still allow a distinction of inactive and active compounds, although the % values as derived, for example, with eqn. (6) will have a increasing uncertainty as $Z' \to 0$. At values of $Z' \leq 0$, the separation band between the $3sd$ limits of the high and low control "variability region" vanishes and the tails of the control distributions start to overlap, making a distinction of active and inactive compounds increasingly error prone and finally impossible. The *screening window* coefficient Z is defined in an analogous way to Z', but reflects the separation and spread of the average sample response S and the low or background control level [26], thus in eqn. (4) replacing H and sd_H by S and sd_S, respectively, resulting in

$$Z = 1 - (3sd_S + 3sd_L)/|S-L| \tag{5}$$

Z is often influenced by the type and composition of the compound collection and by the concentration of the solvent used to prepare the screening library. In practical screening situations, one usually finds $Z < Z'$. Although the Z and Z' values are useful quality measures for antagonist screens, they are less significant for searches for agonists where a significant increase in a signal above the background level may be good enough. In addition to the optimization of the simple assay quality indicators in eqns. (1)–(4), which are typically determined at the readout levels of neutral and positive controls, the reproducibility of the readouts at intermediate values between the 0% and 100% effects should also be carefully checked (for example, with reference compounds of varying potency or by titration of a single reference compound). At the assay optimization stage the derivation of IC_{50} values of such reference compounds and not only a simple comparison of the single point readouts is recommended. For more complex types of assays with multiparameter readouts, also other derived quantities (for example fluorescence polarization values) and their respective variability need to be assessed during this stage.

Usually one calculates normalized values p_i (for example, % inhibition) from the set of raw instrument data x_i to generate more easily interpretable values and to guard against variability between plates, between the days on which experiments are performed, between different cell and/or reagent batches and other not easily controlled factors via

$$^1 p_i = 100(x_i - H)/(H - L) \tag{6}$$

where x_i = signal of sample i. Expression (6) leads to negative % values for inhibitory activity and positive values for stimulatory activity. This is one possible convention and can easily be changed by multiplying eqn. (6) by −1 in order to obtain positive "% inhibition" values. Such transformations of the % scale are trivial and do not affect any of the fundamental aspects of data processing and data correction that are visited in the following sections. Other normalization expressions may be used or may be necessary for certain types of assays (for example, by calculating a ratio to a basal level if there are no known agonists for a specific target) or − and this is one of the important topics of this chapter − when trying to compensate for certain systematic errors in the data (see Sections 7.6 and 7.7)

A useful quantitative tool to assess the degree of reproducibility of a set of experimental values is a graph of the difference $d_i = (x_{i1} - x_{i2})$ versus the average $\mu_i = 0.5(x_{i1} + x_{i2})$, where the indices 1 and 2 denote the replicates of the otherwise identical measurements of the data sets which need to be compared. The mean m and standard deviation sd of all differences d_i can be determined and corresponding horizontal lines at ordinate values $\{m, m + \lambda sd, m - \lambda sd\}$, with $\lambda = 2$ or 2.5, can be added to the graph to depict the approximate normal probability 95 % or 99 % tolerance bands of $\{d_i\}$. This graphical representation is known as dispersion plot or Bland–Altman plot [27] and is also used in assay development [28]. If data are distributed over several orders of magnitude in response values, a logarithmic transformation may be appropriate, especially if the expected error scales roughly in proportion to the measured values. This type of transformation is especially relevant if an assay involves exponential signal amplification [cell growth, polymerase chain reaction (PCR) technology]. The more generally applicable Box–Cox transformation [43] may have to be applied in other cases. Even after such an optional transformation, sd may have a remaining dependence on μ. This type of plot will allow one to assess this dependence very clearly if enough data are distributed over the full range of interest.

Simultaneous optimization of readout values and the derived quality measures described above is most effectively done via rational design of experiments (DOE) methods [29]. DOE methodology can assist the scientist in (1) screening for relevant factors influencing the readout, (2) optimizing an existing assay protocol with respect to the quantities mentioned above or (3) searching for factor combinations that lead to robust readouts when the assay parameters deviate from their respective nominal values. DOE-based data analysis methods even allow one to estimate the influence of so-called uncontrolled or nuisance variables, for example, a non-controllable change of the room temperature, ambient humidity, light levels or other environmental factors, provided that they are recorded in parallel to the execution of the designed experiments (analysis based on randomized block designs [30]). Automated assay optimization (AAO) systems for 96- and 384-well plate technologies can be set up in a laboratory by integrating appropriate DOE software, suitably programmed liquid-handling robots and statistical analysis and graphics software [31]. Such systems are also commercially available [32]. This technology is being successfully applied in many companies for assay development and assay adaptation in the HTS area.

In later sections, a series of fairly powerful methods that allow the correction of systematic background response variations which go beyond the "usual" data normalization step will be presented and reviewed. Nonetheless, it is clearly recommended to work first towards robust assay conditions before starting screening and then having to rely heavily on mathematical correction and result-stabilization methods. Although these correction methods reliably provide much improved data quality, it is obvious that they have limits of applicability. If the necessary data correction factors become "too large", the uncertainties that are inherent in the measured data set can become strongly inflated (see Section 7.7.10), thus adding unwanted sensitivity to small variations of the correction factor and potentially large variability in the final corrected activity values. Under those circumstances, a repetition of the measurements is recommended.

After optimizing the assay to reach an adequate and stable response under HTS or uHTS conditions, one is usually also interested in determining where reasonable primary threshold values can be set. In every high-throughput screen it must be a goal to minimize both the false-positive and false-negative rates. This is important for reasons of economy and logistical effort in the confirmation screen and to maximize the probability of successfully identifying suitable lead compounds or lead series. The detailed investigation of the distributions of the inactive and the active compound populations and estimation the respective contaminations in both directions across the threshold can assist in the search for an optimal value. The particular optimum criteria need to be separately assessed for each screen. Questions related to threshold setting and the assessment of the compound activity distributions are addressed in more detail in the literature [33–35]. Further aspects of this problem will also be readdressed in Sections 7.5.1 and 7.5.3.

7.3
Data Acquisition, Data Preprocessing, and HTS Data Analysis Environment

Several widely used assay readout technologies result in one readout value per well (for example, luminescence, absorbance, fluorescence intensity and radioactivity count values), whereas others result in two numerical values that need to be combined (for example, fluorescence anisotropy or polarization, dual-wavelength HTRF® [Homogeneous Time Resolved Fluorescence]) or in an even higher number of readouts for FCS-based (Fluorescence Correlation Spectroscopy) assay technologies [4, 36]. Multiparameter readouts allow "multivariate" looks into the data which often provide further diagnostic possibilities to discriminate between different types of response effects. In a two-parameter readout the changes in the intensity ratio $I_1/(I_1 + I_2)$ or I_1/I_2 are a measure of the sought effect, whereas changes in the total intensity $(I_1 + I_2)$ can indicate the presence of a response artifact, for example, the attenuation of the signals by a nonspecific mechanism, quenching or the enhancement of the signals due to the presence of autofluorescent compounds. With higher data dimensions and a richer set of information for each measured well, it becomes possible to discriminate, filter out or flag results from wells that exhibit these and other types of unwanted effects (light scattering, dispensing artifacts, cell, toxicity, compound aggregation and so forth). Such artifacts can be detected either in a data preprocessing step before a further reduction and correction to a final % activity value is made or as part of the final quality assessment of the screening data. This is a very valuable approach of well-level QC that helps to eliminate false positives and goes far beyond the possibilities that exist in technologies with single readouts [11, 36].

Preprocessing and data reduction steps to extract the most prominent and most informative features from much larger amounts of raw data are needed in several types of technologies used in HTS. Typical examples of the latter are kinetic measurements where key features of individual curve shapes from fluorescence image plate readers (FLIPR®) [37] for measurements of intracellular Ca^{2+} concentrations or response data derived from images arising from cellular high-content screening (HCS) experiments need to be assembled and extracted before the usual % activity calculations can be made. Derived information from those assay technologies is inherently multiparametric and can be rich in information and in most cases allows the application of the mentioned multivariate data analysis methods to classify and detect unwanted effects. FLIPR curves which usually consist of 100–200 sampling points are often reduced to two to three summary and key feature statistics before normalization and response correction. HCS images and derived data from a limited number of observed cells in individual plate wells provide again a slightly different flavor of HTS data analysis [38]. In each well, suitable ensemble statistics (counts, areas, intensities, quantification of shape changes, nuclear translocation and others) from certain classes of cells, cellular or subcellular features are assembled either as absolute numbers or very often in the form of ratiometric determinations. Such ratios are often less sensitive to systematic variations but factors such as concentration variations and temperature gradients can also generate response trends and edge effects in these kinds of experiments. In HCS, additional well-based quality control quantities besides the ones generally applied in HTS (see Section 7.6) may have to be introduced. HCS data derived from individual wells need to be based on a certain minimum number of cells and this quantity is not necessarily correlated with the global quality measures, such as Z' [eqn. (4)] or an equivalent value, which is based purely on the control wells.

Informatics solutions should allow early format standardization of reader output, automated data set identification and data transport from readers and HTS robots to the data analysis environment. Data sets that need to be analyzed together (for example, for ratio calculations) should be quickly identifiable and retrievable from a common raw data repository. This information needs to be augmented by logistics information about data sets (for example, type of readout in multiparameter experiments or replicate information), plates (plate identification, plate maps), compounds (compound identifiers, chemical structures) and experimental process and meta-data through full integration of the data analysis and visualization environment with the relevant data management components. Although simple on-line data QC systems do not need access to all of this information, a standardized and modular setup will nonetheless allow one to "pull in" additional information or perform additional calculations with little extra effort if the HTS data systems are built on a sound software system architecture. In any case, such information is certainly needed in the main HTS data analysis and visualization software environment in order not only to base decisions on "isolated" data sets, but also to be able to take into consideration any additional information that is directly or indirectly linked to them. An integrated but modular software architecture allows flexible "rewiring" and facilitates the addition of new data processing and algorithmic elements when needs arise. This gives a crucial advantage in today's dynamic environment where measurement technologies and the detailed requirements for data integration and processing are often changing.

7.4
Data Normalization

The usual controls-based normalization step [eqn (6)] applied to plate data may result in misleading % activity data if the majority of the samples are inactive and if there is an appreciable bias b_H between the average of the neutral controls H and the average response level $\bar{x}$ of all wells:

$$b_H = H - average(\mathbf{x}) = H - \bar{x} \qquad (7)$$

where $\bar{x}$ or $average(\mathbf{x})$ is either one of the $mean(\mathbf{x})$, $median(\mathbf{x})$ or $mode(\mathbf{x})$ functionals. The more complicated case of position-dependent response shifts will be treated in later sections. As will be seen further below, the *mean* is often used but sensitive to outliers. Although the *mode* is closest to an intuitive understanding of "average" (that is, the "most probable" value of a data set) one uses much more often the *median* as a more easily calculated value. Some more insight and information on resistant statistics and why it needs be used in the HTS data analysis context will be provided in Section 7.5. Expression (6) can now be suitably corrected to take the response bias [eqn (7)] into account. Such a general background response shift can be generated through different concentrations of solvents in the sample wells compared with the control wells, systematic response errors at the well locations of the controls (edge effects) and solvent effects of compound sample solution on cells. Often the compound samples and the controls are brought on to the plates at different times and are handled by different equipment. Non-robust kinetics, temperature effects and so on can then easily lead to different response behavior.

A serious side-effect of the $b_H \neq 0$ situation is the strong possibility that the false-positive and/or false-negative rates are adversely affected if the hit threshold is not adequately adjusted, especially if the bias varies over the length of the screen or between plates.

Since in HTS one usually has the situation that most of the measured compound samples in a diverse random library are inactive on a given target, it can be assumed that $\bar{x}$ corresponds to 0% normalized activity and with this a modified "sample-based" self-normalization of the data can be defined. With $median(x)$ used to calculate $\bar{x}$, one has a "sample median"-based normalization. It needs to be cautioned, though, that compound libraries that are focused on specific biological targets or target classes may result in many active compounds and then $\bar{x}$ does not necessarily correspond to the 0% activity level. This can lead to erroneous values for b_H and to misclassification of hits and non-hits because the assumptions inherent in sample based normalization are no longer true. In such cases hits should still be identified by the standard control-based normalization scheme. This situation is further illustrated in Section 7.8.

When correcting eqn. (6) for such a background response bias, one arrives at the following second variant of % activity values ^2p_i for the sample-based normalization approach:

$$^2p_i = 100(x_i - \bar{x})/(\bar{x} - L) \tag{8}$$

resulting in values close to 0% for the bulk of the samples. Since there is no experimental information about a similar potential bias b_L of the low controls, we leave L unchanged. If one assumes or, much better, can support with experimental data that such bias is produced through a linear multiplicative response change over the full assay readout scale, then the bias of L (b_L) can be calculated, using eqn. (2):

$$b_L = b_H(L/H) = b_H/\varrho_{HL} \tag{9}$$

and one arrives at the third variant for the normalization expression:

$$^3p_i = 100 \cdot (x_i - \bar{x})/(\bar{x} \cdot (1 - L/H) - L) \tag{10}$$

or

$$^3p_i = 100 \cdot (x_i - \bar{x})/(\bar{x} \cdot (1 - 1/\varrho_{HL}) - L) \tag{11}$$

This type of normalization is clearly only valid for linear responses and multiplicative distortion effects and does not hold for situations where the response is close to saturation levels at either the high or low end of the observation range. These aspects should best be clarified in the assay development phase. For large values of the ratio ϱ_{HL} the resulting % activity values ^2p_i and ^3p_i become practically identical.

The normalizations in eqns. (8) (10) and (11) are still insufficient for situations where there exist appreciable plate position-dependent systematic effects. More refined normalizations that also take such response variations across the plate surface and between different plates into account will be discussed further below, but the basic form of the normalization expressions just derived can be taken over to those situations also. Several aspects of "non-controls"-based normalization are also addressed in the literature [7, 8].

In order to demonstrate the effect of the different types of normalizations in eqns. (6), (8) and (10), representative cases are illustrated in Table 7.1. For all selected cases observed, H (neutral control) and L (positive control) averages of 1000 and 200 are assumed, respectively. All but the first example case in the table exhibit an average sample response $\bar{x}$ of 900 that differs from H by a bias b_H of 100. The mechanism of the response change for the sample area is assumed to be different for all the variant cases: an additive shift (case 2), a multiplicative shift (case 3) and a combination of the two (case 4). Different behaviors of the compound samples compared with the control averages H and L are thus explicitly modeled. In particular, L', the *low* level in the "sample region" of the plate, is only inferred from the change mechanism, whereas H', the *high* level in the sample region, is observed and determined by $\bar{x}$. Depending on the mechanism that creates the systematic differences, the unknown sample response is chosen to correspond to the particular 50% effect level and the resulting % values according to the three different equations for ^1p to ^3p are calculated from the "raw" data numbers. Larger deviations from the correct % value are observed for the control-based normalization ^1p as compared with the two sample-based normalizations, the latter always being closer to the theoretically correct value of 50% and reaching that value for the multiplicative model and the ^3p normalization The normalization expression for ^3p was actually derived for these experimental conditions, so this provides a small consistency check. One can therefore conclude that sample-based normalizations give on average more consistent results than the "classical" control-based normalization, even if the exact mechanism for the response shifts and distortions in a given assay are not known.

Table 7.1 Illustration and comparison of the % activity values derived from the normalization expressions for ^1p, ^2p and ^3p and different types of systematic variations in the sample data (see Section 7.4).

Case No.	Systematic effect for *samples*	Observed values for controls	Unobserved distorted values	Sample x (absolute readout)	Real p (%)	^1p (%) eqn. (4)	^2p (%) eqn. (6)	^3p (%) eqn. (10)
1	No systematic effect	$H = 1000$ $L = 200$ $\bar{x} = 1000$		600	−50	−50	−50	−50
2	Additive effect (shift by -100)	$H = 1000$ $L = 200$ $\bar{x} = 900$	$H' = \bar{x} = 900$ $L' = 100$	500	−50	−62.5	−57.1	−55

Case No.	Systematic effect for *samples*	Observed values for controls	*Unobserved* distorted values	Sample x (absolute readout)	Real p (%)	^1p (%) eqn. (4)	^2p (%) eqn. (6)	^3p (%) eqn. (10)
3	Multiplicative effect (factor 0.9)	$H = 1000$ $L = 200$ $\bar{x} = 900$	$H' = \bar{x} = 900$ $L' = 180$	540	−50	−57.5	−51.4	−50
4	Combined additive and multiplicative effect (factor 0.95 and shift by -50)	$H = 1000$ $L = 200$ $\bar{x} = 900$	$H' = \bar{x} = 900$ $L' = 140$	520	−50	−60.0	−54.3	−52.8

Where to place control samples on a plate and how many of each type are often a topic of debate. Although the placement can be important for the control-based normalization ^1p if an assay exhibits appreciable edge effects or response trends, there are several data correction methods shown and explained later that actually make the placement and, to a degree, also the number of control wells a point of reduced relevance. The methods that rely on the position-dependent sample background response to provide the level of 0% activity (sample-based normalization) are not very sensitive to those choices. Depending on the type of systematic effect and normalization expression used, there can be a remaining small dependence of the corrected % values on the response scale defined by the spread of the neutral and positive controls, as shown, for example, by the dependence of expression (11) on ϱ_{HL}.

7.5
Robust Statistics in HTS Data Analysis

7.5.1
The General Problem

As was noted previously, most of the diagnostic and normalizing calculations need to be protected from the influence of "outlier" data in order to obtain stable results. The methods which are geared towards providing such outlier resistance are at the same time very useful to *detect* such outliers if one needs to identify them for separate treatment. Clearly, if one is applying largely automated data analysis, data reduction and correction procedures – as has to be done in the field of industrial HTS – such resistant methods are actually indispensable.

Robust statistics try to give useful results even when certain assumptions about the behavior of data are violated. One such assumption in classical statistics is that samples are independently and identically distributed (i.i.d.) and can be represented by normal probability densities. Very often real experimental data do not fully correspond to this situation because distributions are not normal, but have heavier tails and outliers (samples far away from the bulk of the data) may be present. In HTS, such "outliers" often appear naturally because there is on average a 0.1–1% and in some (usually cell-based) assays up to a 5% or even higher contribution from active compounds in the screening library. The % activity or potency distribution of HTS readouts is clearly non-normal in the presence of active samples. This situation can be modeled with a mixture distribution which for the zero-effect samples is assumed to be the normal $N(0, \sigma^2)$ for reasons of illustrational simplicity and a second component with a much wider and longer tailed distribution density $g(x|q)$ which describes the "active" sample population and depends on a set of (here unspecified) parameters q. The shape of this distribution component depends on details of compound library composition, assay technology and assay conditions. The modeled activity distribution density is then

$$f(x) \sim (1-p)N(0, \sigma^2) + pg(x|q) \tag{12}$$

with the probability p for the samples to be from the "active" population. The probability p is proportional to the observed hit rate. It needs to be mentioned that the assumption of a normal distribution for the inactive sample population is not fully adequate for most real cases (see Section 7.5.3).

7.5.2
Threshold Setting – a Simple Model

From eqn. (12), it is clear that the standard deviation of f is being inflated above σ for $p > 0$ and that one needs an estimation method to determine the sample standard deviation of the main density peak which is resistant to outliers. This will then also allow one to derive a more reliable hit detection threshold that is based on this sample statistic. Such a threshold can be calculated as

$$\Theta = m + \lambda s \text{ or } \Theta = m - \lambda s \tag{13}$$

depending on the direction of the active samples, where m is an estimate of the average location of the inactive sample population and s an estimate of its scale (spread), for example $s = sd$, the standard deviation. Very often one chooses $\lambda = 3$. Provided that the central distribution is approximately normal, the factor 3 that is typically, but arbitrarily, used to set the lowest possible threshold will limit the false-positive hit probability from the inactive sample population to ~0.13%. In the usual controls-based normalization in eqn. (6), one often just sets $\Theta = \lambda s$ because 0% is the reference level from which sample activity is determined. In assays with high primary hit rates, as often occur in cellular assays, the hit threshold is often adjusted to higher values in order to obtain hit numbers that can still be handled by the downstream process steps (hit compound selection and physical "cherry picking", measurement and analysis of many dose–response curve experiments). An increased level of automation and streamlined sample logistics in this area allow one to go to higher hit numbers for the activity/potency confirmation step and then the statistical limits will provide some guidance of where to expect increasing false-positive contributions. In assays with low hit rates one may loosen the threshold and include more compounds with marginal potency. Since the potency of possible hit compounds alone is not the final selection criterion, this can easily be done as long as the downstream processing can accommodate the load. In this sense, the statistical threshold Θ from eqn. (13) only provides some guidance, but not a hard limit. This is especially true if the distribution of "inactive" samples does not follow the normal, but is intrinsically longer tailed.

7.5.3
Threshold Setting – a Complex Model

Determination and modeling of the distribution density $g(x|q)$ [eqn. (12)] based on suitable experimental data can be of interest for optimal threshold setting when, for example, the balancing the false-positive and false-negative rates is the aim of such an in-depth analysis. It could be of interest to increase the number of true positives while still keeping the number of false positives and false negatives under control. Some possible solution approaches are as follows:

1. Extraction of $g(x)$ from a given measured $f(x)$ when the distribution of the inactive samples (in practice not necessarily a pure normal) is known from measurements [34, 35, 39] or by non-parametric estimation from the activity distribution of the controls.
2. Identification of $g(x)$ by maximum likelihood estimation of a mixture model [40] or by L-moment estimation, a method used in extreme value statistics to determine the distribution parameters [41, 42]. Extrapolation of $g(x)$ below the threshold may allow one to make a rough prediction of the false-negative rate and thus an estimate of whether it would make sense to shift the threshold to a lower value and rescue potential hits – of course, always at the cost of an also increased false-positive rate. However, there may be a net benefit in such a threshold shift.

This problem is not followed any further here because it is not the main topic of this chapter. It needs to be mentioned, however, that analyses along these lines clearly need to be done on data that are *corrected for all positional trends*, otherwise the density distribution of % activity values is easily broadened and skewed by such distorting effects and will mirror a different distribution situation, lead to non-optimal choices for the threshold and to clearly increased false-positive and false-negative rates. This is clearly illustrated by comparing the % activity distributions shown later in Figures 7.14 and 7.15.

7.5.4
The Most Important Robust Estimation Methods for Data Summaries

There are three major classes of robust methods that deal with outliers in the data: (1) methods with *outlier rejection*, (2) bringing *outliers closer to the center* and (3) smooth rejection with weights tending to zero, allowing an explicit zone of "uncertainty" of whether a datum is an outlier or not. The statistics literature describes many resistant measures of location (for example, trimmed mean, Winsorized mean, *median*, Huber M-estimates) and scale [trimmed standard deviation, interquartile range (*IQR*), median absolute deviation (*mad*), Huber M-estimates and others] to summarize data sets [43–48].

For practical applications in HTS data analysis and quality control, the use of the *median*:

$$median = q(x, 50) \tag{14}$$

where $q(x, 50)$ is the 50% quantile of a set of ordered observations $x = \{x_i\}$ is recommended. The interquartile range IQR:

$$IQR = q(x, 75) - q(x, 25) \tag{15}$$

needs to be divided by 1.349 to obtain a consistent equivalent of the standard deviation for the normal distribution:

$$s_R = IQR/1.349 \tag{16}$$

The median absolute deviation or *mad* scale estimator is defined as

$$mad = median_i(|x_i - median(x)|)/0.6745 \tag{17}$$

and again the numerical factor 0.6745 in the expression for *mad* ensures consistency with the equivalent standard deviation of a normal distribution. One can also define the *mmad*:

$$mmad = min(mad_+, mad_-) \tag{18}$$

as an even more resistant version of the *mad*. The values mad_+ and mad_- are the one-sided *mad* estimators for the $\{x_i\}$ with $[x_i - median(x)]$ being >0 and <0, respectively. Particularly in cases where there are considerable mixture contributions from marginally active samples that are strongly overlapping with the inactive population on only one side of the median, *mmad* is anchored on the lighter tailed side of the distribution and is only weakly influenced by the contaminating sample population. It is statistically less efficient (has higher variability) than the *mad*, but this may be a "cost" worth paying when looking for protection against outliers and the effects of strongly asymmetric distributions. The application of other types of robust methods such as trimming, Winsorizing and soft trimming (for example, by smooth residual weighting using Tukey's bi-weight function) [43] and the role of the iterated reweighted least squares (IRLS) procedure [49] in the context of regression problems are addressed in the later section on background response surface modeling and data correction (see Sections 7.7 and 7.7.5).

7.5.5
An Illustrative Example: Performance of Location and Scale Estimators on Typical HTS Data Activity Distributions

The sensitivity of a selection of those estimators for location (average) and scale (spread) on a series of low but typical values for the "contamination" percentage p (0–15%) is illustrated in Table 7.2. The table is based on simulated data sets of 352 samples (a 384-well plate with two columns used for control and reference samples) drawn from the assumed normal mixture distribution:

$$f(x) \sim (1-p)N(0, \sigma_1^2) + pN(50, \sigma_2^2)$$

with $\sigma_1 = 10$ and $\sigma_2 = 20$. From σ_1 and eqn. (13) with $\lambda = 3$, one obtains a theoretical threshold Θ of 30. This value can be compared with the data-based thresholds in Table 7.2 that are derived from the *(mean, sd)* and *(median, mad)* estimators, respectively. The *median* and *mad* are the "estimators of choice" in terms of the degree of outlier resistance and lead to consistently good results even for relatively high mixture contamination percentages of 10 or 15%. The so-called breakdown point (a measure of safety and reliability) of the *median* is 50%, that is, the data set can contain up to 50% "bad" data distributed over both tails without affecting its value. In addition, as seen in eqns. (14) and (17), both *median* and *mad* are simple order statistics and very easily and quickly computable from the data.

Table 7.2 Values of different classical and resistant location and scale estimators as a function of contamination percentage p based on a simulated normal mixture distribution of 352 random sample values. The derived threshold ($\Theta = m + 3s$) has to be compared with the theoretical value of 30.0. Thresholds based on the median and mad estimators are preferred.

p (%)	Location m				Scale s					Threshold Θ	
	mean	5% trimmed mean	Huber	median	sd	s_R	Huber	mad	mmad	mean + 3sd	median + 3mad
0	−1.0	−1.1	−0.9	−1.1	9.6	9.7	9.7	9.8	9.2	27.8	28.3
1	−0.4	−1.0	−0.8	−1.0	11.4	9.8	9.8	9.7	9.4	33.8	28.1
2	0.7	−0.5	−0.4	−0.5	13.7	9.9	10.1	10.0	9.6	42.1	29.5
5	2.0	−0.2	0.0	−0.2	16.4	10.1	10.6	10.1	9.7	51.2	30.1
10	3.9	0.5	1.0	0.5	18.3	10.9	11.8	10.9	10.5	58.8	33.3
15	6.2	1.5	2.5	1.5	20.8	11.5	13.5	11.2	11.0	68.6	35.1

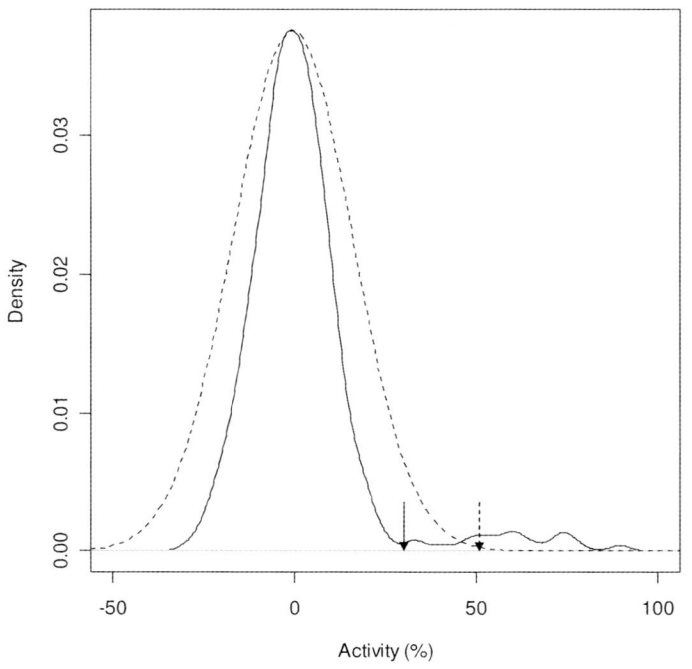

Figure 7.2 Smoothed density distribution of 352 random samples drawn from $f(x) \sim (1-p)N(0, \sigma_1^2) + pN(50, \sigma_2^2)$ with $\sigma_1 = 10$ and $\sigma_2 = 20$ and $p = 0.05$ (solid line). The dotted line indicates the normal distribution density which is based on the calculated sd of the mixture distribution $f(x)$. This distribution is normalized to have the same amplitude at $x = 0$. The arrow at 51.2% indicates the threshold $\Theta = 3sd$ as calculated from the "classical" statistical estimators (*mean*, *sd*), whereas the arrow at 30.1% indicates the threshold as calculated with resistant statistics (*median*, *mad*) and which corresponds very closely to the theoretical value of $3\sigma^1 = 30\%$.

In Fig. 7.2, the $p = 5\%$ mixture distribution example from Table 7.2 is illustrated. The shift of the 3 s threshold Θ as calculated with the classical standard deviation *sd* (dashed overlay of corresponding normal distribution) to the vicinity of the theoretical value at 30% as calculated with the highly resistant *mad* is appreciable.

7.5.6
A Robust Outlier Detection Method

One of the main uses of robust estimators of location and scale (spread) as defined and explored above is the labeling and possibly the exclusion of outliers from further calculation or to treat them separately. Outliers are somehow "anomalous" values in the data that may occur through measurement errors or due to the sample not being from the same parent distribution. For plate control samples one is mostly interested in eliminating "bad" data, whereas for the compound samples the primary interest is to obtain statistical measures that reflect the distribution properties of the inactive samples and are not strongly influenced by the hit population. It has already been indicated that the mean and standard deviation are not appropriate tools for the identification of outliers as they themselves are strongly influenced by the very same outliers they are intended to identify. Replacement candidates which were recommended above are the *median* and *mad*. If they are used instead of the mean and standard deviation, one is led to an outlier identification rule of the form [49]

$$|x_i - median(x)| \geq \lambda_\alpha mad(x) \tag{19}$$

with the parameter λ_α influencing the confidence of the outlier declaration. The constant λ_α can be chosen to standardize the behavior of the outlier diagnostic for normally distributed samples. The concept of an outlier as a datum behaving "differently" to the majority, for example by lying "further away" from the mean or median is not very precise, but it is possible to limit the chance of making false decisions when assuming a known parent distribution density. The constant λ_α is determined here in such a way that with probability at least $(1 - \alpha)$ a single observation $x_i \sim N(\mu, \sigma^2)$ is not identified as an outlier (α for example 5, 1, 0.5 %). This can easily be done via Monte Carlo simulations. The choice of λ_α is especially important for smaller sample numbers which are usually found for control groups on the screening plates. Appropriate values for λ_α for $(1 - \alpha) = 95\%$, 99%, 99.5%, 99.75% and for $n = 3, 4, 8, 16, 32$ as typically used on microtiter plates are shown in Table 7.3 [50]. For very large n, λ_α tends towards the $(1 - \alpha)$ percentile of the normal distribution. Those values are also indicated in Table 7.3.

Table 7.3 Critical values λ_α for the identification of outliers in an otherwise homogeneous dataset of size n when using the statistic $o(x_i) = |x_i - median(x)|/mad(x)$. See Section 7.5.6.

λ_α	$n = 3$	$n = 4$	$n = 8$	$n = 16$	$n = 32$	Asymptotic normal
$1 - \alpha = 95\%$	7.0	3.2	2.6	2.3	2.1	1.96
$1 - \alpha = 99\%$	36	7.8	4.5	3.4	2.9	2.58
$1 - \alpha = 99.5\%$	72	11.3	5.6	3.9	3.2	2.81
$1 - \alpha = 99.75\%$	145	15.6	6.7	4.4	3.6	3.02

In addition to the often used simplistic $3sd$ rule for defining hit thresholds, the "outlier" definition in eqn. (19) can be applied to the identification of hit thresholds after the application of the correction methods for systematic effects that lead to a "normalizing" and "narrowing" effect of the distribution of activity values. An illustration of this desired effect will be shown later (see Section 7.8).

7.5.7
Outlier-resistant Versions of Simple HTS Data Quality Indicators

From the previous explanations, it is obvious that one can define robust equivalents of all the simple indicators of data quality, as for example the coefficient of variation:

$$RCV\,(\%) = 100\,mad(x)/median(x) \tag{20}$$

and the Z' value:

$$RZ' = 1 - (3s_H + 3s_L)/|RH - RL| \tag{21}$$

where s_H and s_L are now calculated via either $mad(x_{control})$ or $s_R(x_{control})$ and the robust high and low control averages RH and RL via $median(x_{control})$ instead of the usual standard deviation sd and arithmetic means. A robust Z value RZ can be defined in a completely analogous way by replacing the high control-related statistics by the compound sample RS values. All other quantities of interest for assessing the quality and stability of an assay or an HTS run can be translated to more outlier-resistant versions in a similar fashion as eqn. (20) or (21).

7.6
Measures of HTS Data Quality, Signaling of Possible QC Problems, Visualizations

After taking the instrument readings, the HTS data should be assessed as rapidly as possible for the presence of systematic effects that could affect the interpretability or validity of the screen. The appearance of process artifacts in the data sets should be diagnosed as fast as possible in order to initiate remedial actions and minimize potential data loss due to insufficient quality. This quality control (QC) and quality assurance (QA) process is best supported by rapid on-line or quasi-on-line calculation of key diagnostic quantities and suitable visualization of data sets and derived diagnostics. Certain data correction algorithms described later in this chapter are well suited for diagnostic calculations based on single plates and they can be usefully employed already at these early stages of HTS data analysis and detection of systematic effects.

Computational and visual assessment of HTS data progresses in three stages, for (1) raw instrument data, (2), normalized data and (3) corrected data. As will be seen, each of these stages provides insight into different aspects of the experiment.

Rapid insight into the behavior of the screen and the quality of the data can be obtained by visualizing key statistical quantities such as (1) plots of individual plate statistics (dot plots, line plots, box plots) [45] or of data summaries over a series of plates, (2) summary plots of trends, that is, summarizing key statistics from the data for individual plates and plotting them as a function of the plate measurement sequence, (3) histograms or smoothed distributions of readout values [43], (4) heat maps (spatial false color images) [51] and (5) whole collections of such heat maps in a trellis arrangement [52].

The following demonstration and illustration of the different graphics and visualization types is based on a simulated, but realistic 384-well plate HTS dataset. In order to illustrate the appearance and diagnosis of typical artifacts, several types of systematic errors are included in the simulation of a 300-plate screen that was "performed" in five separate run batches (boundaries between batches at $k = 50, 80, 150, 200$) The main characteristics are as follows:

- H and L controls are at different initial levels for different plate batches. Readout levels decay over time (plate sequence) in order to mimic the degradation and activity decay of an enzyme.
- The hit rates and the (normally distributed) assay noise are different in the various batches (hit rates between 1 and 2.5%, CV of high controls between 5 and 10%).
- The different plate batches exhibit different position-dependent systematic error patterns (trends created by varying polynomial surface distortions in the simulated data set).
- In batch 2 some of the plates were fed into two readers in an alternating pattern. After a while one of the original readers had to be replaced because of a malfunction. The absolute values of the readouts now differ by a factor of ~5 and the replacement reader produces data with strong positional effects because the intrinsic response distortion correction of the detector was not properly calibrated (plates $k = 60$–80).
- In batch 4 several needles of an eight-tip reagent dispenser were malfunctioning, leading to an interleaved striping pattern of the responses in rows 1–6.
- Extremely high hit rates (~50%) are found in five plates of batch 4 (target specific compound library, $k = 154$–158).

All the process- and data irregularities built into the simulation or present in real HTS data can relatively easily be diagnosed by inspecting the types of plots or displays described in the following sections (7.6.1 to 7.6.4); see also [7, 8] for similar examples.

7.6.1
Trends and Change Points

The following key *trend plots* (or run sequence plots) [53] and *change point graphs* produce a good overview about the time (or measurement sequence) dependence of the assay behavior and possible response differences between different experimental runs:

1. All individual raw data values of high and low control wells are best depicted in a x–y dot or line plot (x direction: plate sequence index k; y direction: raw data values) (Figs. 7.3 and 7.4).
2. As an alternative to graph type (1), the plate-wise summary statistics values of the high and low controls and possibly the compound samples can be a

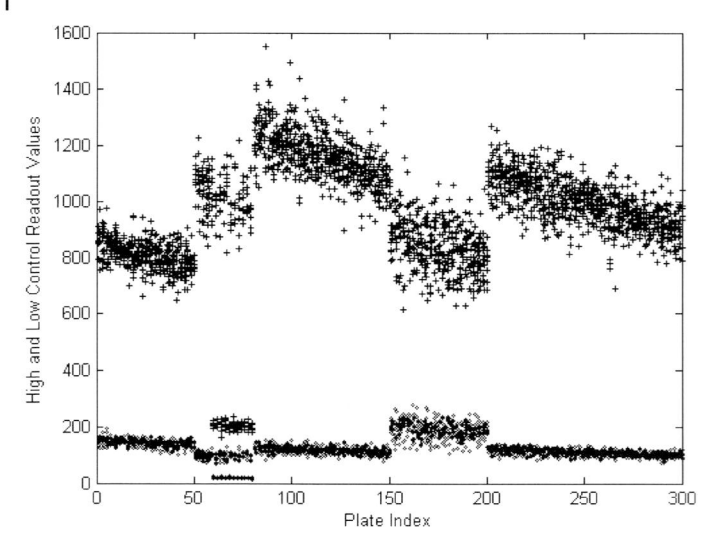

Figure 7.3 Trend plot (sequence plot) of all the individual high and low controls of a series of 300 plates that were measured in different experimental batches (simulated data). Also visible is a decrease in the responses over time within the different experimental runs, which is due to the degradation of an enzyme used in this assay and furthermore certain shifts in the baseline (low control) activity and the signal window.

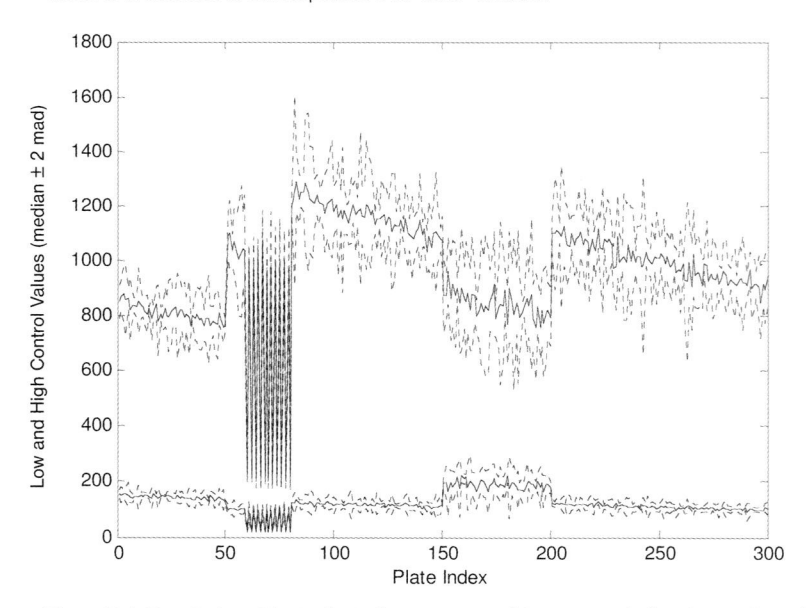

Figure 7.4 Trend plot with similar information content as in Fig. 7.3. The *median* (solid line) and (*median* ± 2 *mad*) values (dotted lines) of the high and low controls, respectively, are depicted. The "erratic" behavior of the high (and low) controls for plate index values from 60 to 80 are due to the data from plates that were measured alternating between two instruments giving different raw data readings levels.

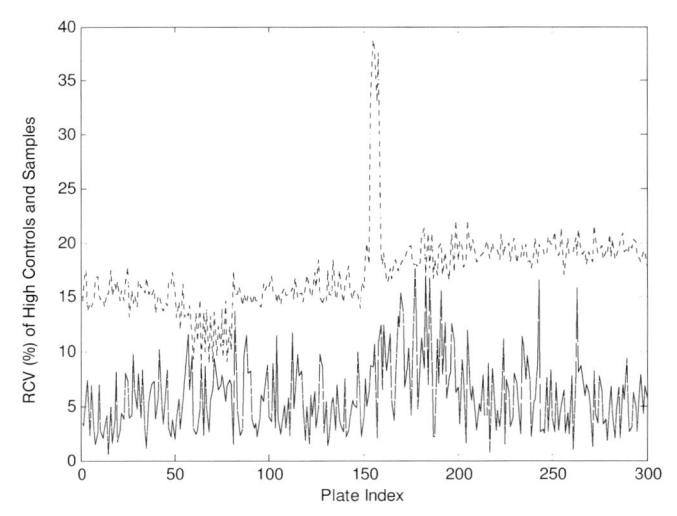

7.6 Measures of HTS Data Quality, Signaling of Possible QC Problems, Visualizations | 173

Figure 7.5 Robust coefficient of variation (*RCV*) of the high control (solid line) and sample data (dotted line) for each of the 300 plates. The large difference between controls and samples is mostly due to systematic response distortions across the plate surface area. The *RCV* peak at plate index ~155 is due to a small series of plates with high hit rates. As described elsewhere, the validity of a sample-based normalization and background response correction for those plates is probably at least questionable. It will have to be carefully checked.

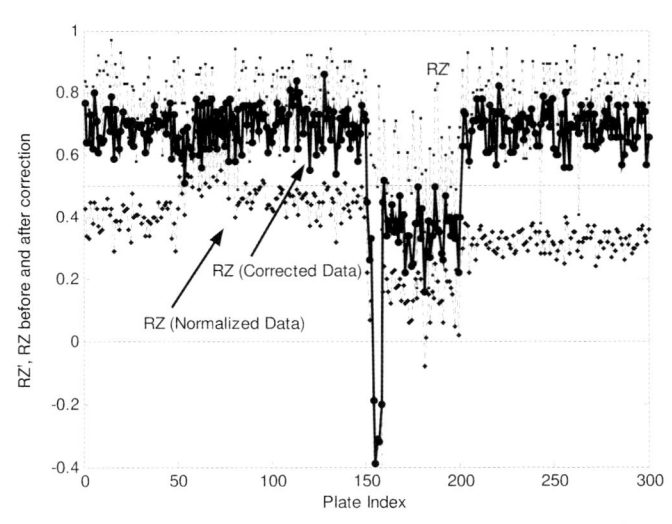

Figure 7.6 Robust *Z'* (assay window) and *Z* (screening window) values (*RZ'*, *RZ*) for the 300 simulated plate data sets. The *RZ* values are indicated for the normalized data (lower dotted curve) and for the equivalent data sets that are corrected for the systematic errors in the responses (solid line). Again, the low (negative) values of the *RZ* for the plates around index ~155 give an indication that a sample-based normalization is inadequate. The good value of *RZ'* (upper dotted line) for those plates (between 0.5 and 0.7) indicates that the assay performed well, but that the "problem" is due to the compound samples.

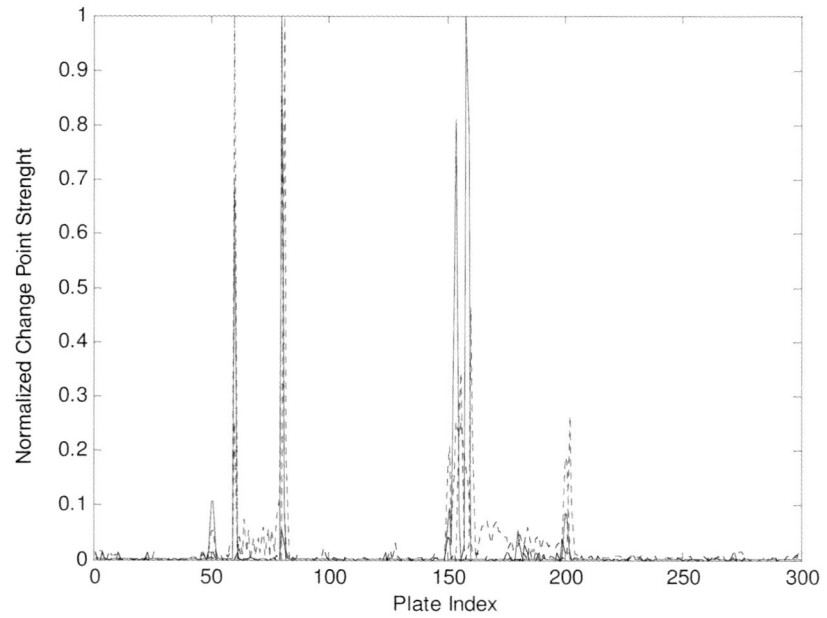

[Figure: Overlay graph showing Normalized Change Point Strenght (y-axis, 0 to 1) versus Plate Index (x-axis, 0 to 300)]

Figure 7.7 Overlay graph of change point determinations $\Delta(k)^2$, each normalized to its own maximum value, for the *median* of the high controls, *RZ'*, *mad* of the compound samples and *well-wise* background response surface *shape change points* (see Section 7.7.4). Measurement batch transition points at $k = 50, 60, 80, 150, 154, 158$ and 200 are clearly discernible. They correspond to a background surface shape change ($k = 50$), beginning of the "alternating" background response difference ($k = 60$), end of this effect together with a background surface shape change ($k = 80$), background surface change ($k = 150$), start and end of the high hit rate plates ($k = 154$ and 158) and another background surface shape change ($k = 200$).

represented as box-and-whisker plots displaying median, 25% and 75% quantiles (*IQR*), minimum, maximum and possibly any outliers.

3. Line plots of variability measures: either *CV* or standard deviations *sd* or preferably their outlier-resistant equivalents *RCV*, s_R or *mad* of neutral (zero effect) controls and the compound samples (Fig. 7.5).
4. Dot or line plots of *Z'*, *Z* or their robust equivalents *RZ'*, *RZ* (Fig. 7.6).
5. Dot or line plots of change point estimators $\Delta(k)^2$ [see expression (29)] of the quantities depicted in (1)–(4) *and* in addition the well-based background response change point counts n_k; see Fig. 7.7. and Section 7.7.4 for more details.
6. Dot or line plots of the extreme values of the response correction factors $a(x, y)$ [eqn. (22)], that is, of $\alpha = max[|a(x, y) - 1|]$ and $\delta = max[a(x, y)] - min[a(x, y)]$ or similarly, the extreme values of the corresponding normalized $^1p(x, y)$% values for each plate, provide a concise measure for the extreme deviations of the background response surface from the "ideal" horizontal plane; see Fig. 7.8 and Section 7.7.1 for more details.

7.6.2
Positional Effects – Summary Statistics and Views

The run sequence plots of types (1) – (5) described in the previous section, which are based on raw or normalized data, obviously cannot tell the complete story of how the zero effect responses deviate from the ideal horizontal plane. Such effects are signaled by the values represented in plot type (6), but obviously only after a series of more involved mathematical operations and not directly from the data. Some systematic errors and effects of position dependent background response shifts can be viewed and diagnosed, for example, by plotting:

7. Row- or column-wise aggregated summary statistics represented as averages per plate as in (1) or as distribution summaries (box plots, box-and-whisker plots) as in (2).
8. Well-, row- or column-wise aggregated counts of hits beyond a certain threshold, for example the $\Theta = 3sd$ average of the neutral controls of an assay run or of the whole screen. A corresponding boxplot representation is

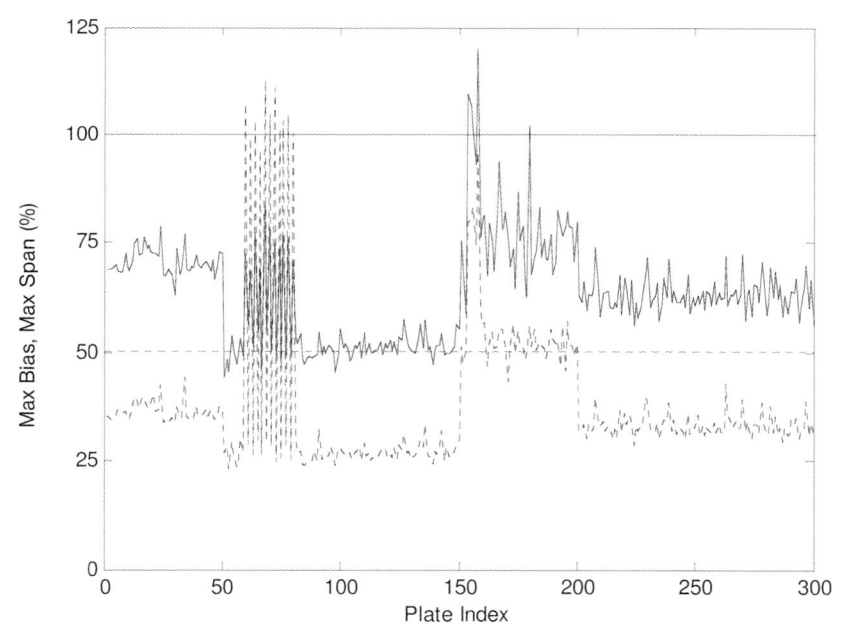

Figure 7.8 The maximal bias (dotted line) and span (solid line) of the "smooth" background response surface give an indication of the strength of the response distortion (see Section 7.7.1 for more details). Relatively high values of this distortion strength can be seen in the region of the experiment where the replaced and miscalibrated instrument was used ($k = 60$–80) and also where high hit rates rendered the background surface determination procedure invalid ($k = 154$–158). Quality control cutoff limits (plate pass/fail limits) can be set heuristically. Here the threshold values for the maximum bias and span are set at 50% and 100%, respectively. Data from plates crossing these thresholds should be inspected more closely.

a)

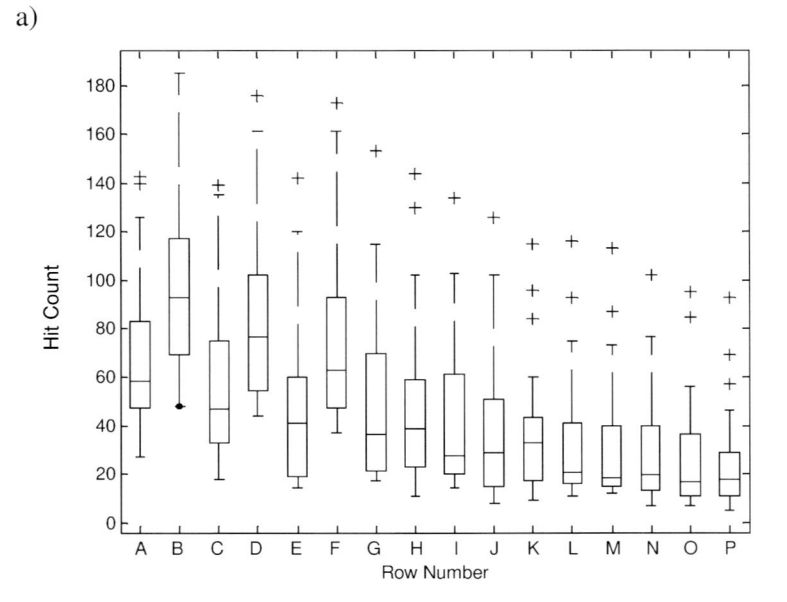

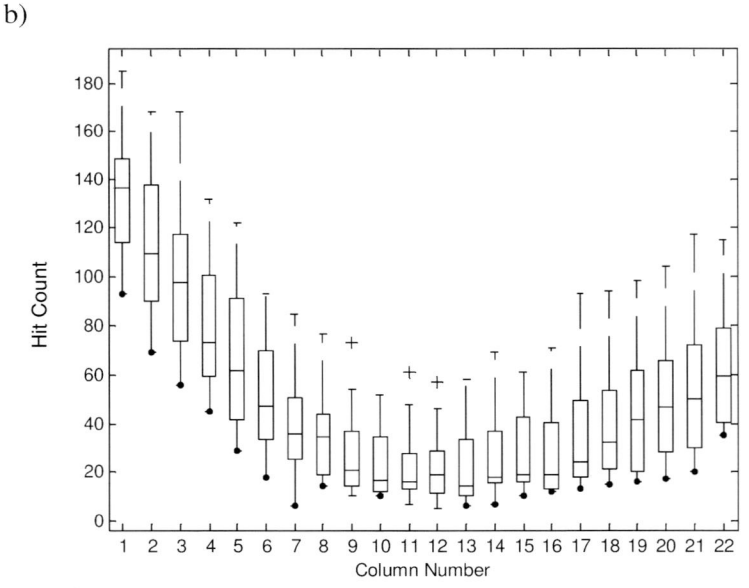

b)

Figure 7.9 The row-wise (a) and column-wise (b) aggregated and summed numbers of hits based on the normalized data indicate a strongly inhomogeneous distribution of hits, especially across the columns. This is highly unlikely in a random compound library and is clearly due to the background response surface distortions in this set of simulated data since the "active" samples were randomly selected from a uniform distribution. Large false-positive and/or false-negative rates can be expected from such data when simple constant thresholds are applied and if the position-dependent background responses are not corrected.

shown in Fig. 7.9. The equivalent hit count distribution after the correction for systematic position dependent response effect with threshold derived from the sample distribution density is shown later in Fig. 7.18. See Section 7.7.10 for further comments.
9. Well-wise aggregated standard deviations (or robust equivalents) of raw and normalized data.

A disadvantage of the aggregation and averaging over plate sets or the whole screen is the possible masking and compensation of systematic row- or column-wise response distortions through the averaging process. Such graphics representations can be adequate for individual screening batches where response distortion patterns are often found to be similar over the whole run.

7.6.3
Positional Effects – Heat Maps, Trellis Plots, and Assay Maps

Much more powerful visualizations of positional effects are provided by heat maps and "trellis" plots of heat maps.
- *Heat maps* are false color images of data values [51], usually in a rectangular arrangement of the two coordinate axes of individual plates or whole screens (see *assay map*, below). Also, the well-wise hit count and well variability values described in plot types (8) and (9) in the previous section are best displayed in this fashion; see examples in Figs. 7.10 to 7.13.
- *Trellis plots* are arrangements of heat maps or other types of graphs in a rectangular grid pattern. The individual elements of the trellis depend on a conditioning variable (here the plate sequence index) [52].

These visualizations allow an image-like representation of individual wells, average row or column effects on each plate or the corresponding values for a complete sequence of all plates in a screening batch or full screen; see Fig. 7.12. Some widely used visualization software programs (e.g. Spotfire [22]) allow one to trellis even higher dimensional representations of the data (spatial coordinates, response values that depend on several factors represented by color, shape and orientation of the glyphs, in addition to the position in the trellis).
- *Assay maps* are special types of heat maps which allow one to represent all well values of a full screen or a screening batch in one single false color image. An assay map is a heat map of all ($n_{rows} \times n_{columns}$) well indices in one coordinate direction (preferably the *y*-direction) and the plate sequence index k for the ($1-n_{plates}$) in the other coordinate direction. They provide a very quick visual insight into the overall behavior of the screen; any systematic patterns and changes in the variability of the response values can be easily discerned because the human eye and brain are very sensitive to changes in color intensity and hue. These types of visualizations are best coupled with an interactive possibility to adjust the color scale in order to "slice" though the data in the most optimal fashion; see Figs. 7.10, 7.11 and 7.13.

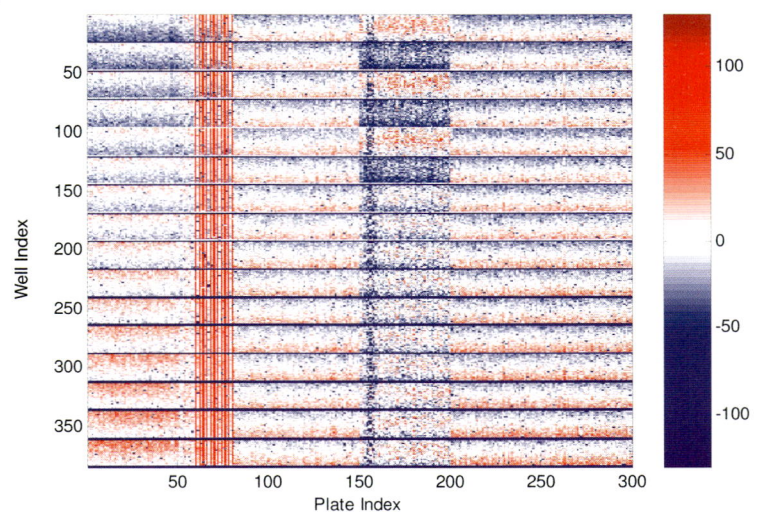

Figure 7.10 Assay heat map ("assay map") of the normalized data of the full run of 300 simulated plates. An assay map provides a very condensed, but at the same time very powerful, depiction of the complete data set. The batch boundaries at k = 50, 60, 80, 150 and 200, and also the high variability (high hit rate) data at k = 154–158 are clearly visible. The staggered pipetting effects in the first six rows of the batch from k = 150 to 200 are also clearly perceptible. The well index is the column-first sequence number of all wells on a plate (i.e. A1 = 1, A24 = 24, B1 = 25, ... P24 = 384). An analogous row-first sequence would produce a different image, but the systematic error features would be visible in a similar fashion.

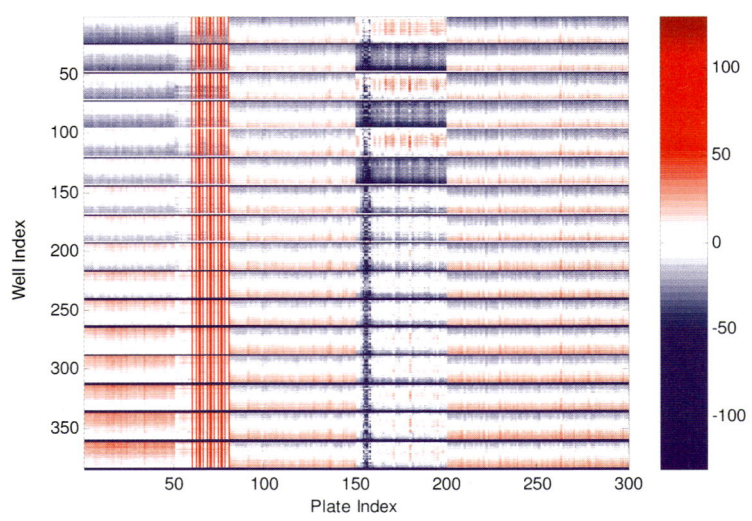

Figure 7.11 Assay map of the systematic background response surface (normalized data) as determined by the EOF analysis procedure (see Section 7.7.8). On comparison with Fig. 7.10, it is obvious that the main systematic (error) features of the data set are well captured.

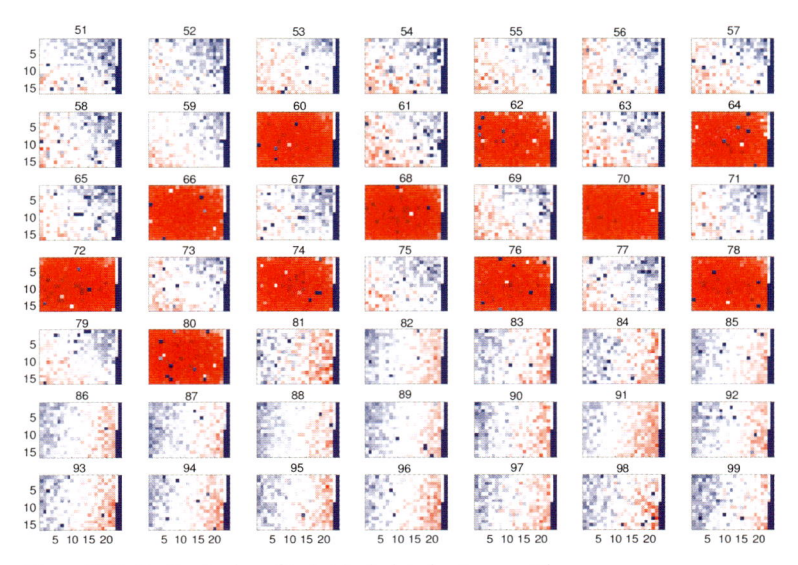

Figure 7.12 Trellis display of individual plate heat maps. The plate sequence from $k = 51$ to 99 is displayed. The alternating response pattern from $k = 60$ to 80 is due to the alternating plate feeding to a miscalibrated reader which was used as an "emergency" replacement. Also visible is a change of the systematic background response pattern which starts at $k = 81$.

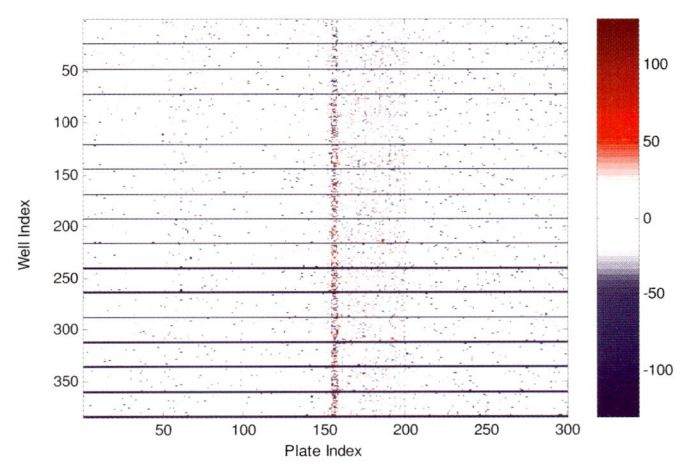

Figure 7.13 Assay map of the corrected data as based on the EOF model of the systematic background response of the complete plate set. The corrected % activity values $^3p'$ depicted here were calculated with eqn. (38). See also Figs. 7.10 and 7.11 for the original data and the EOF response model, respectively. As explained in Sections 7.6.3 and 7.8, the corrected data from the high hit rate plates at $k = 154$–158 look "suspicious" because the frequency of appearance of high positive residuals (red) is relatively large in this area, again pointing to the fact that the sample-based normalization is not appropriate in these cases.

The use of suitable graphics or interactive visualization software which provides heat maps and trellis graphics to display position dependent response patterns in raw, normalized and corrected HTS data is clearly recommended. Visualizing the corrected response data or the residuals in this manner allows one to detect easily and quickly remaining systematic error patterns. Data for plates thus affected should be investigated more carefully. In our example case (Figure 7.13), one can see that the strip which corresponds to the plates with high hit rates (plate indices around 155) also exhibits points with high positive (red) residuals (>+50%). This is a signal that the sample-based normalization may not have worked properly for those plates. Since the assumption that most of the samples of an HTS are inactive is clearly violated for those plates, one needs to review these data more carefully and possibly choose an alternative analysis procedure. This problem will be revisited in Section 7.8.

7.6.4
Distribution Densities – Histograms, Smoothed Distributions

An assessment of the distribution properties of measured or calculated variables is done either through histograms [54] or through graphs of smooth (kernel) density estimations [43, 49, 72]. One needs to be careful with drawing conclusions from

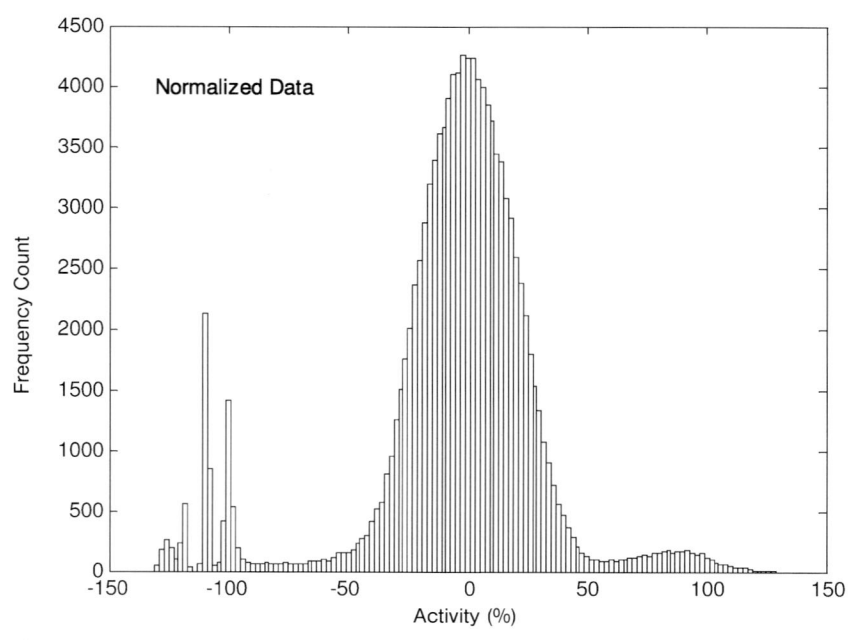

Figure 7.14 % activity distribution histogram of the normalized data.. The peaks at −100% and lower values are due to the low control and background (empty) wells. The second mode of the distribution density around +90% is due to the severe background response distortion of some plates between $k = 60$ and 80.

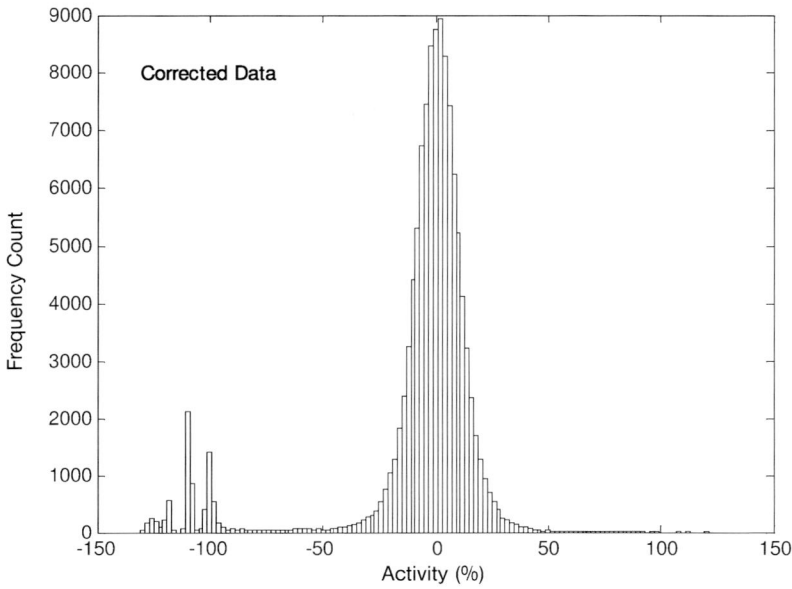

Figure 7.15 % activity distribution histogram of the simulated data set that was corrected for the existing background response distortions. On comparing this distribution with that in Fig. 7.14, the massive improvement in data quality which manifests itself in the much narrower distribution of the % activity values and the disappearance of secondary mode at ~90% can be clearly seen. Again, the peaks at −100% and lower are due to low control and background wells. The remaining (small) tail of the distribution density between +50% and +100% is mainly due to the high hit rate plates (k = 154–158) which are not properly modeled and corrected. This is because the basic assumptions of the sample based normalization procedure are violated for these plates.

histograms with low frequencies and low bin numbers because the position and width of the bins will strongly influence the appearance of the distribution. This is much less the case for average shifted histograms (ASH) [55] or smooth density estimations. In Figs. 7.14 and 7.15 the % activity distribution histograms for the normalized data and the corrected data, respectively, for all 300 plates are shown.

7.6.5
Numerical Diagnostics and Fully Automated QC Assessment

In addition to – and sensibly before – the purely visual data inspection and quality assurance process, it is recommended to have the problems in the data quality and systematic errors diagnosed in an automated fashion via the corresponding numerical representations of the many figures of merit that can be derived from raw and normalized data. The comprehensive set of values which was defined above, that is (robust) Z, Z', CV, b_H, s_H, s_L, α, δ (the extreme deviations of the background response surface as defined in Section 7.6.1) and hit count distribution information can be used to pass or fail individual plate data sets. The data sets that were flagged and failed through such automated procedures should receive priority for further visual and computational assessment and final decision making by the screening scientists. Suitable data analysis software should efficiently and directly guide the scientists to those problematic plates and data sets

7.6.6
Possible Sources of Systematic Errors and Trends

In order to find possible sources of systematic errors. the key diagnostic quantities can be aggregated by various types of process parameters (e.g. pipetting tips, instruments, readers, reagent and/or cell batches and compound libraries) and response differences which are due to either of those factors can be isolated either visually or computationally (e.g. by analysis of variance). Depending on the cause and strength of the effect, tuning and optimization of an experimental or instrumental parameter may be advisable in order to eliminate the systematic variations in the data, even if suitable correction methods are readily at hand for mild to intermediate distortions of the responses.

7.7
Correction of Position-dependent Response Effects

7.7.1
The General Problem

In this section, a series of algorithms that are based on approaches of increasing "sophistication" to model the position-dependent response background effects (trends, edge effects) are described. Such effects are almost always found in microtiter plate-based experiments and can be especially pronounced in cell-based assays. In the following section the question of how to correct or "renormalize" those position-dependent response data to reach the best possible and most homogeneous data quality of an HTS campaign is addressed. A robust representation of the "zero activity" response background surfaces of all plates are needed in order to correct values towards a common reference level. This can be accomplished through several different approaches and mathematical models describing the zero-activity surfaces $z(x, y, t)$, with x and y being the spatial coordinates (corresponding to the microtiter plate grid coordinates or grid indices i and j) and t describing the time of measurement or simply representing the measurement sequence. Some of the possible approaches are described in this section, but further methods from the areas of statistical modeling, signal processing and pattern detection can also be usefully applied.

Following the principle of parsimoniousness, one is searching for a simple linear transformation to correct the systematic positional response variations as follows:

$$z'(x, y, t) = a(x, y, t)[z(x, y, t) + e(x, y, t)] + b(x, y, t) \tag{22}$$

with the observed data being represented as $z + e$, that is, a "smooth" structure z and an added noise component e. If one assumes purely multiplicative distortion effects, one can set $b = 0$. a needs to be determined such that the expectation value $E(z') = \text{constant} = \zeta$ for all x, y and t, which corresponds to a global assay response standardization across all plates. See Section 7.7.10 for further details.

This type of problem can be tackled in various different ways: more direct ones are by estimating the two-dimensional surfaces $z(x, y)$ for each plate or the three-dimensional surface $z(x, y, t)$ in such a way as to represent the screening data set x as $x \sim$ structural model z + noise components e. Some of the modeling procedures can refine this implicitly or explicitly as $x \sim$ smooth structural model z_1 + discrete step-like structural elements (liquid handling artifacts) z_2 + noise components e.

A representation of z can be derived from the data set x by (1) spatial and/or temporal averaging or smoothing, (2) applying parametric or (3) non-parametric regression and optimization via ordinary least squares or robust variants such as LTS (least trimmed squares) or LMS (least median of squares) [46], iterated reweighted least squares (IRLS) [43] or other similar methods, (4) expansion of z into a series of (orthogonal) basis functions (OBF), for example Chebyshev polynomials, Fourier or wavelet series [56] or (5) empirical orthogonal function analysis (EOFA) using principal components/singular value decomposition methods [57, 58]. OBF and EOF expansions need to be followed by a truncation or smooth filtering to a lower number of series expansion terms. The truncation criteria must be chosen such that the structural model of the data is complete, but over-fitting (inclusion of noise components in the structural model parts) is avoided.

Less direct solutions of eqn (22) can be obtained by treating this as a problem of finding the correction function $a(x, y)$ through optimizing certain key properties of the resulting empirical distribution density $g(z')$ as an objective. Such target properties can be, for example: (1) maximum likelihood estimation by expectation maximization (EM) of the density distribution $g(z'|p)$ for a set of system model parameters $p = \{p_i\}$ that directly or indirectly describe the functions a or z, (2) minimization of the variance and skewness of $g(z'|p)$ and (3) minimization of the deviation of $g(z'|p)$ from a normal distribution shape by moment matching. A regularization, that is, imposition of restrictions of the solution manifold, must usually be applied in order to avoid modeling of noise instead of structure, thus restricting the possible forms and values of p. These approaches can be viewed as solving an "inverse problem", namely deriving the "original" data distribution $g(z')$ based on the observed activity distribution and the initially unknown properties of the HTS measurement and distortion "process" $z(x, y)$ or $a(x, y)$. Certain assumptions about the undistorted distribution $g(z')$ that needs to be determined, particularly of the "inactive" portion of the population, need to be made, for example, $g(z') \sim N(\zeta, \sigma^2)$. Bayesian statistical methods and stochastic optimization procedures can be applied to the solution of these problems [59]. These approaches will not be followed any further in this chapter. The focus will now be put on the previously mentioned more direct methods for the determination of $z(x, y)$. In the author's organization there exists long and extensive practical experience with several of the data modeling and correction approaches that are presented below.

They have been proven to be of high practical usefulness for the daily needs for efficient data management in the HTS and uHTS laboratories in an industrial setting.

In the following section headings it is indicated whether the analysis technique being presented and discussed is suited for the assessment and data correction of single plates, for a whole series of plates or both.

7.7.2
Plate Averaging (Multiple Plates)

First impressions on the distortion patterns can be obtained by simply averaging the readout values of individual wells across a whole sequence of plates, the assumption being that environmental and instrumental conditions for a series of plates that were measured over a certain limited time period (in temporal proximity) also lead to similar distortions in the responses. Experience shows that this is actually largely the case, but that there may also be deviations from this behavior between adjacent plates in a sequence, hence local adaptability of the statistical model to accommodate such effects without becoming biased is an important aspect. Possible causes for such non-monotonous behavior can be systematic variations in the properties of the laboratory ware or their pretreatment (e.g. inhomogeneous protein or cell coating, mixing of laboratory ware from different preparation batches in an experiment and variations in the conditions of the microenvironment) and associated "random" placement in the measurement sequence.

For a batch of plates, z_{ij} is the averaged readout value for row index i and column index j across all n_{plates} plate data sets in the measured set x_{ijk} (that is, averaging over the index k). Thus for all k we have an identical average:

$$z_{ijk} = z_{ij} = average_k(x_{ijk}) \tag{23}$$

The averaging operator can again be either the arithmetic mean or – preferably – the median or another resistant location estimator. Similarly, by further averaging over either row or column index, one arrives at column or row summaries for the plate set. Simple averaging is in most cases too crude to serve as an adequately unbiased and robust "zero activity" baseline against which to compare the individual values x_{ij} in all different plates. Although such quantities can serve as a first diagnostic measure for the numeric estimation of the size of systematic effects within an assay run of limited size and over a limited time period (e.g. the daily run set), simple averaging over different runs is discouraged because variations in the distortion patterns will often lead to meaningless average values. Moving averages, that is, running means or running medians [60] or the application of more complex smoothers, for example spline smoothing, loess, kernel smoothing, all in outlier resistant versions [49] as

$$z_{ijk} = smooth_k(x_{ijk}) \tag{24}$$

lead to improved local baselines as compared with eqn. (23) because eqn. (24) has a higher adaptability to variations of responses over time or plate sequence k. A strong disadvantage of this type of "longitudinal" averaging and smoothing is that the local plate context, that is the values of the neighboring area of well (i, j) on the same plate k do not influence the outcome. Intuitively one should assume that readout values of individual wells are on average more strongly correlated with its neighbors on the same plate than with those on other plates, because all wells on the sample plate have the "same" process history (although this is not necessarily true for effects from pipetting, dispensing and contamination owing to liquid handling equipment or process malfunction). Stronger anchoring of the model responses in positional proximity on the same plate needs to be accomplished in order to make the estimates more robust. In that sense, the z estimates above should serve only as diagnostic tools and they should be used carefully when trying to detect hits.

7.7.3
Median Polish Smoothing (Single Plates, Multiple Plates)

An approach recently published by Gunter and coworkers [7, 8] consists of modeling the response of each plate as the additive combination of plate center values T plus row and column effects R and C through the median polish algorithm (basically a resistant version of a two-way ANOVA) [45]:

$$z_{ijk} = T_k + R_{ik} + C_{jk} \qquad (25)$$

for all plates k. An added smoothing term "averages" the remaining systematic effects over nearby plates and helps to stabilize the results. The model background responses z for each well are then

$$z_{ijk} = T_k + R_{ik} + C_{jk} + smooth_k(e_{ijk}) \qquad (26)$$

where e_{ijk} correspond to the residuals of the median polish procedure, ideally the "random noise" of the assay, but often also containing remaining systematic effects. As explained by Brideau et al. [8], smooth( ) needs to be an outlier-resistant operator that is not unduly influenced by hits and at the same time can also reliably model jumps of the response values between plates. Such jumps need to be accommodated and must not be "smoothed" over in a blind fashion. It is therefore important to be able to trace such jumps or change points and to establish algorithmic rules for the value of smooth( ) in the transition region [7, 61].

In addition, it may not necessarily be safe to assume that the plate response can be modeled with independent row and column effects only. In such cases the smooth( ) term also contains implicit interaction terms, nonlinear contributions and localized systematic deviations of the responses from the model in eqn. (25). Straight median polishing of individual plates without additional smoothing is also described and applied in [11].

The residuals

$$r_{ij} = x_{ij} - z_{ij} \qquad (27)$$

around the newly established "zero" activity surface z_{ij} are a measure of the true potency of the compounds. A quantitative relationship of r_{ij} to the corrected % activity scale or a further derived activity score will be explored in Sections 7.7.10 and 7.8.

7.7.4
Change Point Detection (Multiple Plates, Plate Sequences)

Algorithms for the detection of change points [62] in the distortion patterns allow (1) an automated separation of plate batches that may have to be treated separately in the analysis when applying certain types of algorithms (see Sections 7.7.5 and 7.7.6) or (2) an implicit extension of the smoothing methods to take discontinuities safely into account [7, 63].

Change points in a monotone sequence are typically detected through the determination of separate left and right limit values of z_k that is z_k^- and z_k^+ by one-sided smoothing within a window of a certain width h to the left and right of k. Again, different types of smoothers can be applied for this purpose (nonlinear median smoothers, Kernel smoothers and so on). If $k = \varkappa$ is a point of discontinuity, the left and right values differ by the jump size:

$$\Delta(\varkappa) = z_\varkappa^- - z_\varkappa^+ \qquad (28)$$

$\Delta(k)$ is expected to peak near $\varkappa$ and be closer to zero elsewhere. Possible estimators for the change point $\varkappa$ are the maximizers

$$\hat{\varkappa} = arg\ max_k[\Delta(k)]^2 \qquad (29)$$

or

$$\hat{\varkappa} = arg\ max_k|\Delta(k)| \qquad (30)$$

The use of eqn. (29) is preferred because it generally exhibits stronger and more "visible" peaks than eqn. (30). Because there can be several such change points in an observed data sequence, the maximizing rule needs to be applied locally within a certain window of index values around k. A lower threshold for $\Delta(k)$ that depends on its mean (or median) m_Δ and standard deviation (or mad) s_Δ needs to be set such that jump or change points can be automatically detected. In order to account for variations of those quantities over a longer plate series, resistant smoothing of the local averages $m_{\Delta k}$ and $s_{\Delta k}$ is recommended. From these values one can define local thresholds $\Theta_k = m_{\Delta k} + \lambda_\alpha\ s_{\Delta k}$ with λ_α chosen such that the frequency of randomly occurring threshold crossings from $k-1$ to k with $\Delta_k > \Theta_k$ remains below a certain (small) probability α for a stationary series with the equivalent average, variance and autocorrelation structure [64].

This procedure can be applied separately to different quantities characterizing the assay behavior and will help to partition the plate sets if analysis algorithms need to treat them separately or if different hit threshold conditions should be applied to the distinct plate subsets (see further examples below). One would typically investigate the behavior of the neutral and positive control raw data levels and/or variance measures with an algorithm as described above and can then identify changes in the raw data or normalized response distortion patterns by applying the change point detector for each well location across the plate sequence. An illustration is given in Fig. 7.7. For practical use, the well-based change point information (in practice 384 or 1536 numbers per plate times the number of plates in the investigated set) has to be condensed further to one value per plate. This is most easily done by counting all well change threshold transitions for each plate k according to a "minimal jump size" rule Θ_k and arriving at a vector $n = \{n_k\}$ of number of transition counts for each plate. Subsequently one can apply a threshold N_{min} to identify those plates that exhibit simultaneous discontinuous pattern changes towards neighbors with $n_k > N_{min}$ wells. As mentioned before, this is a useful automated procedure for partitioning the data sets for further analysis if algorithms are employed which are sensitive to such groupings or in order to test such sensitivity by comparing the results of a global assay analysis with those of the partitioned data sets. For all methods that need to rely on the simultaneous analysis of a whole plate series, it is recommended to eliminate clearly erroneous (e.g. with $Z' < 0$) and problematic plates (e.g. with $Z < 0$) from the final analysis step (see also Section 7.8).

7.7.5
Parametric (Polynomial) Surface Fitting (Single Plates, Multiple Plates)

The simplest choice in this category is to fit a two-dimensional nth-degree polynomial surface of the form

$$z(x, y) = a_0 + a_{11}x + a_{12}y + a_{21}x^2 + a_{22}xy + a_{23}y^2 + \ldots a_{n\,1}x^n + \ldots + a_{n\,n+1}y^n \quad (31)$$

to the observations $z(x, y) = z_{i\,j}$ on *each separate plate*. Within the ordinary least squares (OLS) minimization framework, the linear parameters a_{ik} of polynomial (31) are easily calculated with standard algebraic methods [20, 65, 66]. The degree of the polynomial and the number of terms to include can be assessed by the R^2 and F statistics [65] and by the cross-validation Q^2 [67]. The lowest possible polynomial degree with adequate R^2 and Q^2 is the preferred model. OLS fitting is plagued by the equivalent outlier and robustness problems as outlined and illustrated for the simple location estimates in Section 7.5.5, so the application of resistant fitting procedures is again recommended. The easiest method to create a more robust result z is (1) trimming, that is, eliminating a specific percentage of the smallest and largest observations $z_{i\,j}$ or all observations outside the limits calculated as in eqn. (19) with λ_α corresponding to the asymptotic normal, (2)

Winsorization, that is setting the response values beyond the $\lambda_\alpha s$ cutoff limits to that very limit [43] and (3) soft trimming by applying a weight function w to the residuals r (e.g. Tukey's biweight) [43]:

$$w(r) = [1 - (t/R)^2]_+^2 \tag{32}$$

with $t = r/s$ and $R = 4.685$; $[\;]_+$ denotes the positive part (that is, for all $|t| \leq R$, where the expression value inside the brackets is ≥ 0); $w(r)$ is zero outside this range.

Because even the resistant initial estimates for $m = median(z)$ and $s = mad(z)$ as used in eqn. (19) can be biased or inflated if background response distortions across the plate(s) are present, it is recommended to iterate the cutoff and fitting procedures with a recalculated s that is now based on the residuals of the initial fit. Soft trimming and weighting of residuals require in any case an iterative procedure because it is obvious that the weights depend on residuals that depend on the function values based on certain parameters that again depend on the influence of the individual data points or rather the weights of the respective residuals. The iterated reweighted least squares (IRLS) procedure is a very versatile tool for resistant linear regression as in our particular two-dimensional polynomial case, but can also be used for nonlinear regression problems.

Applying eqn. (31) to a whole *plate series* (or batch) requires some data preconditioning and also some caution. First, one should have some confidence that the distortion patterns on all the plates have a "similar" shape, and second, one needs to be careful about differences in the levels of the absolute readout values z_{ijk}. Polynomial fitting to determine the background surface distortion of a whole HTS plate data set can be applied after standardization, that is, normalizing the data through mean centering (subtracting the mean response m of a plate from all values) and scaling to unit variance (division of all residuals by s) [20]. In order to lessen the danger of introduction of correction bias, it is recommended to run the change point detector previously described across the normalized plate data set. One can then automatically partition the plate sets if discontinuities are being detected. Although simultaneously fitting a polynomial to the whole data set certainly stabilizes the model, there is also a danger of introducing bias if individual plates deviate from the average behavior. As described in [20], partitioning can also be done through a cluster analysis and classification of plates into families of "similar" response patterns. *K*-means clustering was used in this particular example. This is certainly a necessary step if the measurement sequence is not known. The surface fit values and the residuals on the original variable scale can be obtained by reversing the data standardization step that needed to be applied for the simultaneous determination of the polynomial coefficients. Diagnostics and visualization of residuals in order to look out for plates and surface areas with remaining systematic spatial patterns should be applied in a routine fashion [65].

Polynomials of low degree can only model smooth surfaces, hence liquid handling artifacts which usually manifest themselves in step-like patterns in one coordinate direction cannot be modeled and will either contribute to the *noise component* or bias the polynomial coefficients or both. Trimming, Winsorizing or residual downweighting will make the fit less sensitive to strong step and stripe patterns, but effects of lower response amplitude will remain in the data and lead to highly structured systematic patterns in the residuals.

It is possible to detect such stripe and step-like artifacts in the data with specially designed pattern detection algorithms and flag appropriate plates and wells for separate analysis [50] and possible repetition of the measurements in an added HTS run (preferably after repair or adjustment and standard quality check of the faulty liquid handling equipment).

7.7.6
Nonparametric Surface Fitting, Local Regression, and Smoothing (Single Plates, Multiple Plates)

The idea of nonparametric fitting, smoothing and local regression is not to specify a parametric model for the surface $z(x, y)$ as for example in eqn. (31), but rather to allow the data to determine the appropriate functional form. In loose terms, one assumes only that the function is smooth. Different approaches are again possible and some have been applied to HTS data analysis: Kernel smoothing [69], local regression [50, 70–72], smoothing splines [68, 73, 74] and orthogonal series expansion (e.g. Fourier series) [75]. Some aspects of the latter will be discussed in following subsections. The application and performance of local regression to HTS activity background surface modeling were described in [9, 76] and have been productively applied in a highly automated HTS data processing pipeline in the author's laboratories since 1997. The underlying principle is that any smooth function can be approximated by a low-degree polynomial in the neighborhood of any grid point (i, j). Linear and quadratic expressions are most often used in practice. This local approximation can then be fitted by weighted least squares with a weight function that depends on a certain bandwidth h. In the two-dimensional case and a local quadratic approximation around the grid point (i, j), one has

$$z_{ij}(x, y) = a_0 + a_1(x - j) + a_2(y - i) + a_3(x - j)^2 + a_4(x - j)(y - i) + a_5(y - i)^2 \quad (33)$$

and an appropriate weight function $W(d, h)$ [71] that reduces the contributions from points (x, y) at increasing distances d from the grid point (i, j). The local regression estimate $\hat{z}$ at the grid point (i, j) is then

$$\hat{z} = \hat{z}(i, j) = \hat{a}_0 \quad (34)$$

An important adjustable parameter in this procedure is, of course, the bandwidth h of the smoothing kernel $W(d, h)$. It can be set to an appropriate "best" average value by generalized cross-validation (GCV) [77] and evaluation of the residuals for a comprehensive suite of data sets that exhibit various different types of response distortions and noise contributions. Artificial data sets that are generated through Monte Carlo simulation and that exhibit a wide variety of random variations in the surface shape can be used fairly well in such a search for a global optimum of h. It is impractical to adjust h for each new assay or assay run, but experience with a very large number of real assay data sets shows that the procedure with an average h behaves very well. The smoothing bandwidth needs to be chosen differently for different plate densities, however. The procedure is capable of modeling highly irregular and varying surface shapes on each separate plate.

As with ordinary polynomial regression described in the last subsection, one has to take considerable care about the effects of "outliers" and especially the contributions from the population of active samples. Soft trimming and iteration via the IRLS scheme lead to very flexible and resistant estimations of the smoothed zero activity surface $z(x, y)$ via two-dimensional local regression. By automatically applying *a priori* knowledge about the rough position of this surface (namely being "close" to the level of the neutral controls H) and an initial soft trimming step around this level, it is possible to reach breakdown points of >50% even for cases where there are many highly active samples leading to a distinct density peak at higher activity levels. The breakdown point is smaller if many actives with marginal activities are found in the tail of the density distribution.

For simultaneously fitting a whole *plate series*, the same data preparation steps as for the parametric polynomial case need to be made. After mean centering and unit variance standardization of the data from each plate, the robust local regression procedure as just described can be applied to the complete data set at once. Also, as previously recommended, it is advisable to partition the data set and perform separate surface determinations if global change points, that is, changes in the general shape of responses between different plate batches, are being detected. The background surface values for each separate plate can again be obtained by back-transformation to the original variables. As above, the resulting residuals should be checked for areas of remaining systematic effects. The determination of the robust smooth background response surface by performing separate fits for each plate has clearly a higher adaptability and one usually finds a smaller number of areas with remaining local bias than for the global fit that treats all plates at once (see also Section 7.7.9).

7.7.7
Expansion into Orthogonal Basis Functions (Single Plates)

The expansion of $z(x, y)$ into an orthogonal set of basis functions (OBF) has already been mentioned as another useful method to model a smooth surface. There are many possible choices for such basis functions, but Fourier expansion into a series of sine and cosine terms or the expansion into a series of Chebyshev polynomials [56] is often used. In Fourier smoothing, one expands the discrete spatial data into a series of sine and cosine terms in each of the variable directions, creating a frequency representation of the original data, and then performs a low-pass filtering and reverse transformation step to obtain a smoothed estimate of the original data [75].

In practice, FFT (fast Fourier transform) and DFT (discrete Fourier transform) algorithms are employed to perform the forward and backward transformations. In the area of HTS data analysis and quality control, Root et al. [19] applied this technology for diagnostic purposes, mainly to detect "hidden" spatial patterns and for occasional correction of the response surface in specific cases. In their approach, the power density spectrum (periodogram) is calculated from the two-dimensional DFT (2D DFT) and the frequency components that have a higher amplitude than those expected from a spatially uncorrelated data set can be extracted and transformed back to the original spatial data representation on the plate coordinate grid via reverse DFT. Those components with excess amplitude represent certain spatial periodicities that can easily be visualized after back-transformation. The respective amplitude contributions for those correlated patterns for each well in relation to the original data amplitude give an indication of the strength of the pattern at that particular plate location, but cannot be used directly for subtraction of the systematic pattern. As proposed by Root et al. [19], a specific utility of this method can also be found in the setup and periodic quality control of robotic liquid handling equipment. Uniform test plates processed by the equipment under investigation can be investigated with the described periodogram pattern detection method to check whether regularly appearing systematic errors originate from there. Steps to improve the process performance can then be initiated if the source of the regular patterns is clarified. In contrast to the smooth low-pass filtering mentioned above, the reverse transformation of the DFT where the high-power frequency contributions are eliminated through a hard cutoff can in most cases not be safely used to apply direct corrections to the original data. This is due to the well-known Gibbs phenomenon of Fourier approximation theory that manifests itself in a spatial "ringing" effect. Root et al. [19] gave an example where direct correction was found to be appropriate (strong checkerboard pattern).

7.7.8
Empirical Orthogonal Function (EOF) Analysis, Singular Value Decomposition (SVD) (Multiple Plates)

In the previous subsection, the plate data were expanded into a predetermined series of orthogonal basis functions. The analysis of individual plate response surfaces with those methods is fairly straightforward, provided that the computational procedures that are used are adequately resistant to outliers and data originating from the "active" sample population. Applying those analysis methods to a whole plate series needs some precautions that were elaborated on in the relevant sections. In contrast, the empirical orthogonal function (EOF) analysis [57] is a very natural and purely data-driven procedure to model simultaneously the responses of a whole plate series. The method actually *only* works reasonably well with a certain minimal number of plates (≥10) and *not at all* with single or very few plates. Any spatially continuous stochastic function $f(x, y, t)$, in our case the three-dimensional plate background response surface, can always be expanded into an infinite series of deterministic spatial functions $u(x, y)$ and stochastic coefficients $c(t)$ that depend only on time t (or, in our case, plate sequence number k):

$$f(x, y, t) = \sum_{m=1}^{\infty} u_m(x, y) c_m(t) \qquad (35)$$

An approximate (smoother!) representation of f can be obtained by truncating the series at some index n and retaining only the leading terms. It can be shown that the EOF decomposition of a finite data set is equivalent to principal component analysis (PCA) [78] or singular value decomposition (SVD) [79]. The functions u_m depend only on spatial (plate grid) coordinates and the coefficients c_m are functions of time or sequence number only and describe the strength or "importance" of the spatial patterns or modes at that particular time point or sequence index k. The computationally fast SVD algorithm decomposes the two-dimensional $(n_{row}n_{col}) \times n_{plates}$ data matrix $\{x_{ij,\ k}\}$ into $n_{max} = min[(n_{row}n_{col}), n_{plates}]$ eigenmodes u_m (or eigenpatterns, principal plate patterns, eigenplates) and n_{plates} principal component scores c_m used as expansion coefficients for all n_{max} modes.

SVD ranks the EOF–PC pairs by their importance to the overall variability of the whole dataset. The relative importance of each mode is determined by its associated eigenvalue. The scores c_m plotted as "time series" over k quantify the global strength of the mth pattern. As mentioned above, the series (here written in the equivalent form used with the discrete plate coordinate indices) is truncated after n terms:

$$f_{ijk} = z_{ijk} = \sum_{m=1}^{n} u_{mij} c_{mk} \tag{36}$$

The n retained eigenmode matrices u_m should represent the main systematic effects and trends in the data and the noise contributions should remain in the truncated part of the expansion. Standard diagnostic criteria used in PCA or SVD to determine the number of components to include in the expansion can usually be employed [78], but see the caveat further below. Although it was stated above that the retained terms will create a "smooth" representation $\{z_{ijk}\}$ of the data set $\{x_{ijk}\}$, this is only to be understood in the sense that experimental noise can be separated from the systematic behavior. The EOF decomposition can efficiently model repeated step patterns in a series of plates. The analysis can be made with the original data matrix x or with its centered and variance standardized equivalent. A cautionary remark needs to be made here because it may happen that the centering and standardization step already removes most of the "pattern" structure in the data and then the SVD is essentially left with very little structure above the noise. One of the overly simplistic "rules of thumb" in PCA to include at least 75% of the data variance with the retained PCs will not work under those circumstances, as this would lead to excessive addition of noise components, unless the contribution of the mean centering is included in the total variance calculation. Diagnostics based on investigating the magnitude of the n_{max} ordered eigenvalues (EV) and identifying the last step change (kink) before flattening out when going to the smaller EVs will more safely identify the transition from "structure" to "noise". In most cases only a very limited number of terms (<10) need to be retained to obtain a noise-free "smooth" representation of the data set $\{x_{ijk}\}$. When explicitly visualizing the detected eigenpatterns of such centered and standardized data, one should always include the mean center matrix and look for patterns there also (see Fig. 7.16).

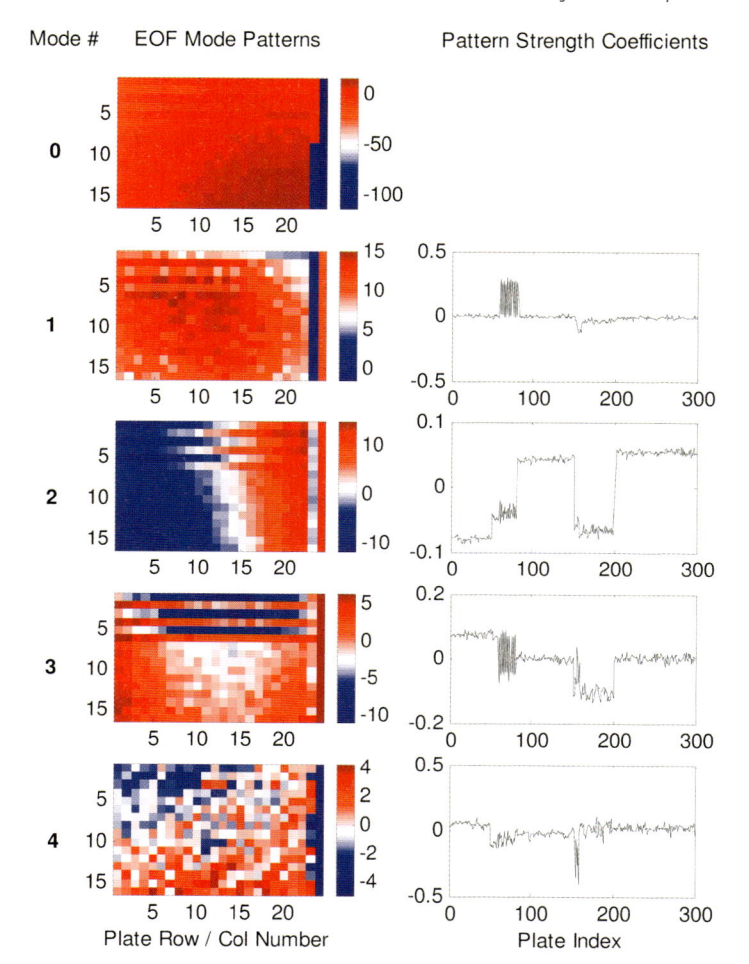

Figure 7.16 The first four modes and expansion coefficients (pattern strength coefficients) and the overall average (mode 0) of the EOF model of the simulated data set for all 300 plates. The systematic error patterns which can be identified in the original data as shown in the assay map of Fig. 7.10 are reflected in the mode patterns. To correlate the modes finally with the original plate data, the corresponding expansion coefficients also have to be taken into account. It can also be seen that the average contribution of mode 4 as measured by the product of its amplitude (−4 to +4) and the EOF expansion coefficient (mostly close to zero apart from higher amplitudes for plates around index 60–80 and 154–158) is already becoming small and that the mode pattern is less structured and noisier than the lower modes. This indicates that the truncation limit of the EOF series is close. An $n = 5$ cutoff was used.

It is sometimes advantageous to create separate EOF models of data from clearly separate batches with distinct systematic error patterns that occur only on a smaller number of plates or that have clearly different variances (see also Section 7.7.9). The partitioning of the data set through change point detection or through the separation into the daily batches can be done automatically and the separate models can be handled transparently in such an analysis system. The user can still be provided with the information about the response surface model(s) and the corrected sample activities and their distribution for the whole plate series, irrespective of partition boundaries. In many cases the detected change points correspond to experiment batch limits that are visible in the assay heat map display of raw and/or normalized data.

In order to make the EOF/SVD model more resistant to outliers and to the distracting contributions from the active compound population, it is recommended either (1) to trim or Winsorize all variables of the original dataset (individual plate data) [80] or (2) to use robust PCA algorithms as described, for example, in [81–83]. Also, the computational methods developed for independent component analysis (ICA) can be used in the decomposition eqn. (36) [84, 85]. The latter method is able to extract and represent the dominant features in the observed data that are as independent from each other as possible. The EOF/PCA method does not guarantee that a certain systematic effect in the original data is only reflected in a single mode u_m, only that the u_m are uncorrelated. This is nicely illustrated in Fig. 7.16, where the prominent striping pattern that is present in some of the plates manifests itself in several consecutive modes. In the plates that are not affected by the striping pattern, the effect exhibited in the various modes cancels out because the relevant contributions from the different products $u_m c_m$ differ in sign and magnitude. The derivation of independent components may be of interest if the aim of the data modeling is not only to create a "smooth" representation for the purpose of data correction, but also to detect, recognize and classify the different kinds of systematic errors which are present in the screening data set. Further features of the EOF analysis and diagnostics applied to HTS data analysis are described in [86].

7.7.9
Some Remarks on the Correction of Individual Plates versus Complete Plate Sets

Correction algorithms for individual plates that do not rely on data from neighbors in temporal proximity are especially important for on-line and near-real-time assessment of systematic effects. In these cases, data from the other plates of the screening batch may not be available. Single plate correction methods also have their merit in the higher adaptive power for modeling plate-specific distortions without being biased by temporal or sequential averaging. Every smoothing procedure has to deal with the bias (remaining systematic error) versus variance (random error or estimation variance) tradeoff and since one dimension of influence (t, sequence) is eliminated, there is potentially less distorting influence from systematic response variation in neighboring plates. Of course, this requires that the basic modeling procedure to represent the position-dependent systematic effects $z(x, y)$ across a single plate is adequate and efficient. Multi-plate EOF models can also end up with remaining small systematic errors if the series (36) is truncated "too early" and very localized structured effects are not adequately represented in the retained modes of the series. If certain kinds of systematic patterns appear on only a limited number of plates, then the model variance of methods that rely on the presence of multiple occurrences (such as EOF and related procedures) can become quite large. This is illustrated in Fig. 7.17, where the considerable residual roughness of the EOF model for plates with a very specific response pattern that appears only a limited number of times in the measured sequence is easily discernible (Fig. 7.17 a). The single plate local regression model is much smoother and provides a much more stable basis for the analysis of the residuals (that is, finally the "true" sample activities), whereas the absence or presence of one or a few plates with the same "companion" pattern in EOF analysis can change the respective residuals in an appreciable way. Both the single-plate local regression model and the multi-plate EOF model produce a stable background response surface for the second example (Fig. 7.17 b). The systematic pattern of this case has a much broader support which is provided by a higher number of plates. This results in models of roughly the same quality for both modeling approaches. From a practical point of view, the "single-plate" algorithms are also very useful during assay development or hit confirmation where smaller numbers of plates will be run.

In the *best of all worlds*, a series of complementary data modeling and correction approaches can be made available in the data analysis and QC software components for the HTS scientists to cover all possible needs. The combined methods can then provide reliable results for (1) single-plate on-line diagnostics, (2) smooth and discontinuous response effects spanning a larger series of plates, (3) "singular" patterns which have a distinctive shape only shared by very few other plates in a larger series and need to be represented in a robust and stable manner and (4) generally for a small number of plates where some multi-plate modeling methods do not work reliably.

7.7.10
Position-dependent Correction of Background Response Surface

Once a faithful representation of the "smooth" response surfaces $z(x, y)$ [or z_{ij} for all well coordinate points (i, j)] has been created by using any of the data-driven methods previously described, the corrections to the data original HTS data set $\{x_{ij}\}$ can be applied and an improved representation of the activity data can be obtained.

Two equivalent correction approaches are (1) renormalizing $z(x, y) = z_{ij}$ via eqn. (22) to the global constant average response level (horizontal plane) $E(z') = $ constant $= \zeta$ (assuming a multiplicative change effect), where ζ can be chosen arbitrarily to be the overall average of the model response or of the original data: $\zeta = average(\{z_{ijk}\})$ or $\zeta = average(\{x_{ijk}\})$. One then arrives at the position-dependent raw data correction factor $a(x, y) = \zeta/z(x, y)$; see eqn. (22). In practice, there is little difference between the two choices for ζ.

The alternative approach (2) is to perform the renormalization of the plate responses by interpreting the $z(x, y)$ values as a more refined position- and plate-dependent version of $\bar{x}$ in the previously established normalization equations to

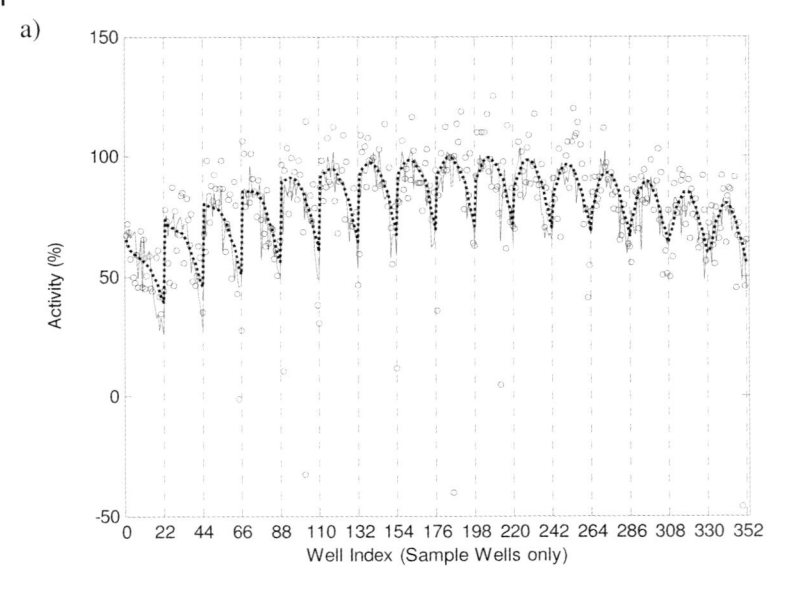

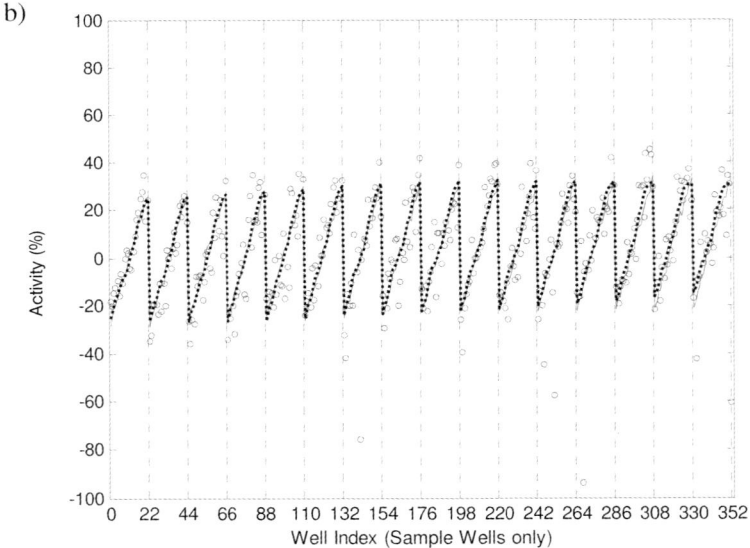

Figure 7.17 Sample response data for plate $k = 64$ (a) and plate $k = 265$ (b) with the local regression model representation (dotted lines) and the EOF model representation (solid lines). The roughness of the EOF model in (a) is appreciably higher than that in (b) because the pattern represented in plate $k = 64$ appears on only a very limited number of plates. The smooth local regression model is derived from the respective single plate data sets only; there is no dependence on data from other plates, as is the case for the multi-plate EOF analysis method. For systematic patterns that do not occur very frequently in the data set, the estimated efficiency of the EOF model is much lower than for the local regression model. See Section 7.7.9 for further explanations.

arrive at modified expressions for the % activity measures denoted by ^2p and ^3p [eqns. (10) and (11)]. One thus obtains the corresponding corrected $^2p'$ and $^3p'$ % activity values where the inhomogeneous plate responses at each separate well coordinate point (i, j) of the plate are eliminated. These quantities can also be expressed in terms of the residuals r_{ij} [eqn. (27)], that is

$$^2p'_{ij} = 100(x_{ij} - z_{ij})/(z_{ij} - L)$$
$$= 100 r_{ij}/(z_{ij} - L) \tag{37}$$

and

$$^3p'_{ij} = 100 \cdot (x_{ij} - z_{ij})/(z_{ij} \cdot (1 - 1/\varrho_{HL}) - L)$$
$$= 100 \cdot r_{ij}/(z_{ij} \cdot (1 - 1/\varrho_{HL}) - L) \tag{38}$$

As was mentioned earlier in Section 7.2, other types of normalization equations may have to be chosen for certain types of assays. Modifications of those expressions to obtain corresponding versions for the *sample-based normalization* approach are easily made by applying similar substations as above. As was mentioned previously, large values of the ratio ϱ_{HL} will result in practically identical % activity values $^2p'_{ij}$ and $^3p'_{ij}$, whereas appreciable differences can arise for smaller ratios ($\varrho_{HL} < 5$–10).

In practice, one needs set acceptance limits for the correction factor $a(x, y)$ in eqn. (22) because of the introduction of additional uncertainties into the final result. Often the variability of the responses of compound samples with activities around 50 % is found to be larger than that of the controls. This can be explained by added error contributions due to compound dispensing errors and the increased sensitivity of a sigmoidal dose–response relationship to variations in the concentration in the region of 50 % activity between high and low controls. This larger variability of the compound samples is, in the case of sample-based normalizations, further changed by the correction factor a. Reasonable practical limits for the average correction factor are 0.5–$0.7 \leq a \leq 1.7$–2. The exact limit setting needs to take the level of the full activity controls into account, and therefore depends on ϱ_{HL} [eqn. (2)]: Smaller H/L ratios require a much tighter limit setting than higher ratios. Often one is willing to tolerate excursions of a to slightly more extreme values in small areas of the plate. It should be mentioned that corrections outside those boundaries can be accepted if, for example, the dispensing of the zero activity controls was flawed owing to an accident during an experimental run (e.g. an undetected reservoir blockage). The whole batch of data can still be completely rescued by sample-based normalization and calculation of the corrected $^2p'$ values according to eqn. (37), since those values do not depend on H or ϱ_{HL}. In such an event, all other quality criteria need to be within their respective acceptance range. Such a rescue procedure has some relevance for screens utilizing expensive or difficult to obtain reagents.

The extreme deviations $\alpha = max[|a(x, y) - 1|]$ and $\delta = max[a(x, y)] - min[a(x, y)]$ or equivalently the corresponding deviations of normalized $^1p(x, y)$ % values calcu-

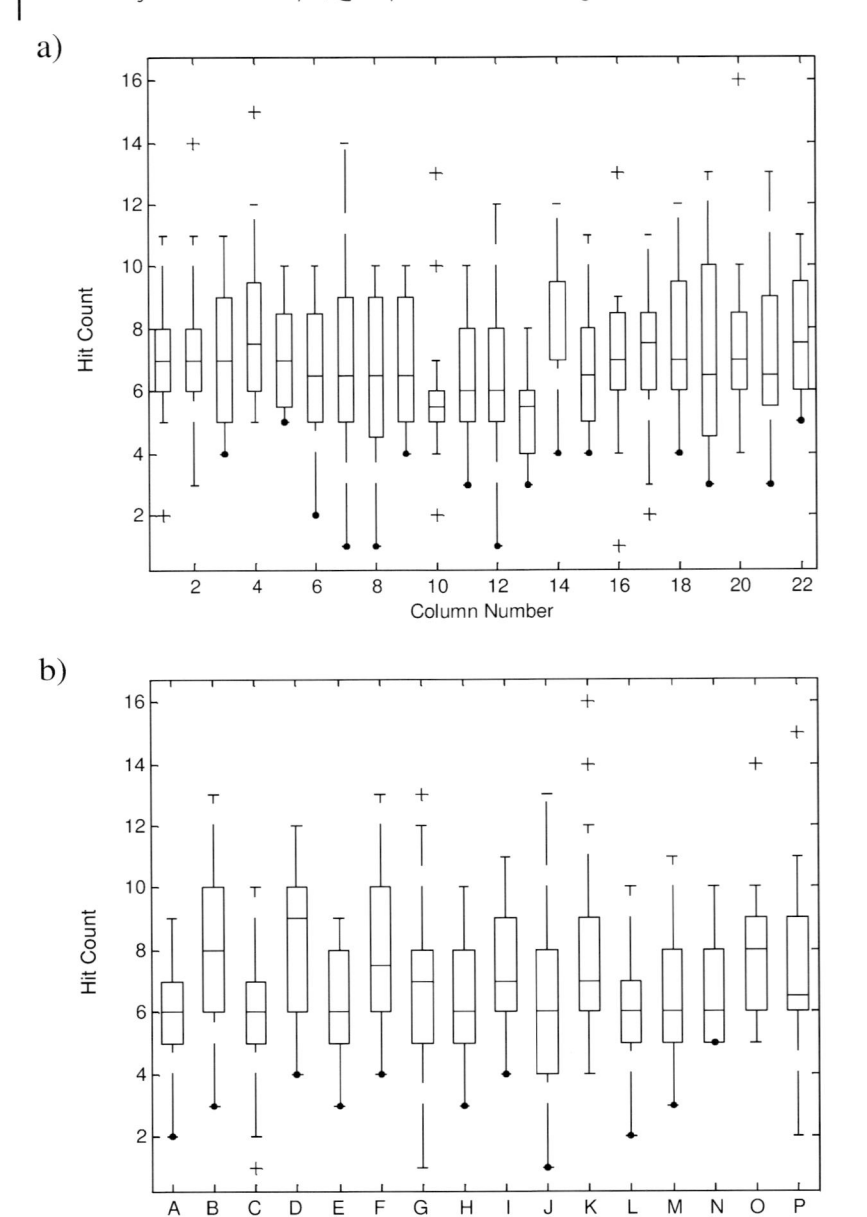

Figure 7.18 The row-wise (a) and column-wise (b) aggregated and summed numbers of hits based on the corrected data. This should be compared with Fig. 7.9. After the correction (renormalization) of the systematic background response distortion, the hit counts are fairly evenly distributed across the rows and columns and the average value of 6–7 hits per well below the −30% threshold corresponds very closely to the average value that can be calculated based on the simulation parameters (~6 hits per well across the 300-plate "screen").

lated from the background response surface representation $z(x, y)$ are themselves data quality criteria that need to be assessed and used for a pass/fail decision on each plate (see also section 7.6.1).

The % activity distribution density which is corrected for systematic response variations with eqn. (37) or (38) is narrower than the uncorrected normalized results with eqn. (6). In Figs. 7.14 and 7.15 the histograms for the normalized data and the corrected data for all 300 plates of the simulated example data set are shown. The considerable narrowing is clearly visible. Also, the distribution of the well-specific hit rate is becoming very homogeneous in both the row- and column-wise projection with values close to the average hit number or hit rate per well as corresponding to the simulation input (Fig. 7.18). Another clear sign of the improvement in the screening data quality after the background response correction is the fact that the RZ values for all the plates are much larger than with the original control-based normalization and are now roughly the same as the corresponding RZ' values, thus signaling a very similar variability and signal window for both the compound samples and the controls (Fig. 7.6). As can be seen for this simulated data set, an average improvement of RZ of between 0.3 and 0.35 units results from the background response correction procedure.

7.8
Hit Identification and Hit Scoring

A direct result of the application of the data corrections is that the average response values now form a horizontal plane centered at 0% and that the density distribution of the inactive samples is narrower and more normal (Fig. 7.15). One can therefore calculate new scale parameters (sd or resistant equivalents s) for the distribution width of the inactive sample population. Z-scores [not to be confused with the Z factor in eqn. (5)] measure the deviation of an individual value x_i from the average value $\bar{x}$ of the whole data set in units of its standard deviation sd_x [13, 28]:

$$Z_i = (x_i - \bar{x})/sd_x \tag{39}$$

and thus attribute to each x_i a level of "significance" of the deviation from $\bar{x}$ ("Studentized residuals"). As outlined earlier (Section 7.5.2), sample activities with $|Z_i| > 3$ (or $|R_i| > 3$, see below) are usually considered for selection into the assay's primary hit list. Economic and logistic considerations about the absolute number of hits or considerations about the shifted balance of false-negative versus false-positive rates beyond a certain threshold may impose different settings. Further important aspects of selection, deselection or even enrichment of HTS compound hit lists which go beyond the direct data-related statistical considerations and which are based on physicochemical and biological properties, structural similarity, chemogenomic classification and biological annotation information are treated elsewhere in this book (Chapters 1 "Chemical Genetics", 9 "Chemoinformatic Tools for HTS Data Analysis and 11 "Data Mining using Cell-Based Assay Results").

For the many reasons already explained earlier, it is preferable to switch from the classical Z-score to resistant location and scale variables and thus arrive at one possible definition of an R-score (R for "resistant", "robust") as

$$R_i = [x_i - median(\mathbf{x})]/s_x \qquad (40)$$

where x_i and $\mathbf{x}$ are the corrected response values, or alternatively (relevant for the corrected % activity scale p' which is centered at 0%) as

$$R_i = p'_i/s_{p'} \qquad (41)$$

Similar R-scores can be defined with the *mode* and/or the *IQR*-based s_R as location and scale estimations (see Section 7.5.4 on robust estimation). Such R-scores (also termed B-scores in [8]) are very useful for comparing the activities of different primary assays because, for example, a 40% activity value in an assay with $s = 7\%$ is highly significant ($R = 5.7$), whereas the same activity value in assay with $s = 20\%$ corresponds at best to a "borderline" significance ($R = 2.0$). In order to perform relevant searching and data mining in primary screening databases and corporate data warehouses, the "significance" scale s for each plate should always be stored together with the corrected % activity values [eqns. (37) and (38)] and should be taken into account when retrieving data and performing further calculations and comparisons based on them.

As was also mentioned in [8], one needs to be careful with hit identification via R-scores if they are based on sample-based normalization and the underlying assumption is invalid. This happens when an appreciable percentage of the samples on a single plate exhibit strong activity. The final sensitivity depends on the degree of "outlier" resistance of the background surface modeling method that is being applied. As was seen earlier in the data and visualization examples, those problematic cases can be easily spotted and flagged for special treatment in the hit analysis phase. Also, an appropriate annotation of the compounds and plates in the screening deck (for example, of target specific compound libraries and "frequent hitters" for specific screening technologies) can provide discrimination information for the hit analysis phase. In the scatterplot of R-scores versus the uncorrected % activity of all screening samples one can clearly identify the benefits of applying a case specific hit identification strategy (Fig. 7.19): The compounds considered to be inactive by both hit selection criteria can be found in quadrant (1). All compounds with an R-score <–3 (remember the negative scale choice for "inhibition" results!), that is, points in quadrants (3) and (4), are the selected hits following the $|R_i| > 3$ criterion. All samples in quadrant (2) are only identified by the –55% activity threshold without any correction of the systematic response errors. It was mentioned above that the sample-based correction and normalization cannot be applied for plates with high hit rates and it is those cases that can be identified in quadrant (2). Although these data are not identified as hits with the

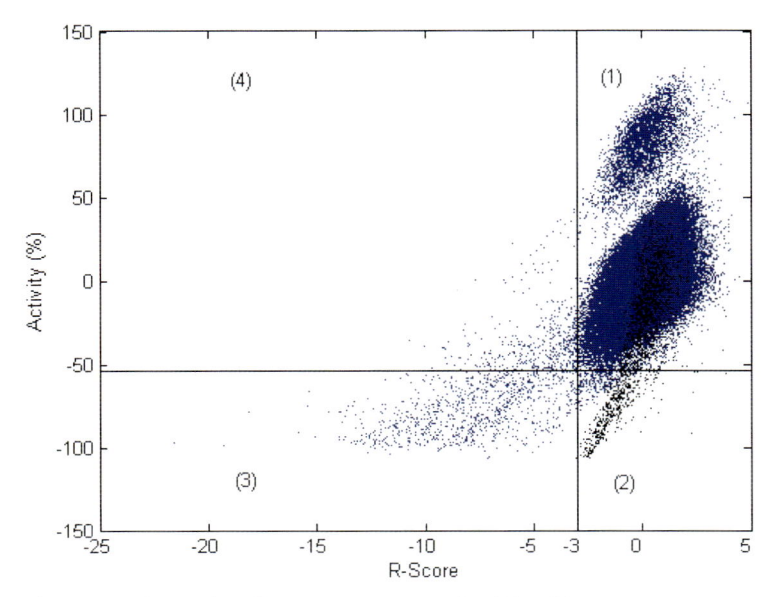

Figure 7.19 Scatterplot of R-score versus normalized % activity data of the set of 300 plates. See Section 7.8 for explanations of the quadrants (1)–(4) and the comments about the data points depicted in black (corresponding to the data from the high hit rate plates $k = 154$–158)

R-scores, it is clear that the data correction cannot be applied for the corresponding plates and potential hits have to be alternatively identified by the % activity criterion. The points corresponding to the simulated plate data with high hit rates are shown as black dots in Fig. 7.19 and they make up the largest portion of the data in this quadrant of the graph. Besides those special cases in quadrant (2) which indeed need to be analyzed in the "classical" fashion, it is obvious from the figure that not many other samples would need to be considered "false negatives" if only the R-score hit selection criterion were to be followed. The safest hit selection strategy is, of course, to propagate all samples from quadrants (2)–(4) to the next confirmation screening or at least to the next *in silico* hit "filtering" stage.

7.9 Conclusion

A comprehensive set of computational tools is available to assess and to correct reliably primary HTS data in a fully automated fashion. In combination with the "right" types of graphics and visualization possibilities, scientists can very quickly focus on the problematic data sets and process areas of the HTS experiments by identifying possible sources of artifacts and improve the measurement process as needed. The data assessment and correction algorithms that were discussed in this chapter can be assembled into a fully automated and efficient data analysis pipeline, which – coupled with suitable visualization components – allows for highly improved data quality in the screening data repositories. This strong improvement in the data quality of primary HTS data collections through the elimination of response distortions and some noise components across the full screen clearly benefits the downstream data mining, cross-assay analysis and knowledge creation processes.

Acknowledgments

The author would like to thank Sigmar Dressler for several stimulating discussions which often triggered further investigations, algorithm development and numerical experimentation to clarify and illustrate certain features of the mathematical and statistical methods presented in this chapter. He also contributed to the continued evolution of the HTS data quality control and analysis platform in the Discovery Technologies department of the Novartis Institutes for BioMedical Research which has "assessed" and "cleaned" all HTS data in a fully automated fashion since 1997. The author also thanks Christian Parker and Martin Beibel for careful reviews of the manuscript.

References

1 S.J. Fox (ed.), *High Throughput Screening 2002: New Strategies and Technologies*, High Tech Business Decisions, Moraga, CA, **2002**.
2 J.P. Helfrich, Data management in high-throughput screening, Bio-IT World, Framingham, MA, 9 September **2002**. http://www.bio-itworld.com/archieve/090902
3 P. Gribbon, S. Schaertl, M. Wickenden, G. Williams, R. Grimley, F. Stuhmeyer, H. Preckel, C. Eggeling, J. Kraemer, J. Everett, W. W. Keighley, A. Sewing, Experiences in implemeting uHTS – cutting edge technology meets the real world, *Current Drug Discovery Technologies*, **2004**, *1*, 27–35.
4 C. Eggeling, L. Brand, D. Ullman, S. Jaeger, Highly sensitive fluorescence detection technology currently available for HTS, *Drug Discovery Today*, **2003**, *8*, 632–641.
5 J. Woelcke, D. Ullmann, Miniaturized HTS technologies – uHTS, *Drug Discovery Today*, **2001**, *6*, 637–646.
6 B. Tulsi, Balance technologies to overcome HTS bottlenecks, *Drug Discovery and Development*, **May 2004**. http://www.dddmag.com/CrtlssueArch.aspx
7 B. Gunter, C. Brideau, B. Pikounis, A. Liaw, Statistical and graphical methods for quality control determination of high-throughput screening data, *Journal of Biomolecular Screening*, **2003**, *8*, 624–633.
8 C. Brideau, B. Gunter, B. Pikounis, A. Liaw, Improved statistical methods for hit selection in high-throughput screening, *Journal of Biomolecular Screening*, **2003**, *8*, 634–647.
9 H. Gubler, M. Girod, S. Dressler, R. Bouhelal, J. Ottl, HTS data analysis in the real world: practical experience with HTS data quality assurance systems and recent integration of the GeneData Screener software, in *Proceedings, Lab-Automation 2004*, 1–5 February 2004, San Jose, CA, Association for Laboratory Automation, Geneva, IL, **2004**, p. 65.

10 Spotfire User Conference Presentations, Life Sciences, http://www.spotfire.com/events/userconference/. Spotfire, Somerville, MA

11 P. Gribbon, R. Lyons, P. Laflin, J. Bradley, C. Chambers, B. S. Williams, W. Keighley, A. Sewing, Evaluating real-life high-throughput screening data, *Journal of Biomolecular Screening*, **2005**, *10*, 99–107.

12 S. Reimann, M. Lindemann, B. Rinn, O. Lefevre, S. Heyse, Large scale, comprehensive quality control and analysis of high-throughput screening data, *European BioPharmaceutical Review, Applied R&D*, Spring, Samedan, London, **2003**.

13 S. Heyse, Comprehensive analysis of high-throughput screening data, *Proceedings of the SPIE*, 2002, 4626, 535–547.

14 *Genedata Screener® Software*, http://www.genedata.com/screener.php, Genedata, Basel.

15 C. Heuer, T. Haenel, B. Prause, A novel approach for quality control and correction of HTS based on artificial intelligence, *Pharmaceutical Discovery and Development 2002/03*, PharmaVentures, Oxford, **2002**.

16 *CyBi® SIENA Software*, http://www.cybio-ag.com/, CyBio, Jena.

17 N. Haan, A Bayesian approach to array image variability, microarray technology, *Pharmaceutical Discovery*, **2004**, Sep 15, 2004

18 *BlueFuse for Microarrays Software*, http://www.cambridgebluegnome.com/, Bluegnome, Cambridge.

19 D. E. Root, B. P. Kelley, B. R. Stockwell, Detecting spatial patterns in biological array experiments, *Journal of Biomolecular Screening*, **2003**, *8*, 393–398.

20 D. Kevorkov, V. Makarenkov, Statistical analysis of systematic errors in high-throughput screening, *Journal of Biomolecular Screening*, **2005**, *10*, 557–567

21 B. P. Kelley, M. R. Lunn, D. E. Root, S. P. Flaherty, A. M. Martino, B. R. Stockwell, A flexible data analysis tool for chemical genetics screens, *Chemistry and Biology*, **2004**, *11*, 1495–1503.

22 *Decision Site and Decision Site for Lead Discovery Software*, http://www.spotfire.com/products/decisionsite_lead_discovery.cfm, Spotfire, Somerville, MA.

23 *Partek Pattern Visualization System® Software*, http://www.partek.com/html/products/products.html, Partek, St. Charles, MO.

24 *ActivityBase Software*, http://www.idbs.com, ID Business Solutions, Guildford.

25 *MDL® Assay Explorer, MDL® Assay Explorer Visualizer Software*, http://www.mdli.com, Elsevier-MDL, San Leandro, CA.

26 J. H. Zhang, T. D. Chung, K. R. Oldenburg, A simple statistical parameter for use in evaluation and validation of high throughput screening assays, *Journal of Biomolecular Screening*, **1999**, *4*, 67–73.

27 J. M. Bland, D. G. Altman, Statistical methods for assessing agreement between two methods of clinical measurements, *Lancet*, **1986**, *i*: 307–310.

28 C. L. Hurbert, S. E. Sherling, P. A. Johnston, L. F. Stancato, Data concordance from a comparison between filter binding and fluorescence polarization assay formats for identification of ROCK-II inhibitors, *Journal of Biomolecular Screening*, **2003**, *8*, 399–409.

29 G. E. P. Box, W. G. Hunter, J. S. Hunter, *Statistics for Experimenters*, Wiley, New York, **1978**.

30 *NIST/SEMATECH e-Handbook of Statistical Methods*, http://www.itl.nist.gov/div898/handbook/, NIST, Gaithersburg, MD, Chapter 5, **2005**.

31 P. B. Taylor, F. P. Stewart, D. J. Dunnington, S. T. Quinn, C. K. Schulz, K. S. Vaidya, E. Kurali, T. R. Lane, W. C. Xiong, T. P. Sherrill, J. S. Snider, N. D. Terpstra. R. P. Hertzberg, Automated assay optimization with integrated statistics and smart robotics, *Journal of Biomolecular Screening*, **2000**, *5*, 213–225.

32 *Sagian™ Automated Assay Optimizer*, http://www.beckman.com, Beckman, Fullerton, CA.

33 G. S. Sittampalam, P. W. Iversen, J. A. Boadt, S. D. Kahl, S. Bright, J. M. Zock, W. P. Janzen, M. D. Lister, Design of signal windows in high throughput assays for drug discovery, *Journal of Biomolecular Screening*, **1997**, *2*, 159–169.

34 J. H. Zhang, T. D. Y. Chung, K. R. Oldenburg, Confirmation of primary active substances from high throughput screening of chemical and biological populations: a statistical approach and practical considerations, *Journal of Combinatorial Chemistry*, **2000**, *2*, 258–265.

35 P. Fogel, P. Collette, A. Dupront, T. Garyantes, D. Guedin, The confirmation rate of primary hits: a predictive model, *Journal of Biomolecular Screening*, **2002**, *7*, 175–190.

36 S. Turconi, K. Shea, S. Ashman, K. Fantom, D. L. Earnshaw, R. P. Bingham, U. M. Haupts, M. J. B. Brown, A. Pope, Real experiences of uHTS: a prototypic 1536-well fluorescence anisotropy-based uHTS screen and application of well-level quality control procedures, *Journal of Biomolecular Screening*, **2001**, *5*, 275–290.

37 K.S. Schroeder, B. D. Neagle, FLIPR: a new instrument for accurate, high throughput optical screening, *Journal of Biomolecular Screening*, **1996**, *1*, 75–80.

38 V.C. Abraham, D. L. Taylor, J. R. Haskins, High content screening applied to large-scale cell biology, *Trends in Biotechnology*, **2004**, *22*, 15–22.

39 Z. Li, S. Mehdi, I. Patel, J. Kawooya, M. Judkins, W. Zhang, K. Diener, A. Lozada, D. Dunnington, An ultra-high throughput screening approach for an adenine transferase using fluorescence polarization, *Journal of Biomolecular Screening*, **2000**, *5*, 31–37.

40 D.M. Titterington, A. F.M. Smith, U. E. Makov, *Statistical Analysis of Finite Mixture Distributions*, Wiley, New York, **1985**.

41 B.V. M. Mendes, H. F. Lopes, Data driven estimates for mixtures, *Computational Statistics and Data Analysis*, **2004**, *47*, 583–598.

42 J.R. M. Hosking, L-moments: analysis and estimation of distributions using linear combinations of order statistics, *Journal of the Royal Statistical Society, Series B*, **1990**, *52*, 105–124.

43 W.N. Venables, B. D. Ripley, *Modern Applied Statistics with S-Plus*, Springer, New York, **1994**.

44 P.J. Huber, *Robust Statistics*, Wiley, New York, **1981**.

45 J.W. Tukey, *Exploratory Data Analysis*, Addison-Wesley, Reading, MA, **1977**.

46 P.J. Rousseeuw, A. M. Leroy, *Robust Regression and Outlier Detection*, Wiley, New York, **1987**.

47 F.R. Hampel, The breakdown points of the mean combined with some rejection rules, *Technometrics*, **1985**, *27*, 95–107.

48 D.M. Hawkins, *Identification of Outliers*, Chapman and Hall, New York, **1980**.

49 J.E. Gentle, W. Härdle, Y. Mori (eds.), *Handbook of Computational Statistics*, Springer, Heidelberg, **2004**.

50 H. Gubler, Novartis Institutes for BioMedical Research, Basel, unpublished.

51 M.B. Eisen, P. T. Spellman, P. O. Brown, D. Botstein, Cluster analysis and display of genome-wide expression patterns, *Proc. Natl. Acad. Sci. USA*, **1998**, *95*, 14683–14868.

52 W. S. Cleveland, *Visualizing Data*, Hobart Press, Summit, NJ, **1993**.

53 J. Chambers, W. Cleveland, B. Kleiner, P. Tukey, *Graphical Methods for Data Analysis*, Wadsworth, Belmont, CA, **1983**.

54 D.W. Scott, On optimal and data-based histograms. *Biometrika*, **1979**, *66*, 605–610.

55 D.W. Scott, Averaged shifted histograms: effective nonparametric density estimators in several dimensions. *Annals of Statistics*, **1985**, *13*, 1024–1040.

56 W.H. Press, B. P. Flannery, S. A. Teukolsky, W. T. Vetterling, *Numerical Recipes in C: The Art of Scientific Computing*, Cambridge University Press, Cambridge, **1992**; http://www.numerical-recipes.com/.

57 R.W. Preisendorfer, *Principal Component Analyses in Meteorology and Oceanography*, Elsevier, Amsterdam, **1988**.

58 H. von Storch, F. W. Zwiers, *Statistical Analysis in Climate Research*, Cambridge University Press, Cambridge, **1999**.

59 J.A. Scales, L. Tenorio, Prior information and uncertainty in inverse problems, *Geophysics*, **2001**, *66*, 389–397.

60 P.F. Velleman, Definition and comparison of robust nonlinear data smoothing algorithms, *Journal of the American Statistical Association*, **1980**, *75*, 609–615.

61 I. Gijbels, P. Hall, A. Kneip, On the estimation of jump points in smooth curves, *Annals of the Institute of Statistical Mathematics*, **1999**, *51*, 231–251.

62 G. Gregoire, Z. Hamrouni, Two non-parametric tests for change-point problem, *Journal of Nonparametric Statistics*, **2002**, *14*, 87–112.

63 P. Hall, D. M. Titterington, Edge preserving and peak-preserving smoothing, *Technometrics*, **1992**, *34*, 429–440.

64 P.J. Brockwell, R. A. Davis, *Introduction to Time Series and Forecasting*. Springer, New York, **1996**.

65 N.R. Draper, H. Smith, *Applied Regression Analysis*, Wiley, New York, **1981**.

66 G.A. F. Seber, *Linear Regression Analysis*, Wiley, New York, **1977**.

67 M. Stone, Cross-validatory choice and assessment of statistical prediction, *Journal of the Royal Statistical Society, Series B*, **1974**, *36*, 111–147.

68 P. Craven, G. Wahba, Smoothing noisy data with spline functions, *Numerical Mathematics*, **1979**, *31*, 377–403.

69 T.J. Hastie, C. R. Loader, Local regression: automatic kernel carpentry (with discussion). *Statistical Science*, **1993**, *8*, 120–143.

70 J. Fan, I. Gijbels, *Local Polynomial Modelling and Its Applications*, Chapman and Hall, London, **1996**.

71 W.S. Cleveland, Robust locally weighted regression and smoothing scatterplots, *Journal of the American Statistical Association*, **1979**, *74*, 829–836.

72 C. Loader, *Local Regression and Likelihood*, Springer, New York, **1999**.

73 P.J. Green, B. Silverman, *Nonparametric Regression and Generalized Linear Models: a Roughness Penalty Approach*, Chapman and Hall, London, **1994**.

74 G. Wahba, Data-based optimal smoothing of orthogonal series density estimates, *Annals of Statistics*, **1981**, *9*, 146–156.

75 S. Efromovich, *Nonparametric Curve Estimation*. Springer, New York, **1999**.

76 H.G. Zerwes, J. C. Peter, M. Link, H. Gubler, G. Scheel, A multiparameter screening assay to assess the cytokine-induced expression of endothelial cell adhesion molecules, *Analytical Biochemstry*, **2002**, *304*, 166–173.

77 D.A. Girard, Asymptotic comparison of (partial) cross validation, GCV and randomized GCV in nonparametric regression, *Annals of Statistics*, **1998**, *26*, 315–334.

78 W. Härdle, L. Simar, *Applied Multivariate Statistical Analysis*, Springer, Heidelberg, **2003**.

79 A. Björck, *Numerical Methods for Least Squares Problems*, SIAM Press, Philadelphia, PA, **1996**.

80 N. Kettaneh, A. Berglund, S. Wold, PCA and PLS with very large data sets, *Computational Statistics and Data Analysis*, **2005**, *48*, 69–85.

81 M. Hubert, S. Engelen, Robust PCA and classification in biosciences, *Bioinformatics*, **2004**, *20*, 1728–1736.

82 G. Boente, A. M. Pires, I. Rodriguez, Influence functions and outlier detection under the common principal components model: a robust approach, *Biometrika*, **2002**, *89*, 861–875.

83 M. Hubert, P. J. Rousseeuw, K. Vanden Branden, ROBPCA: a new approach to robust principal component analysis, *Technometrics*, **2005**, *47*, 64–79.

84 C. Jutten, J, Herault, Blind separation of sources. Part I: an adaptive algorithm based on neuromimetic architecture, *Signal Processing*, **1991**, *24*, 1–10.

85 A. Hivärinen, E. Oja, Independent component analysis: algorithms and applications, *Neural Networks*, **2000**, *13*, 411–430.

86 H. Gubler, EOF and REOF analysis for assessment and correction of systematic errors in HTS data, manuscript in preparation.

8
Chemoinformatic Tools for High-throughput Screening Data Analysis

Peter G. Nell and Stefan M. Mundt

8.1
Introduction

Over the past decades, drug discovery strategies have changed dramatically and adopted many new concepts. High-throughput screening (HTS) evolved from the ability to screen libraries of moderate size. More recently, ultraHTS (uHTS) utilizing high-density microtiter plate formats allowing one to screen more than a million compounds for a single target in a reasonable timeframe has emerged. The pharmaceutical industry has incurred great expense in designing and synthesizing large and diverse libraries of small organic molecules via combinatorial chemistry. When these compound libraries are tested against possible drug targets applying HTS techniques, an-ever increasing amount of data is generated, which needs to be handled in a clever and inevitably automated way to identify and prioritize new hits and lead candidates. Today, chemoinformatic techniques assist the medicinal chemist and molecular pharmacologist in viewing and organizing data and understanding relationships between measured biological data and properties of compounds. Selection of the most promising lead compounds and an early and systematic removal of suspected "false positives" together with the identification of "false negatives" are vital to the success of HTS. Ideally, the characteristics of compounds chosen for further lead-finding steps will be drug-like in several pharmacokinetic, physicochemical and structural parameters. Chemoinformatic methods have a great impact in the identification and application of these parameters to narrow down data sets [1–3] as *in vitro* and *in vivo* screening cannot cope with the typically large number of primary HTS hits. More restricted parameters for drug/lead likeness can be applied for the design of a screening library. Applying further chemoinformatic methods can reduce cost and optimize use of human resources in the area of synthesis and pharmacological testing. Sequential screening, including virtual screening (VS) techniques, are an alternative option to random screening. This chapter will give an overview of the chemoinformatic techniques and methods that are applied particularly in the

High-Throughput Screening in Drug Discovery. Edited by Jörg Hüser
Copyright © 2006 WILEY-VCH Verlag GmbH & Co. KGaA, Weinheim
ISBN: 3-527-31283-8

area of HTS data evaluation and will focus on providing key references rather than giving a complete survey.

8.1.1
Definition of Chemoinformatics

The term *chemoinformatics* (or "cheminformatics") was introduced by Frank Brown [4, 188]. To this day there is no clear definition, but a consensus view appears to be emerging. Chemoinformatics is the method of applying information technology to the drug discovery and optimization learning cycle. It integrates all data, information and derived knowledge that are needed for decision support. Central to any chemoinformatics system is the graphical user interface (GUI) that allows easy access to integrated biological data including bioinformatics and pharmacological data and also chemical databases. The tools applied for data analysis and data mining differ from chemometrics, computational chemistry and molecular modeling, even though they share the same underlying algorithms [5–7].

One prerequisite for the general applicability of software tools was the evolution of computer power, that following "Moore's law" [8], has doubled every 12–24 months, keeping pace with the explosion in the rate of generation of chemical, biological and other data.

In the early days of computational chemistry, giant server-computers had to be used and computation time was very expensive. Nowadays, even the desktop and laptop computers that are available to research scientists are very powerful and can perform many operations in a short time and with less financial investment. This development also had an impact on software-developing companies. Most of the new software packages are available for personal computers or Linux systems, making it possible to provide these packages to every research scientist within an organization and not limit their use to specialists.

Integrated systems that combine all tools, starting from visualization of the HTS results in an appropriate form, organizing data, applying tools for filtering out non-lead-like compounds, identifying lead series in comprehensive data sets using various chemoinformatic tools and finally resolving an initial SAR within these lead series will bring these tools to more scientists and speed up and reduce costs of HTS screening methodology. Examples of commercially available software packages will be described in the last section of this chapter.

8.1.2
High-throughput Screening

Owing to past investments and acquired expertise in the fields of combinatorial chemistry and assay technologies, along with the steadily growing pressure on discovery units to improve their productivity, screening has developed into high-throughput screening. Currently there are two different screening paradigms used in pharmaceutical companies: (i) random screening, "mass random" or "primary library" screening, and (ii) sequential screening, also called "focused library"

Random Screening

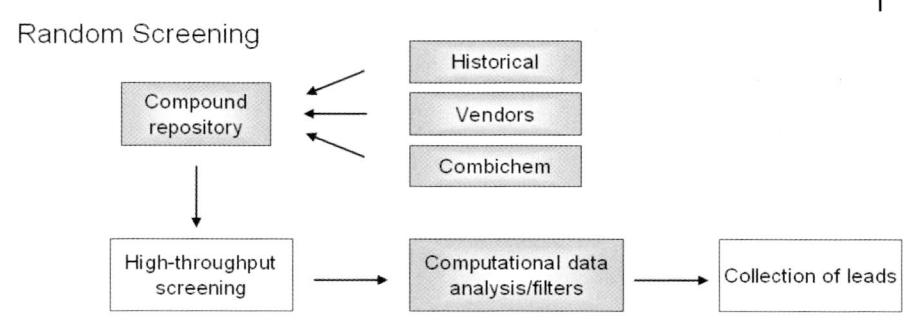

Figure 8.1 Random screening workflow.

screening. These paradigms have different benefits and limitations. Both kinds of workflow are supported by chemoinformatics tools.

8.1.2.1 Random Screening

In *random screening*, the entire corporate compound archive is tested in an assay (Fig. 8.1). These libraries can be built from historical collections, natural products, combinatorial synthesis and third-party collaborations. Currently, some drug discovery companies have corporate compound libraries of one to two million compounds available for screening, although routine screening of more than one million compounds is still rare. When applying a random screening approach, no pharmacophore knowledge of the target is needed. The assay is used as a filter to identify the promising hits that justify further analysis for lead finding (e.g. clustering and data mining). Data quality in HTS efforts has improved in recent years such that normally all data, including those associated with weakly active and with inactive compounds, can be used in subsequent analysis. Functional tests can monitor interactions with different binding sites [9, 10]. A major advantage of screening an unbiased and diverse large library in this way is the increased probability of finding biologically active compounds with a new and unexpected structural scaffold and potentially yielding a major advantage regarding the intellectual property and patent situation of a resulting lead candidate.

In a review process, the initial paradigm to screen a complete company compound archive for all pharmaceutical targets in search of drugs, different rational approaches based on chemoinformatics were developed to design combinatorial libraries and to populate the screening library on the basis of properties that have proven to be associated with successful therapeutics, especially physicochemical and ADMET properties [11, 12]. These properties can be predicted using chemoinformatic tools that will be described in Section 8.3.

8.1.2.2 Sequential Screening

The second major screening paradigm is referred to as *sequential screening* (Fig. 8.2). It is characterized by the selection of an initial compound sample set prior to synthesis or biological testing [13, 14]. The main feature of this approach is that only a small set of compounds is assayed and the results are statistically analyzed to produce a SAR model. This model is then used to select further promising candidates for additional screening. Screening of these compounds will then be used to refine the model and generate an improved version. This process may be repeated several times to identify the most promising compounds for the lead optimization team. Chemoinformatics support this approach at multiple levels. Methods applied include pharmacophore-based virtual screening, diversity selection, recursive partitioning, prediction and use of filtering for "drug likeness" or "lead likeness" properties based on *in silico* prediction methods [15–17].

Chemoinformatic tools first come into play when the initial screening set needs to be defined [18]. If there is knowledge about the target, e.g. an X-ray structure or a pharmacophore model, the compounds can be scored using computational chemistry methods such as docking and modeling. Also, chemoinformatic tools evaluating similarity features can be applied comparing compounds with known active inhibitors [14]. If there is no information on the structure of the target, a subset should be chosen that provides the greatest random diversity for a given sample set size. Again, chemoinformatic methods such as similarity searching based on descriptor methods can be applied. These can be either fragment-based descriptors that count the presence/absence or frequency of various fragments or continuous descriptors such as calculated physicochemical parameters including $clogP$ and polar surface area.

In the next stage, sequential screening also makes use of chemoinformatic algorithms for building the statistical model after each sequence of screening and hit evaluation including clustering of chemical compounds [19].

The compounds of the initial screening set might not only be a selection of a pre-existing compound repository but may also contain compounds from newly designed and generated focused libraries. Many such examples have been described in the literature (e.g. kinase [20] or GPCR focused libraries).

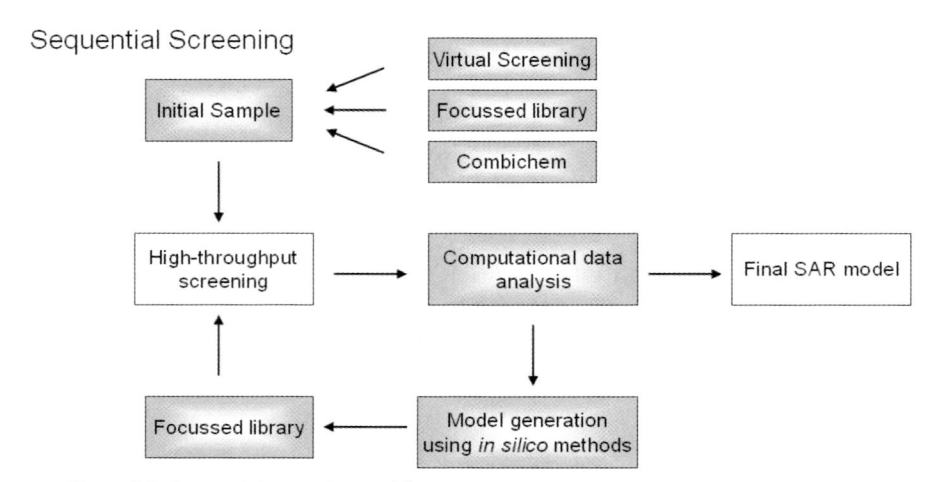

Figure 8.2 Sequential screening workflow.

Although the hit rate for sequential screening might sometimes be higher than when applying the random screening approach, the risk of missing active compounds and therefore potential lead series, so called false negatives, might also be much higher. In addition, random screening has the advantage of identifying new classes of active compounds, which have not been pre-selected.

The following sections will deal with those chemoinformatic tools mainly used in random high-throughput screening.

8.2
Workflow of High-throughput Screening and Use of Chemoinformatics

Random screening starts with a large screening library. In the initial high-throughput *in vitro* assay, primary hits are identified (Fig. 8.3). Owing to technical reasons described in other chapters of this book, the identified compounds need to be re-tested to confirm activity. In general, the resulting set of confirmed hits would be in the range of 0.1–1 % of the tested compounds. At Bayer HealthCare, this is typically in the range of 50–2000 compounds.

The confirmed hits are explored using various chemoinformatic tools. For example, compounds with undesired structural fragments that have been associated with toxicity or chemical reactivity can be filtered out using substructure search or similarity search algorithms. There are further reasons why some compounds should not be considered for further investigation. Some of the compounds might be identified as "frequent hitters" by using profiling tools. Frequent hitters are defined as molecules generating hits in many different assays covering a wide range of targets and therefore are not specific enough to be suitable drug candidates. Two general categories describe this behavior: (1) the hit compound is not selective in its activity for the target; these compounds are also called "promiscuous inhibitors" [21]; and (2) the compounds are disturbing the assay or detection method; examples of this class are often colored or fluorescent compounds or are lipophilic compounds interfering with membranes. This effect might be confirmed using experimental examination. Using chemoinformatics, the activity profile of each hit from historical screening results may also identify unselective compounds. Obviously this profiling is only helpful for compounds that have already been synthesized and tested in several assays, therefore as a possible alternative ensemble methods [22] and virtual screening methods have been developed and tested for the identification of frequent hitters [23]. Using ensemble methods, a combination of support vector machines and prior removal of redundant features (see below for explanation of these methods), Merckwirth et al. [22] achieved ensembles exceeding a classification rate of 92 % in a cross-validation setting. The misclassification rate regarding the problem of frequent-hitters was only 4–5 %, an acceptable rate regarding the technical background, although one should be aware that this might still be too high, as important hits might be excluded using this method. In a second approach, Roche et al. [23] elaborated a scoring scheme from substructure analysis using multivariate linear and nonlinear

High-throughput Screening Workflow

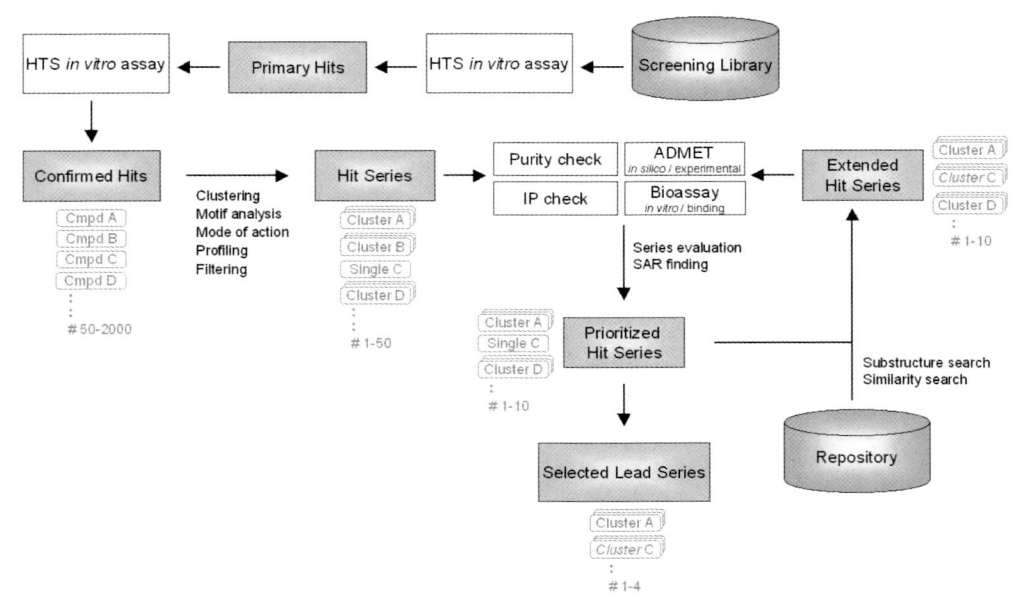

Figure 8.3 Typical high-throughput screening workflow.

statistical methods applied to several sets of one- and two-dimensional descriptors. The resulting model was based on a three-layered neural network. The system correctly classified 90% of the test set molecules in a 10-times cross-validation study. The method was also applied to database filtering, identifying between 8% (for a compilation of drugs) and 35% (for the Available Chemicals Directory) potential frequent hitters.

Grouping of the confirmed hits according to their structural similarity by using clustering algorithms makes it possible to order and explore the data obtained. A typical hit series will consist of approximately 1–50 clusters and singletons depending on target and assay type.

Next, the hit series need to be evaluated for whether it meets the desired characteristics of a potential drug and can be designated a lead series in the "hit to lead" process [24]. To reduce the risk of failure, more than one lead series should be identified. To balance cost and time, the number will normally be limited to four or five series, depending on the available resources.

The purity of compounds in the hit series can be examined using well-known analytical techniques such as LC–UV–MS, LC–MS–MS, and, in some cases, NMR or, if available, LC–NMR [25].

Some pure HTS hits show activity in the biochemical assay but still are not desired for further investigation. Reasons can be chemical instability, chelate or polyionic binding or the compounds can be protein-reactive, resulting in covalent binding to the target [26, 27]. In most cases, it is best to identify these structures by inspection of the hit list by an experienced medicinal chemist or molecular pharmacologist. Chemoinformatic methods can support them in this filter process. For example, if a compound is binding covalently and irreversibly to the assayed protein owing to a chemical reaction, it is a reactive compound. As this reactivity of the compound can often be assigned to a specific fragment, a screen using substructure search compute engines can assist the scientist. Moreover, it is possible to take the structure of known reactive compounds and perform a similarity search to find similar compounds within the hit list [28]. The same algorithms can also be applied to questions regarding intellectual property by identifying similar compounds that have already been patented or revealing that there is freedom to operate.

An impractical chemical synthesis or optimization can be another good reason to stop further investigation on a specific compound. The experienced medicinal chemist remains the best judge of this compound property.

Filtering out impure compounds, frequent hitters and false-positives is an important part of the HTS workflow as starting the lead optimization process with these risky compounds can be time consuming and costly. Early detection of problematic compounds is crucial for the success of an HTS program [29]. Ideally, these undesired compounds would be removed from the screening library or corporate compound collections prior to screening, especially when HTS capacity is limited, but in most cases this is impractical or impossible. On the contrary, keeping a few frequent hitters and false positives might be desirable to serve as internal controls to monitor the assay and process performance.

Some active compounds might not be identified during the initial screening process owing to technical problems. These compounds are called false negatives. An approach to automatically identify these false negatives in HTS data was developed by Engels et al., making use of logistic regression to set up a model that can be employed to screen a data set and predict potential outliers [30]. The method is based on the assumed relationship between structures of screened compounds and their respective biological activity on a given screen expressed on a binary scale. By means of a data mining method, an SAR description of the data set is developed that assigns probabilities of being a hit to each compound in the screen. Then, an inconsistency score expressing the degree of deviation between adequacy of the SAR description and the actual biological activity is computed. This inconsistency score allows the identification of potential outliers that can be selected for validation experiments. This approach might only be meaningful for assays with a low rate of systematic unspecific compounds (biochemical tests with direct readout). Applying this method in HTS campaigns at Johnson and Johnson, the initial hit rate varied from under 0.05% to more than 2%. In proportion to the initial hit rate, between several hundred and 2000 compounds were identified as possible outliers and re-tested after being assigned as potential false negatives. Engels et al. stated that in almost all cases, a significant enrichment over the original hit rate had been achieved. Owing to the consistently valuable results, this method has been automated and integrated into the HTS workflow at Johnson and Johnson.

In the final step of the HTS process, the biological activity determined under HTS conditions for the lead series must be confirmed in a pharmacologically relevant *in vitro* or binding assay, including selectivity assays. Parameters for absorption, distribution, metabolism, excretion and toxicology (ADMET) may also be used as criteria for prioritization of hit series. If possible, experimental data will be used which can be complemented by *in silico* data. Chemoinformatic methods used here will be described below. All these properties and parameters will be used to prioritize the hit series and deprioritize unpromising compound classes with identified liabilities.

To expand the structural classes and to include predicted false negatives and represent an environment for the singletons under evaluation, compounds from a repository might be added based on similarity or substructure searches. The resulting data sets are again tested in biological assays and reprioritized following the workflow described above. This cycle is run until suitable lead series have been identified [189], which will not always be the case. The hit series to be followed up will typically consist of one to four structural classes, depending on the available resources for the lead optimization process.

8.3
Chemoinformatic Methods Used in HTS Workflow

In general, there are five steps that are important for discovering knowledge from information and data: (i) data integration, searching and retrieval, (ii) data transformation, e.g. format conversion, data normalization, descriptor assignments and calculation of fingerprints for chemical structures, (iii) data reduction, e.g. clustering, (iv) data mining, e.g. pattern discovery, classification of mode of action, SAR hypothesis generation, and (v) assessment of virtual compounds based on a prediction model to prioritize molecules for synthesis.

The following section is focused on steps ii, iii and iv, the data transformation, data reduction and data mining steps, in addition to the chemoinformatic methods mentioned in the workflow for HTS, including some specific example tools.

8.3.1
Substructure Search/Similarity Search

Substructure searching and 2D similarity searching are among the most widely used chemoinformatic methods for mining structural data. The basic principle of these methods is that compounds with structural features in common will have similar properties. This reflects the controversial underlying assumption: structurally similar molecules have similar biological properties [31]. Another application of similarity searching based on properties has been the discovery of compounds with similar biological activity but different chemical structure, this has been termed *scaffold hopping* and will be explained in Section 8.3.8.

Recently, a debate started about the validity of the assumption that structurally similar compounds behave alike, initiated by the observation that some compounds similar to a probe by means of calculated similarity may not have the expected activity [32]. This observation should not really come as a surprise when considering that often the molecular descriptors do not take into account the stereochemistry, salt forms, conformers or spatial flexibilities of compounds. In addition, structure–activity and structure–property relationships underlying different mechanisms or properties can be significantly overlapping in the structure space, adding further complexity, Therefore, structural similarity can only be one component of pharmacological similarity.

The term *substructure* has been used widely in the literature. Generally, a substructure is a fragment of a 2D representation of a chemical structure. An example of substructure-based searching methods is MDL's Molecular ACCess System (MACCS) [33, 34], which includes atom types and counts, functional groups and ring types. Generic (fuzzy) searches can be performed using Markush definitions, named after Eugene Markush, the first inventor to include them successfully in a US patent in the 1920 s. Markush definitions are lists of groups of atoms and bonds which can provide constrained variability in a structure by replacing one or more atoms in an exact substructure with pseudo-atom symbols that correspond to lists of atoms or functionalities [35, 36].

Processing of structural chemical data differs greatly from other numerical data analysis. It is a simple thought process for a trained medicinal chemist to recognize which chemical groups are present or to compare two chemical structures including, e.g., homomorphism, isomorphism and maximum common substructures. In contrast, to make structure and substructure searching feasible for computer programs, special approaches for structural representation with substructural fragments, so called "descriptors", were required. In this way, chemical structures can be stored and handled by an atom-by-atom search algorithm. From a mathematical point of view, a chemical formula can be regarded as a graph with nodes (atoms) and edges (bonds). Using this level of abstraction, a molecule can be described by enumerating all bond paths through it. In addition, various 3D structural conformers and hybridization can also be handled by computational algorithms, but these are computationally costly: these algorithms are so called "NP-complete problems", meaning that the required computation time and memory will grow exponentially with the number of atoms within a structure [37].

8.3.1.1 Structural Descriptors/Fingerprints

Descriptors are used to allow for data reduction and to enhance computational performance. There is an infinite variety of potential descriptors and new molecular descriptors are continuously being developed.

Structural descriptors can be classified according to their level of spatial complexity: (i) one-dimensional (1D) descriptors, derived from compound properties, e.g. atom counts, molecular weight, volume and log P; (ii) two-dimensional (2D) descriptors, encoding molecular fragments, e.g. UNITY fingerprints [38], MACCS

A

Chemical Structure

Fragment List

Fingerprint Descriptor

fp₁ | 1 | 0 | 1 | 0 | 0 | 1 | 0 | 0 | 1 | 0 | 0 | 0 | 1 | 1 | 0 | 1 |

B

Binary Descriptor Comparison

fp_1 | 1 | 0 | 1 | 0 | 0 | 1 | 0 | 0 | 1 | 0 | 0 | 0 | 1 | 1 | 0 | 1 |

fp_2 | 1 | 0 | 1 | 1 | 0 | 1 | 0 | 0 | 1 | 1 | 0 | 0 | 1 | 0 | 0 | 1 |

$fp_{(1\&2)}$ | 1 | 0 | 1 | 0 | 0 | 1 | 0 | 0 | 1 | 0 | 0 | 0 | 1 | 0 | 0 | 1 |

Tanimoto Distance

$$\tau_{(1\&2)} = \frac{Nfp_{(1\&2)}}{Nfp_1 + Nfp_2 - Nfp_{(1\&2)}} = \frac{6}{7 + 8 - 6} = 0.66$$

Euclidean Distance

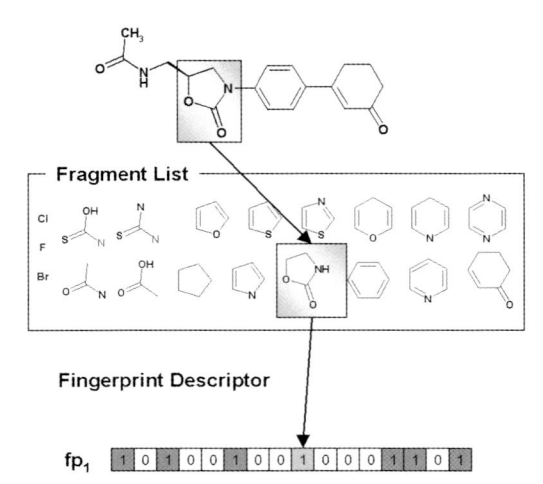

$$D_{(1\&2)} = \sqrt{\Sigma (fp_1 - fp_2)^2} = \sqrt{3} \quad D^{norm}_{(1\&2)} = 0.57$$

Figure 8.4 (A) Molecules are split up into fragments. 2D structural fingerprints encode the absence or presence of specific small chemical fragments in a vector by assigning binary "0" or "1" values. (B) Encoding molecular structures in a bit string allows quantification of molecular similarity of two structures by functions such as Tanimoto distance (τ, number of bits in common) or Euclidean distance (D, straight line distance in a Cartesian coordinate system, can be normalized to the range 1 to 0). A distance of zero indicates that both structures have no bits in common whereas a value of one indicates that the fingerprints are identical.

structural keys or Daylight fingerprints [39]; (iii) three-dimensional (3D) descriptors or spatial pharmacophores, e.g. solvent-accessible surface area, CoMFA (comparative molecular field analysis). Comparative overviews of various descriptors have been given; see Bayada and coworkers for a more detailed discussion [40, 41].

Most commonly, 2D description is used for encoding of compounds in bit strings [42]. Each molecule is split up into substructural fragments. If a particular fragment is present, then a corresponding bit is set to 1 in the bit string (Fig. 8.4A). Some algorithms use predefined fragments (e.g. Unity fingerprints). others generate lists of all fragments dynamically.

Hashing is often used to limit the length of fingerprints. Instead of using only a single bit to encode the occurrence of a certain feature, multiple bits are set to 1. For a string of 100 bits, this means the possibility of encoding up to 2^{100} (= 1.3^{30}) instead of 100 different structural attributes. Although two structures may have a different number of atoms and bonds, they may share the same fingerprint bits because the fingerprints encoding limits the structural information [43].

Fragment-based descriptors such as UNITY fingerprints demonstrate efficient performance on large, structurally diverse compound sets whereas clustering of structurally homogeneous compound sets is preferably performed using CATS descriptors, as these provide a more intuitive grouping of compounds, comparable to medicinal chemists' intuitive experience. CATS descriptors belong to the category of atom-pair descriptors and encode topological pharmacophore information [44].

8.3.1.2 Measures of Similarity

In classical quantitative structure–activity relationship (QSAR) analysis, a common free-energy scale relates independent variables to each other. Concepts of similarity are then possible by simple arithmetic difference of values. Concepts of similarity of chemical structures are inherently more complex and less intuitive because a structure needs to be described in terms of a descriptor, e.g. 2D fingerprints, and therefore comparisons cannot be based on a common free-energy scale.

Many different types of similarity measures, so-called "similarity coefficients", have been introduced as a quantitative measure of the degree of descriptor-based relationship. Similarity coefficients are measures of the distance between objects

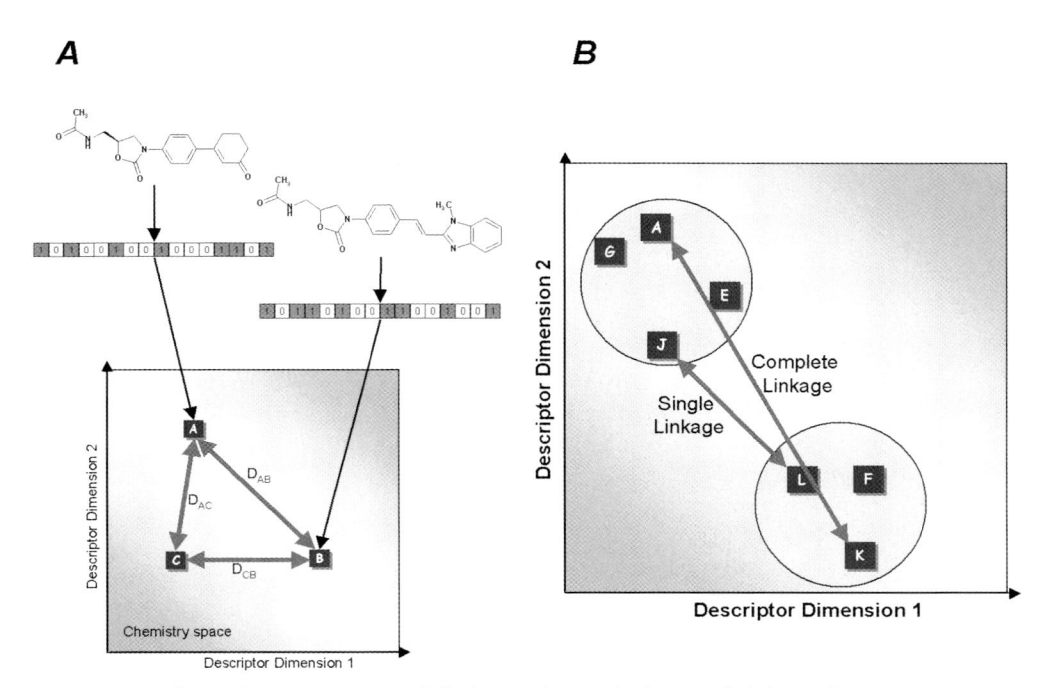

Figure 8.5 (A) Chemical structures are encoded by fingerprints (e.g. 16-bit string, points in a 16-dimensional space). The concept of similarity relations can be illustrated as points in a simplified two-dimensional chemistry space. Distance comparison of three structures must obey the triangular inequality $D_{AB} \leq D_{AC} + D_{CB}$. (B) There are several methods for defining distance (or similarity) between clusters. One of the simplest methods is single linkage, also known as the nearest neighbor technique. The distance between groups is defined as the distance between the closest pair of compounds. The complete linkage, also called furthest neighbor, clustering method is the opposite of single linkage. Distance between groups is now defined as the distance between the most distant pair compounds.

and, in most cases, the values are normalized to values between 0 and 1. All types of similarity coefficients share the following properties: (i) distance values must be zero or positive and the distance from an object (e.g. a chemical structure) to itself must be zero, (ii) the distance values are symmetrical, (iii) the values obey triangular inequality (Figure 8.5) and (iv) distance of non-identical objects are greater than zero [45].

The Tanimoto coefficient is the most common similarity coefficient for binary descriptors and the Euclidean distance is used for non-binary descriptors (Figure 8.4B). The Tanimoto coefficient is a particularly intuitive similarity measure, as it is "normalized" to account for the number of bits that *might* be in common relative to the number that *are* in common. That is, it is a measure of the number of common substructures shared by two molecules. Euclidian distance is a measure of the geometric distance between two fingerprints, where each fingerprint is thought of as a vector in the total number of bits (the fingerprint's size). Often the Euclidian distance is normalized to a range of 0.0 (identical) to 1.0 (all bits are different).

A compound that may be identified as similar or dissimilar to another molecule is only such in the descriptor space that is used for the selection. However, descriptors cannot adequately predict similar binding patterns to a protein from pairs of compounds that are structurally different. Therefore, there is no "universal descriptor" for all bioassays, despite the mathematical possibility of deriving one in a particular chemical descriptor space [46–50].

8.3.2
Clustering

The technique of clustering has been addressed in many contexts and by researchers in many disciplines. This reflects the broad appeal and usefulness of clustering as one of the steps in exploratory data analysis. Cluster analysis is a technique for grouping data by a common characteristic and finding themes in data. The most common application of clustering methods is to partition a data set into clusters or classes, where similar data are assigned to the same cluster and dissimilar data belong to different clusters. Clustering methods can be used for any numerical data such as pharmacological observations or calculated properties, but also for chemical structures using descriptors.

Clustering methods provide unsupervised classifications of objects determining "natural" groupings according to similarities in patterns, e.g. pharmacological observations, predicted values or calculated descriptors into groups (clusters).

For the stepwise aggregation of objects (e.g. chemical structures), a similarity measure (e.g. Tanimoto distance) of their bit-string features is used. A shorter distance between objects means higher similarity [51–53].

A major disadvantage of clustering methods is that it is not possible to add new compounds to an existing cluster analysis; instead the clustering procedure must be rerun from scratch. Clustering methods can be classified into hierarchical aggregation or nonhierarchical aggregation.

8.3.2.1 Hierarchical Clustering

The characteristic feature of hierarchical clustering (Fig. 8.6) is that the objects are combined according to their similarity distance with respect to each other. Hierarchical clustering is a bottom-up approach: iteratively two nearest elements or clusters are joined per step to form a single, larger cluster. The aggregation of clusters can be carried out by different weighting of linkage variants; most common are single linkage and complete linkage (Figure 8.5). Initially, each molecule of the set is treated as an individual cluster and the procedure is repeated until all molecules are in one cluster. Hierarchical clustering yields a dendrogram representing a nested grouping of patterns and the similarity levels between the groups. Assigning compounds to clusters can be achieved with different amounts of diversity from coarse, where all compounds are joined in one cluster at the root, to fine, where all compounds are singletons at the leaves. Although statistical tests that measure the probability of the existence of any particular number of clusters have been developed, frequently no clear-cut optimum can be determined.

8.3.2.2 Nonhierarchical Clustering

Nonhierarchical clustering methods (Fig. 8.7) do not generate a dendrogram structure. The clusters are nonhierarchically ordered and may be partitioned

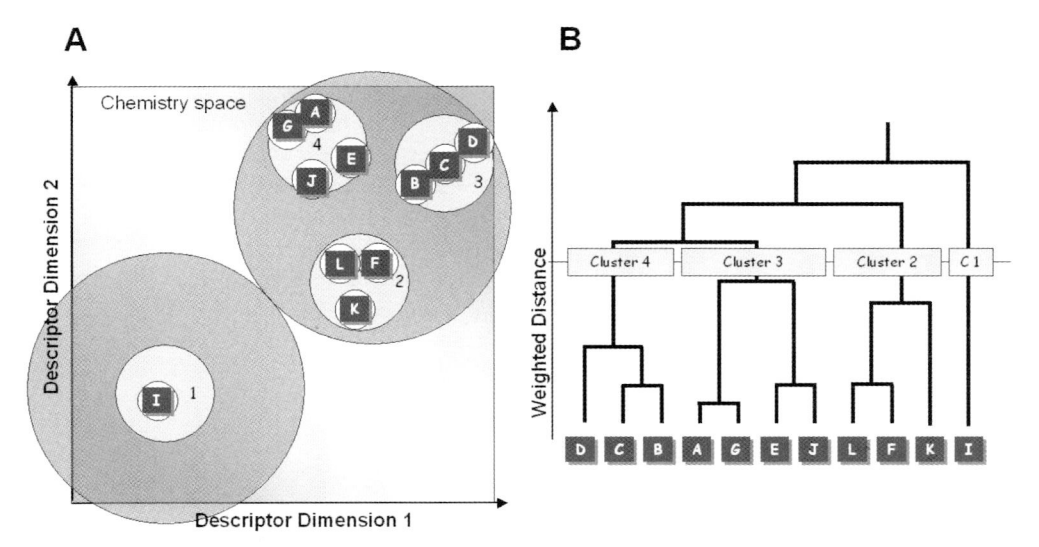

Figure 8.6 Hierarchical clustering. (A) The two most similar compounds form the first cluster with minimal distance in descriptor space (A–G), compounds are grouped by increasing dissimilarity: either a new cluster is formed (E–J) or a single compound is added to an existing cluster (CB–D) or two clusters are merged (AG–EJ). (B) The course of the analysis can be displayed in a distance-weighted dendrogram. The individual compounds are displayed as leaves, branches show the merging into clusters and the root of the dendrogram joins all compounds into one cluster. Assignment of compounds to clusters can be achieved at different horizontal levels.

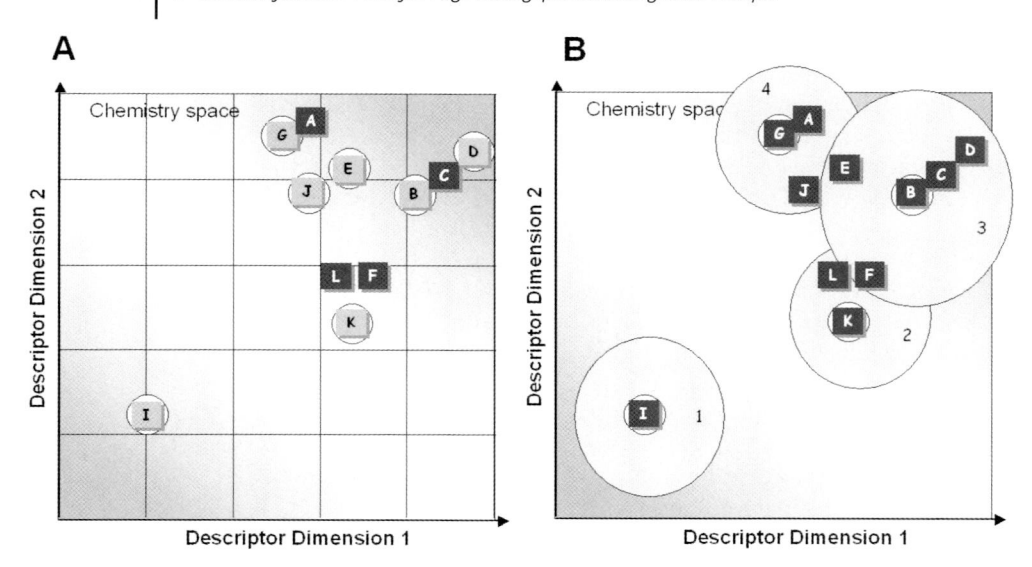

Figure 8.7 (A) Partitioning methods establish a reference frame that subdivides the descriptor space. Grid-based selection extracts only one molecule of each cell (G, E, D, J, B, K, I). (B) Non-hierarchical clustering methods may assign structures to clusters independently: clusters may overlap (E in clusters 4 and 3, F in clusters 3 and 2).

independently of each other, with the result that they may overlap. A non-trivial parameter setting is usually required that reflects some knowledge of the data set, e.g. number of desired/meaningful clusters. Compared with hierarchical methods, automatic determination of cluster boundaries is a major advantage of these methods. The most common method used is based on nearest neighbors. The method initially takes the number of components of the population equal to the final required number of clusters. The final desired number of clusters is chosen such that the points are mutually furthest apart. Next, each component in the remaining population is examined and is assigned to one of the clusters depending on the minimum distance. The centroid's position is recalculated every time a component is added to the cluster and this continues until all the components are grouped into the required number of clusters.

8.3.2.3 Partitioning Methods

Partitioning methods represent a natural procedure of portioning the chemical space. Each dimension of a chemical space is a descriptor and the individual compounds are points in this space whose coordinates are the values of each of the different descriptors. Each dimension is divided into bins according to the range of the property. This binning further defines a grid of bins or cells within the chemistry space. Several cell-based diversity measures have been proposed. The smallest, most diverse subset is then selected by choosing one molecule of each cell. Partition-based methods are especially useful for comparing different compound populations and useful for identifying diversity voids, i.e. cells not occupied. New compounds can be easily added in the analysis of the set. However, in some cases the arbitrariness of cell boundaries can limit the usefulness of these methods [54, 55].

8.3.2.4 Principal Components Analysis

Principal component analysis (PCA) is widely used in all forms of data analysis because it is a simple, nonparametric method of extracting relevant information from complex large data sets that are often confusing, clouded and even redundant.

In the HTS workflow, it is used for data reduction for a large number of measured or calculated pharmacological activities, chemical properties or structural fingerprints [56, 57]. The number of variables under investigation is called *dimensionality* of the data. Reduction of variables' dimensionality is very important for graphical inspection of the data. PCA transforms a number of (possibly) interdependent or correlated variables into a smaller number of significant, independent and uncorrelated, orthogonal variables or the *principal components*, by a linear transformation. The first principal component (PC_1) accounts for as much of the variability in the data set as possible and each succeeding component accounts for as much of the remaining variability as possible.

To explain the results of PCA, an example data set of 1600 compounds with antibiotic activity on eight different bacteria strains was analyzed (Fig. 8.8.). The

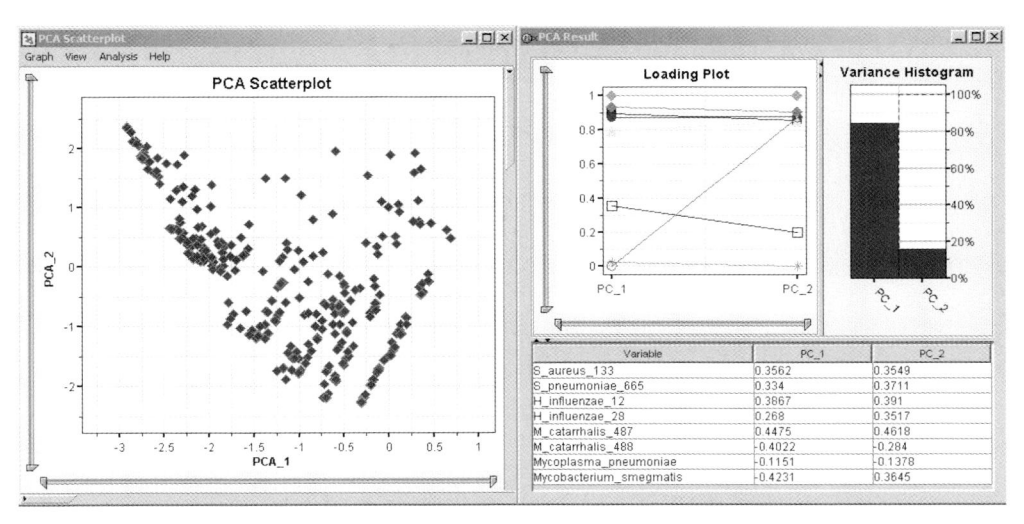

Figure 8.8 Principal component analysis is applied as a method for dimension reduction. An eight-dimensional activity space of different bacteria strains is transformed into an orthogonal two-dimensional space of the new principal components (PC_1 vs. PC_2). The Loading Plot shows the influence of all observed activities for each principal component. The Variance Histogram explains the variability and importance of each principal component.

"variance histogram" of this "2D" PCA containing the relative importance of each principle component revealed that PC_1 accounts for 85%. A "loading plot" shows which activities of the bacteria strains were most influential for each principal component; their weights are used to calculate the linear transformation of the observed activity to PC_1: *Mycoplasma pneumoniae* (lacks a cell wall which leads to osmotic instability) shows the least variance in the observed activity and has the smallest influence; *Moraxella catarrhalis* (non-motile, Gram-negative aerobic bacterium) shows the largest influence.

Interpretation of the grouping of the compounds is usually carried out by visual inspection groupings in the "PC scatterplot". Compounds with a similar activity profile for the different bacteria (inactive, active or highly active) show up as neighbors.

8.3.3
Maximum Common Substructure – Distill

Within a structure cluster, automatic identification of a common core structure by chemoinformatic tools may be of interest. The search can be performed by a maximum common substructure (MCS) algorithm. The MCS is the largest common part between two or more structures regarding non-hydrogen atoms. By reducing the structures into atoms and bonds, the algorithm analyzes the dataset pair wise. MCS algorithms make use of graph-based encoding of chemical structures, starting with an initial match between atoms and developing to fragments increasing in size. Owing to the NP-complete nature (see Section 8.3.1) of these searches, the effort to find the maximum common substructure increases for larger molecules [58, 59]. One application of MCS is to define the core for a subsequent automated R-group deconvolution.

Distill is an implementation of the MCS method applicable to structural heterogeneous data sets [60]. Distill is a structure-based hierarchical clustering routine that classifies compounds according to their common substructures (Fig. 8.9). The algorithm yields a hierarchy in which each node contains a group of compounds that share a common core, a maximum common substructure (MCS). In contrast to other clustering methods where only one tree is created, Distill may create a "forest" of hierarchical trees. The root or top node of each tree is represented by the largest molecular fragment that all of the nodes below have in common. Each tree has several "leaf" nodes, which are the MCS of one or more compounds. Going up the tree from the leaves to the root, the MCS fragments at each node become smaller. The use of fuzzy atoms and bonds based on Markush representation allows the identification of larger skeletal backbones. By coloring the dendrogram nodes by property values (e.g. average activity), Distill can visually relate substructures to activity. Summary statistics of properties are provided for the compounds at each node. The visualization of activities (e.g. by coloring) combined with the clustering by common substructures shows how incremental change in structure adds or detracts from activity. It aids structure–

Figure 8.9 Distill provides a graphical user interface to navigate within a dendrogram of Maximum Common Substructures (MCS). The MCS can be used for R-group deconvolution analysis.

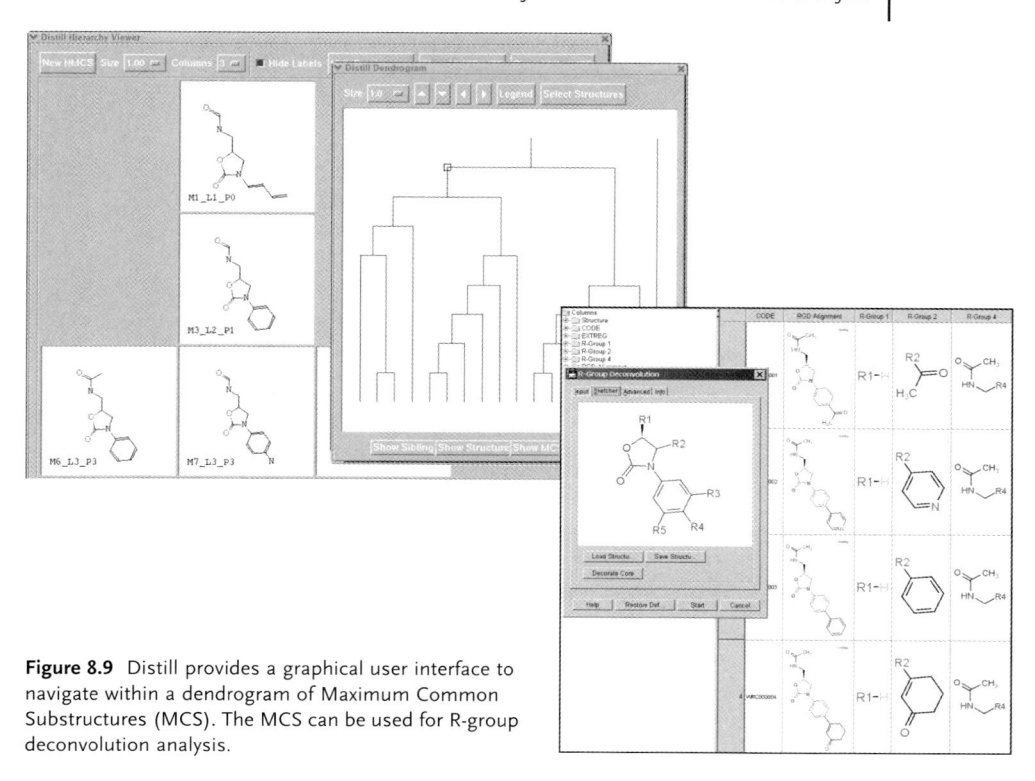

activity relationship analysis by providing a pathway to enhance activity by addition of structural elements and helps formulate a SAR hypothesis.

8.3.4
Mode of Action – Profiling

Structure–activity relationships (SAR) are obtained by relating the biological activities of molecules to numerical descriptions of whole or part of the structure. Similarities based on conventional fingerprints have been used to explore biological space in applications such as virtual screening [61] and substructural analysis [62]. In contrast, molecules can also be characterized by its activity profile measured in pharmacological biochemical high-throughput assays or receptor-targeted cell-based *in vitro* assays. Here, each molecule is characterized by a pharmacological fingerprint, which is a vector containing activity values in a standardized response. It is possible to discover new and hidden pharmacological effects of structural scaffolds by mining the properties that the assays have in common.

Heatmaps of chemical clusters and assays, clustered by target or readout principle, are a powerful tool for mode-of-action analysis. These analyses are widely used in genomics expression profiling [177, 178, 200] and are also applied in the novel fields of chemical genomics [179, 180] and high-content screening [201]. Columns and rows are ordered independently by clustering or by simply sorting by known properties. The neighboring cells of the matrix showing similar properties correspond to SPR–SAR combinations and reveal possible areas with a new mode of action. Further statistical validation such as statistical test models are usually applied beyond this quick exploratory overview analysis.

In the example in Fig. 8.10, compounds containing the DNA topoisomerase inhibitor core show weak to strong activity in all promoter assays and in most of the GPCR assays using aequorin and in some GPCR assays using firefly luciferase. Compounds containing the phosphodiesterase core show graduated activity on PDE assays only.

The combined analysis of chemical clusters within a whole group of assays allows a more confident assessment of activity profiles and can help to prioritize HTS hit classes.

8.3.5
Artificial Neural Networks (ANNs)

Problems involving routine calculations are solved much faster and more reliably by computers than by humans. Nevertheless, there are tasks in which humans do perform better, such as those in which the procedure is not strictly determined or the procedure cannot be split into individual algorithmic steps. One of these tasks is the recognition of patterns. ANNs constitute a method which models the

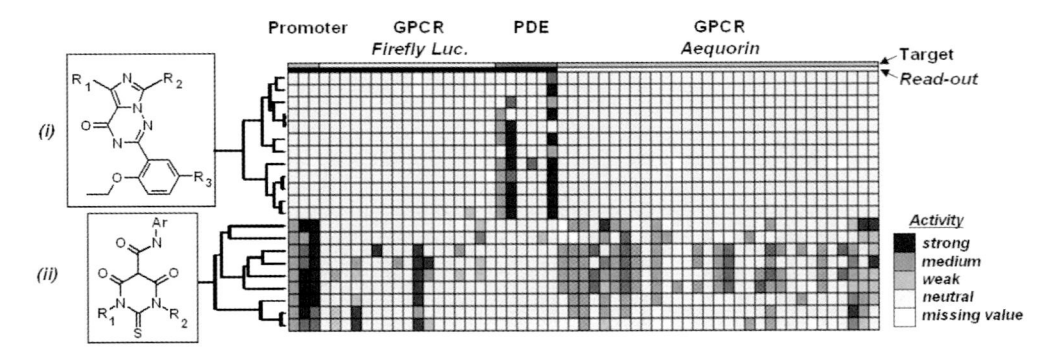

Figure 8.10 Heatmap of combined structural and pharmacological data for analysis of desired activity profiles: The two pharmacophores phosphodiesterase inhibitors (i) and DNA topoisomerase inhibitors (ii) show distinct pharmacological activity profiles. Structural fingerprint clustering was used for a similarity grouping of the compounds (left dendrogram, map-rows). Biochemical and cell based assay are grouped by target class and readout principle (map-columns): promoter, whole cell bacteria, targeting an essential metabolic promoter; GPCR, mammalian whole cell expressing transgenic G-protein coupled receptors coupled to bioluminescent readout; PDE, biochemical assays targeting phosphodiesterase coupled to bioluminescent readout. Besides firefly luciferase, aequorin was used as an alternative readout for GPCRs. Normalized activity values are depicted in grayscales.

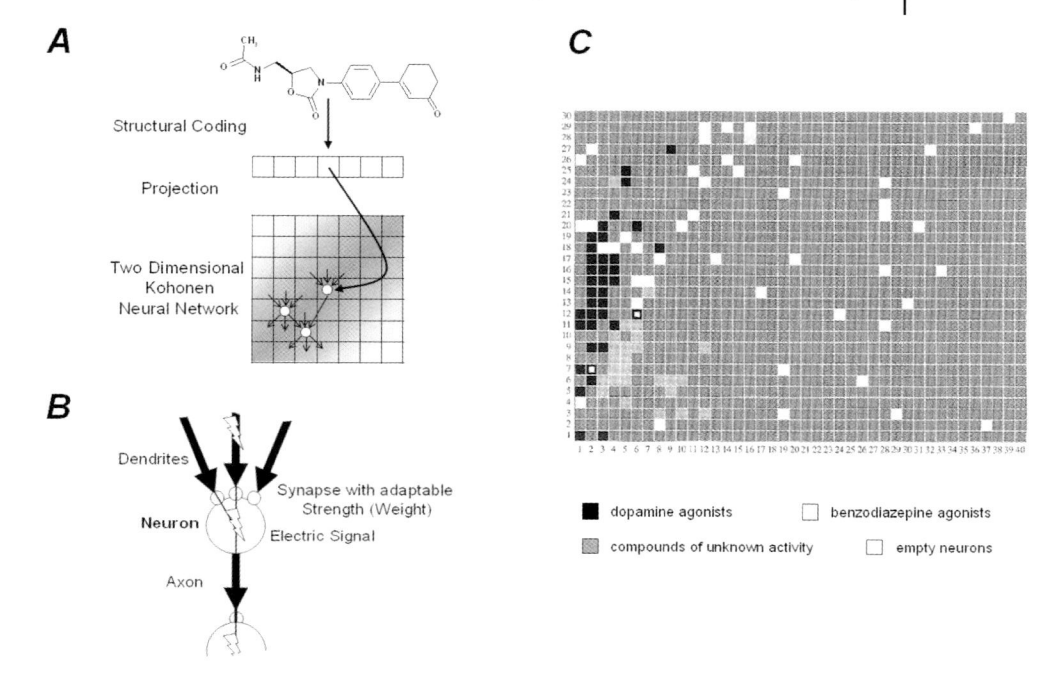

Figure 8.11 Kohonen networks/self-organizing maps (SOM). (A) Architecture; (B) schema of neurons; (C) application example from Bauknecht et al. [65]. Classification of biologically active compounds by a map of 40 × 30 neurons: 112 dopamine agonists (DPA), 60 benzodiazepine agonists (BPA) in the bulk of 8323 structures of unknown activity. Only the activity type of the compound mapped into the individual neurons is used for coloring. DPA and BPA only occupy limited areas of the map; compounds assigned to neighboring neurons are similar.

functionality of a vertebrate's brain. Just as the brain consists of neurons which are connected and interchange signals via axons and branching dendrites with one another, an ANN comprises interrelated artificial neurons. The transmission of information from the axon to the dendrites of other neighboring neurons through a synapse is gated by its variable strength. The building of synapse strength is the essence of learning. Just like humans, ANNs learn from examples (training): the synaptic connections between the neurons are adapted during the learning process. Two types of learning strategies are used in ANNs: *unsupervised* and *supervised learning*.

Neuronal networks based on *unsupervised learning* algorithms do not need training; the network tries to group data points by similarity by assigning them to the same neuron within the network. Examples are the Kohonen networks, so called self-organizing maps (SOMs) (Fig. 8.11). In *supervised learning,* the minimization of the deviation from a known correct result for a well-known training set is used to adapt weights of the network. Examples are back-propagation networks [63, 64].

8.3.6
Decision Trees/Recursive Partitioning

Decision tree learning is a very common method used in data mining [66–68]. Recursive portioning was originally referred to as the decision tree method [69]. A decision tree depicts rules that divide data into groups. The first rule splits the entire data set into subsets. Another rule may be applied to each subset, different rules to different pieces, forming a second-generation split. The goal of this method is to divide the dataset up using a series of simple rules (e.g. compounds with or without a certain chemical group, derived from predefined molecular substructural descriptors). Using a series of rules, each generating an additional branch in the decision tree, it is possible to classify compounds into groups with similar structural and biological features. Recursive partitioning has been used in many studies combining structural descriptors and measured pharmacological data [70–75].

8.3.7
Reduced Graph-based Methods

As has been shown in the previous section, clustering and similarity methods can be very powerful to support the hit to lead process. However, there are also

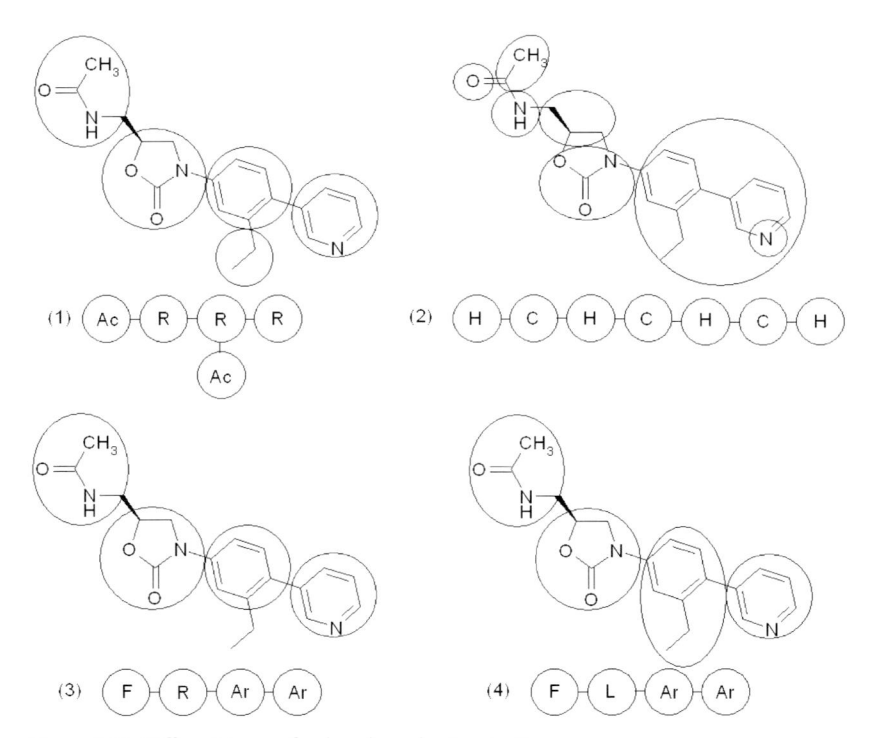

Figure 8.12 Different types of reduced graphs (see text).

limitations. For example, opportunities for "scaffold hopping" will most likely not be identified using normal clustering methods. Scaffold hopping describes the process of replacing a central fragment within an active compound by another fragment, thereby changing the scaffold of the structure, without losing activity. Various algorithms have been developed to address this feature. These include analyses based on reduced graphs and fragmentation methods in combination with statistical methods. Examples will be described in the following section.

Two-dimensional fingerprints were originally developed for substructure searching; they are now also commonly used for similarity searching. However, they are inherently limited to the structure from which they were derived, making them particularly effective at identifying close analogs and less effective at finding compounds that share similar activity but are based on different structural series. Fingerprints usually distinguish atoms on the basis of the element and therefore different ring systems with different heteroatoms cannot be correlated with equivalencies in pharmacological activity or other properties.

Applying reduced graphs can facilitate scaffold hopping [76, 78], by summarizing the features of molecules while maintaining the topology between these features. There are different approaches to reduced graphs. One version was described by Gillet's group [77]. They introduced a graph where each node represents a group of connected atoms and an edge exists between two nodes if there is a bond in the original structure between an atom contained within one node and an atom in the other node. Different types of reduced graphs following this approach are possible (Fig. 8.12): (1) nodes correspond to ring systems and connected acyclic components; (2) nodes correspond to connected carbon components and connected heteroatom components; (3) nodes correspond to aromatic rings, aliphatic rings and functional groups; (4) nodes correspond to aromatic rings, functional groups and linking groups [77].

8.3.7.1 Reduced Graph Theory Analysis: Baytree
Another option to reduce structures is based on the omission of atom information [78, 79]. Bayer HealthCare adapted this methodology and introduced a tool,

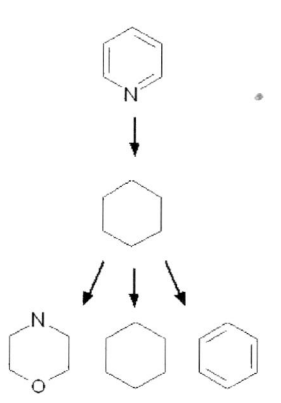

Figure 8.13 Scaffold-based grouping; examples for six-membered ring systems.

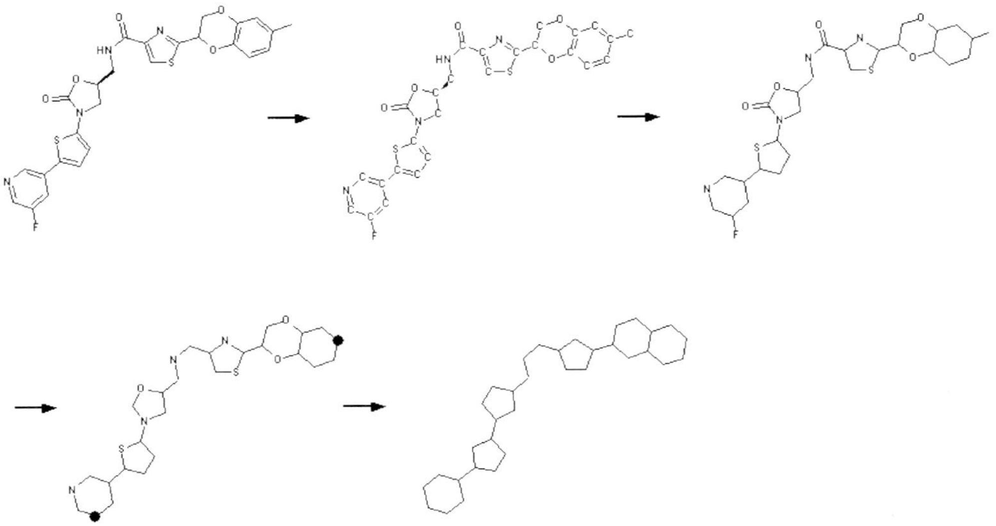

Figure 8.14 Framework generation process.

"Baytree" [181], that is based on the hierarchical ordering of compounds by their scaffold topology [80]. The compounds are classified according to the following schema: (1) the input structure is split into the main framework and its substituents; (2) all bond types of the main framework are set to single and all heteroatoms are converted to carbon; (3) the framework is split into rings and linkers; for example, the analysis would group together a phenyl group and a morpholinyl group as a six-membered ring (or a thiazole and a thiophene as a five-membered ring) (Fig. 8.13); (4) a "molcode" is generated based on the number and size of rings and linkers present in the compounds. The compounds are then sorted according to these molcodes. The framework generation process is shown in Fig. 8.14.

In order to make the result visible to the user, a set of hierarchical trees is generated and the compounds are sorted into the leaves of their respective trees. A node represents a group of compounds with a common ring and linker makeup.

An example of a section of a dendrogram representing the Baytree result is given in Fig. 8.15.

Three example structures are shown. Using the Baytree approach, compounds B and C are located next to each other, in contrast to most other similarity search methods. The resulting tree can further be refined by using a color by value function that highlights regions of more active compounds or compounds with common properties such as physicochemical parameters. In this case, all three compounds were active in a similar range against a specific target. It can be seen that this method enables the research scientist to quickly find possible sites for scaffold hopping as all three examples (A–C) show different scaffolds.

8.3 Chemoinformatic Methods Used in HTS Workflow

Applying reduced graph analysis, initially developed by Bemis and Murcko, to 5120 compounds from the Comprehensive Medicinal Chemistry Database (CMC) [81] revealed that there were 1179 different scaffolds and 306 acyclic compounds. Within this set of scaffolds 783 were unique, being present only in a single compound each. Looking at the frameworks that exist in at least 20 substances, it was found that this resulting set of only 32 frameworks accounted for 50% of the 5120 total drug molecules. In comparison, the 306 acyclic compounds accounted for only 6%. This finding supports the assumption that drugs might be characterized by their framework and that this can be used for relation to drug likeness or unlikeness. A list of the 32 frameworks is given in the original reference.

Other groups have described similar systems, e.g. Jenkins et al., who included feature point pharmacophores (FEPOPS) [82].

All these examples show that reduced graphs can be a good alternative to clustering and at the same time assist the discovery of new lead classes by applying scaffold hopping. The reduced complexity for visual inspection makes these tools easily usable by all research scientists.

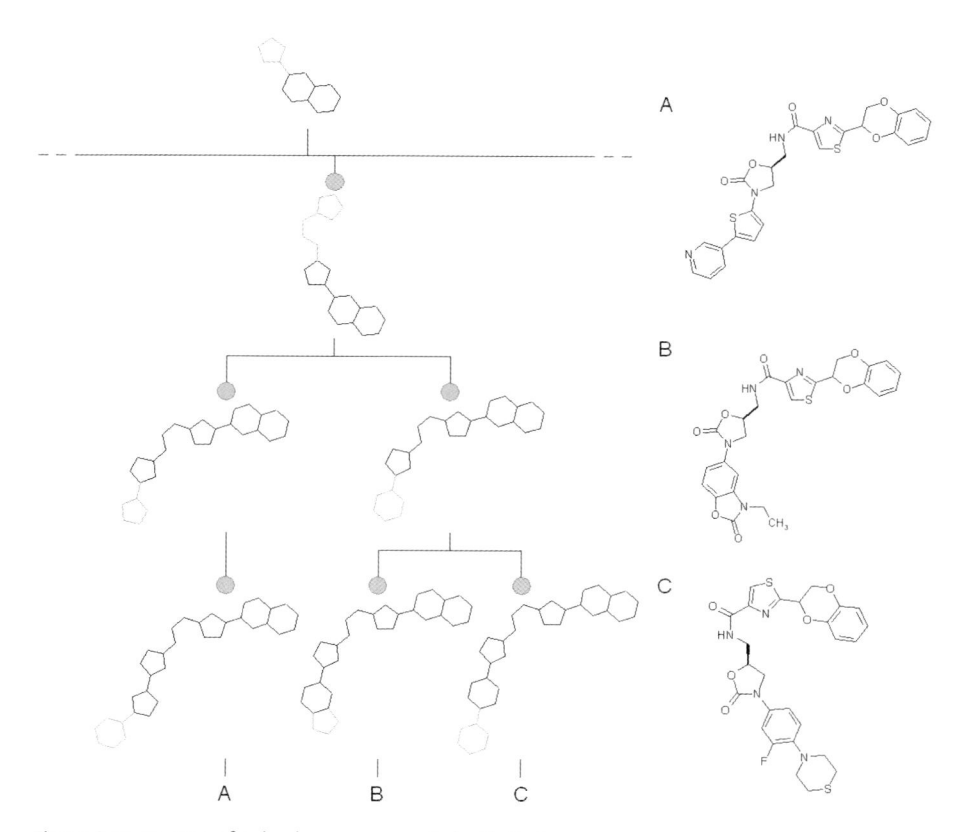

Figure 8.15 Section of a dendrogram generated with Baytree.

8.3.8
Fragment-based Methods – Structural Units Analysis

Reduced graphs are one alternative tool to clustering. Another approach is based on the fragmentation of compounds into small fragments. The contribution of the given compounds to activity is then evaluated regarding the presence of the fragments in active or inactive compounds.

One of the tools following this approach is "structural units analysis" (SUA), designed to assist the research scientist in rapidly mining HTS or lead optimization data to identify structurally related compounds that are enriched in active compounds. A data set is analyzed for preferred connected fragments that appear in active compounds more often than other fragments. First, all structures are divided into small fragments at points connecting rotatable and rigid parts of the molecule. These structural units are then used by the algorithm [83] to explore the data set for combinations of units that disproportionately result in active compounds. Specific combinations of structural units or rules, reflecting a structure–activity relationship, are generated and provide an automated method to determine groups of compounds that are the basis for chemical series that show a preferred activity. The rules provide the structural units that contribute significantly to activity, the order of the connectivity of these fragments and an example compound. An issue in creating SAR models that can be difficult to address with similarity-based clustering methods is that a set of active compounds might not be just a series of homologous derivatives, but might differ by replacement of the main core or a major fragment with, e.g., a bioisosteric group [84–86]. This change is often significant enough that the similarity of those compounds based on fingerprints is very low. By introduction of a wildcard concept, the SUA algorithm is able to take into account such close relationships of actives which are not reflected in the atom-by-atom similarity of molecules. Resulting rules that contain a user-defined maximum number of wildcards may then include compounds that reflect scaffold hopping or bioisosteric replacement within that series without losing activity. For SUA to give statistically validated rules, both compounds found to be active in a particular screen as well as similar compounds which were not should be examined [202].

Figure 8.16 shows an example of an SUA-generated rule using a Bayer screening dataset containing approximately 1600 compounds with antibacterial activity. As

Figure 8.16 SUA result: rule and examples.

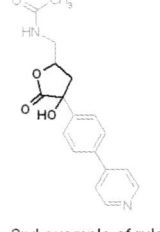

mentioned, a set of data must contain inactive compounds in order to obtain meaningful results. Therefore, the user provides an activity cutoff. In this example, the activity cutoff was set so that about 15 % of all compounds were in the "active" range. One of the 11 rules obtained is shown in Figure 8.16. It can be interpreted that a compound containing the fragments that are shown in the "series" column have a high probability of being active. The connectivity of these units is given in the next column. One example of the series is shown in the first column highlighting possible wildcards in bold. In this case the oxazolidinone is only a scaffold and can be replaced by different fragments without losing activity. A second active example of this series is shown next to the rule to illustrate this point: compounds having a hydroxy–lactone are active keeping all other fragments constant (activity data for both compounds were in the same range).

The rules from structural units analysis can be used to search a database for similar compounds, including bioisosteric fragments, to replace the wildcards and also for the optimization of lead compounds by medicinal chemists.

8.4
Chemoinformatic Methods in the Design of a Screening Library

As HTS technologies became available and have dramatically shortened the time it takes to screen, there was much debate as to the value of library design to screen every available compound in the library so that there is no chance of missing any potential lead compound [87]. However, this may not always be feasible as the number of compounds that could be synthesized is astronomically large, with estimates ranging from 10^{18} to 10^{300} [88–90, 190]. Even using conservative estimates, the number of possible pharmaceutically relevant small molecules has been proposed to be about 10^8 compounds. For practical reasons, a number of different pharmaceutical companies have chosen to make selections of compounds for screening from their internal or vendor compound collections and compounds that could be synthesized around combinatorial scaffolds, in order to maximize the efficiency of their screening efforts. The selections of compounds are made using a variety of methods that evaluate important aspects of drug and lead likeness, ADME parameters and diversity [91]. It is important to stress that assembling a screening compound collection is aimed at finding hits that can be developed into leads rather than identifying compounds with perfect drug-like properties. Therefore, the major goal is not only to increase the number of hits being produced in each HTS, but rather to ensure that as many as possible of the hits produced are compounds suitable for optimization [182].

8.4.1
Drug and Lead Likeness

A large, chemically diverse library such as a compound repository of a pharmaceutical company will contain compounds that do not have drug-like properties.

Over the past several years, the topic of drug likeness has been discussed intensively in the literature and many technologies have been developed to recognize or design drug-like compounds [92–101]. One of the most prominent filter rules is Lipinski's rule of five [102–104]. Aiming at the prediction of oral bioavailability, it takes into account the lipophilicity (calculated octanol–water partition coefficient of <5), number of H-bond donors (<5) and H-bond acceptors (<10) and the molecular weight of a molecule (<500 Da). These parameters can be easily calculated by various *in silico* methods that are now part of many commercially available software packages. The risk of cutting out active compounds that can still be optimized in some of the above parameters might be very high when applying these rules. For example, those compounds with a molecular weight slightly higher than 500 Da might still be important for building up a structure–activity relationship or identifying new scaffold classes even though these compounds will probably never be drugs themselves. Application of this type of filter for prioritization seems to be reasonable but should be made with this understanding.

Another concern is that hits may not be suitable for optimization owing to the complexity of the structures, high molecular weight or ADMET properties (absorption, distribution, metabolism, excretion and toxicity) in the "just acceptable" border range. A paradigm shift was proposed to identify more lead-like rather than drug-like compounds. This gives the medicinal chemist some "freedom" for the lead to drug optimization that is normally characterized by an increase in molecular weight, complexity of chemical structure (number of rings, heteroatoms and rotatable bonds) and improvement of ADMET parameters [105] Some of these criteria might not be restricting for natural products (e.g. molecular weight) [106, 107].

Also, the prefiltering of hits based on lead-like libraries should be used judiciously. Compounds with reduced structural features might result in lower activity and therefore be hard to detect with the assay conditions used in HTS; false negatives might be the result.

On the other hand, drug/lead-like features can be used successfully for the design of screening libraries.

An interesting simplified approach to the identification of drug-like compounds introduced by Muegge et al. [108] uses a simple pharmacophore point filter. Based on the observation that non-drugs are often underfunctionalized, a minimum count on well-defined pharmacophoric features is required to pass the drug-like filter. The reliability for several test sets was lower than with other more sophisticated methods such as neural network approaches [109] or decision tree methods, but there are advantages such as the selection of the pharmacophore points using "chemical wisdom" and the simplicity of the algorithm.

As has been mentioned, non-lead likeness is often defined by inadequate physicochemical parameters (e.g. solubility), pharmacokinetic parameters (mainly ADMET) or structural properties. Prediction tools for ADMET parameters and solubility [110] are on the market or have been developed in academia. The next section will give an overview of the prediction of these parameters.

8.4.2
ADME Parameters

Although it has increased drastically in the recent years, the throughput of *in vitro* ADMET (absorption, distribution, metabolism, excretion and toxicology) assays still remains low compared with the throughput capacity of HTS and combinatorial chemistry. Therefore, these *in vitro* assays can only be applied to a small selection of compounds in an HTS data set or combinatorial library, fueling the need for computational or *in silico* models to compensate for throughput limitations. This is reflected by the numerous reviews that can be found in the literature [111–118, 191].

According to Segall and Beresford [119], the currently available computational ADMET models can be classified into two main categories – empirical and mechanistic. Empirical models are characterized by the use of linear or nonlinear statistical tools to explore relationships between structural descriptors and measured ADMET parameters. Mechanistic models, in contrast, apply quantum mechanical models to calculate atomic interactions between substrates and enzymes or transporter proteins, which are involved in ADMET processes. As 3D structures of ligands and of target proteins or enzymes are needed and involved, the required computational power for mechanistic models is much higher than for the empirical models.

Although many different *in silico* ADMET models are available and are refined in an ongoing process, predictability is still a limitation [111] owing to the many different underlying mechanisms that result in the complicated nature of *in vivo* ADMET processes.

Once again, it should be mentioned that even if a cluster or single hit from HTS might not fulfil all ADMET parameters, a medicinal chemist or molecular pharmacologist might learn from these structures for the optimization for other hit series. Prediction of ADME parameters makes more sense for library design.

8.4.2.1 Absorption

A promising lead compound has to be sufficiently soluble [120–122, 151] in the gastrointestinal fluids of the stomach and gut. Membrane affinity and the molecular volume of the compound are further critical parameters that influence absorption. Some degree of lipophilicity is required for a molecule to permeate through the lipid membrane of the gut wall. On the other hand, compounds that are too lipophilic might remain in the membrane or stick to other biological materials in the gut. Caco-2 cell monolayers are most often used for the prediction of human intestinal permeability. A recent review on the effectiveness of these artificial membranes as well as computational-based predictions has been given by Ungell [123]. Lipophilicity [124] can be measured as the logarithm of the partition coefficient of a compound between n-octanol and water ($\log P$). Several *in silico* tools for the prediction of $\log P$ are available [125, 126].

The polar surface area can also be taken into account for predicting the degree of absorption [127, 128].

A recent overview on the prediction of absorption was given by Hämäläinen and Frostell-Karlsson [129]. They list commercially available databases and software systems modeling absorption. They summarize that in *silico* methods can provide a high-throughput means of screening compounds but their accuracy and predictive value depend critically on the reliability of database information. Dose dependency and inter-individual variability in drug response, together with the differences in experimental protocols applied in various laboratories, can lead to widely varying results, which can compromise the reliability of the conclusions reached by using *in silico* methods.

A bioavailability score has been developed by Martin [130] covering the dependency of bioavailability on different physical properties of compounds contingent on the predominant charge at pH 6–7. For compounds that are negatively charged, the polar surface area seems to be predictive whereas for compounds that are uncharged or positively charged, the rule of five seems to be most predictive [130].

8.4.2.2 Distribution

The distribution in the body of a compound from blood and the penetration of the respective target tissue are characterized by the plasma protein binding. Human serum albumin (HSA), a serum protein, is viewed as a model component of the blood. The binding strength of compounds to this protein is an important factor in the distribution from the circulatory system to target tissues in the body. The parameters that best describe the potential of binding to HSA involve hydrophobicity together with some modulating shape factors [131].

Central nervous system (CNS)-active drugs need to satisfy another important criterion, namely that they need to cross the blood–brain barrier (BBB). Prediction of brain penetration using various descriptors has been discussed in the literature [132–134].

8.4.2.3 Metabolism

The main factors controlling metabolism are active processes such as the binding to and degradation of drug molecules by liver enzymes [135, 136]. Cytochromes P450 oxidize a wide range of endogenous compounds and xenobiotics, transforming them into active, inactive or even toxic metabolites. The microsomal, membrane-associated P450 isoforms CYP3A4, CYP2D6, CYP2C9, CYP2C19, CYP2E1 and CYP1A2 are responsible for the oxidative metabolism of more than 90% of the marketed drugs [137].

Apart from studying metabolism by *in vitro* or *in vivo* approaches, *in silico* methods have been developed [138, 139]. Several methods based on QSAR and pharmacophore models [140, 141] and methods using support vector machines are available [142]. Protein models have been introduced that allow modeling of the active site of the CYP enzyme [143–145]. Recently the X-ray structures of human CYP2C9 [146] and CYP3A4 [147] have been solved, both unliganded and bound to either an inhibitor or a substrate. This might fuel the use of pharmacophore docking methods to predict metabolism induced by the CYP family.

Expert systems have also been reported such as the metabolic fingerprint concept, METAPRINT, introduced by Keserüu and Molnár [148], for the assessment of metabolic similarity and diversity of chemical libraries.

CYP induction is undesirable since there might be some impact on the efficacy of coadministered drugs. In contrast to CYP inhibition, there are only a limited number of *in silico* methods known to predict CYP induction. These are based on PXR (pregnane X receptor) binding [149–151]. More sophisticated, but also most demanding, is the approach of modeling the entire process of cytochrome-mediated hydroxylation quantum chemically [139, 152, 153]. Owing to the enormous computational expense needed for this method, it is currently not applicable to HTS.

8.4.2.4 Excretion

Excretion is the physical elimination of a compound through hepatic and renal clearance. Owing to the multi-faceted nature of drug excretion and elimination, it is difficult to make good predictions and only a few methods have been reported [154].

8.4.2.5 Toxicity

For a drug to reach the market, it must be safe regarding, for example, carcinogenicity and developmental toxicity. Prediction techniques are desirable in order to avoid costly problems at a late stage of the discovery process. SAR and QSAR techniques have been applied to a wide variety of toxicological end-points from the prediction of LD_{50}s and maximum tolerated dose to AMES assay results, carcinogenic potential and developmental toxicity effects [155]. A review of prediction systems available has been given by Greene [156].

Another critical issue is the potential for QT interval prolongation that has been linked to incidences of sudden cardiac death [157]. Compounds with this cardiotoxic potential should be identified in an early stage. I_{Kr} potassium channels encoded by the human ether-à-go-go related gene (hERG) are responsible for the repolarization in the heart. Therefore, the potential for inhibition of hERG K^+ channels is used as an *in vitro* predictor for QT prolongation. The only reliable and precise method for the determination of K^+ channel inhibition is patch-clamp electrophysiology, which is time consuming and labor intensive, making it impracticable for high throughput. Many groups have therefore investigated the potential of *in silico* prediction of hERG channel blocking. Some of the resulting methods give reasonable results at least for supplying a qualitative filter [158–160].

8.4.3
Diversity

As has been shown in the previous sections, HTS can be an effective approach to lead discovery especially to find new chemical entities. However, to take full advantage it is necessary to start from screening libraries with high diversity. Only such an approach can ensure that one might find a variety of different leads that will improve the success rate of finding a suitable lead that can result in a development candidate, a major milestone on the way to a marketed drug [183, 184]. Several hits within one structural class might help in setting up an initial SAR. If the number of these hits gets too high, this might mean redundancy and an inefficient use of resources. On the other hand, Nilakantan et al. showed in a case study that it may be necessary to test at least 100 compounds from a particular group of related molecules in order to be confident that actives of a given series can be identified [161]. The measurement of diversity depends on the correlation of parameters with a chosen reference. For example, if two compounds are compared regarding their structural similarity, this can be classified as "structural diversity" [192], whereas if one takes into consideration the biological effects of two compounds, diversity can be classified as "functional", bringing back the discussion of whether structural similar compounds in general show similar pharmacological profiles.

There are two major classes of compound libraries, natural product [193] and synthetic libraries [194, 195], in addition to collected libraries (from vendors) and internal legacy (from in-house medicinal chemistry projects). Although the generation of synthetic libraries is not a new topic, it has gained much importance during the past 15 years owing to the recent developments in combinatorial chemistry (see Chapter 9). In principle, combinatorial chemistry starts from a defined scaffold as a building block for combinatorial enumeration by variation of the different R-groups, depending on the synthetic feasibility (e.g. standard reactions that can be performed under the given conditions). Usually, this results in compound collections with moderate structural diversity. In two reviews [106, 196], the current status of natural product chemistry as a source for screening libraries is discussed. Advantages include the complexity of structures, high diversity and in some cases a "target focus" is discussed. However, many disadvantages can be listed: (i) costs of extraction or for supply by vendors; (ii) a possibly time-consuming elucidation of molecular structure; (iii) when identifying hits the structure and its novelty might not be known; (iv) owing to the sometimes high complexity, chemical optimization and scale-up during the process to a development candidate might be infeasible; and (v) intellectual property (IP) issues might occur, since more countries are now actively protecting their biodiversity rights. Some researchers combined the advantages known from natural products with synthetic combinatorial chemistry as guidelines for the design of synthetic libraries.

In principle, there are two problems regarding diversity that can be addressed by chemoinformatic methods [197]. First, from the given virtual chemistry space which would result in too many compounds to be synthesized, one needs to select compounds that represent the highest degree of diversity, e.g. methods for measurement of diversity are needed. Second, for a given defined compound library, diversity voids or missing diversity need to be identified [55, 185–187]. These diversity calculations can be carried out at the level of either reagents or reaction products. Reagent-based diversity selection is extremely fast but not as accurate as product-based enumeration of compound libraries, because diversity in reagents does not necessarily translate into diversity of products.

One of the simplest approaches for subset selection of a library is to select compounds randomly. Such algorithms are intuitive and inexpensive in computation time, as these are linearly dependent on the number of compounds to be selected for the desired library. In order to maximize diversity of the selected compound set, each newly selected compound is compared with the already previously selected ones. This is done using the descriptor distance, resulting in a similarity cut-off value, e.g. 2D fingerprints and Tanimoto distance. By this procedure, compounds that are to closely related or redundant are filtered out. Such distance-based diversity algorithms appear to give a satisfactory result for simple diverse subset selection. Matter and Pötter showed that 2D fingerprint descriptors allow the handling of global diversity and these can be applied to design a general screening library with optimally diverse compounds [183, 184]. Clark et al. developed "OptiSim"/"OptDesign", a stochastic dissimilar selection algorithm [187]. This allows adjustment of the diversity criteria between maximum spread and minimum dissimilarity-based selection and includes the possibility to bias the selection of candidate reagents of a library for inclusion in subsets.

A disadvantage of this plain distance-based diversity approach is that only pairwise inter-compound distances are considered. The absolute location of compounds in chemistry space cannot be calculated by this procedure and therefore it is unlikely to identify voids in this way. In contrast, using cell-based diversity algorithms chemistry space is partitioned by dividing each axis of the multidimensional structural descriptor space into bins. For a given library the absolute position of compounds in the chemistry space is easily available, in addition to diversity voids by the detection of empty cells. However, whereas distance-based algorithms can be applied in either a high- or low-dimensional descriptor representation of chemical space, cell-based algorithms can only be applied in low-dimensional chemistry-spaces of typically 5–10 dimensions. This might limit the number of properties that are considered for structural diversity.

Pearlman and Smith developed the "Diverse Solutions" software, a cell-based partitioning method with nonlinear binning combining atomic and molecular property-based descriptors to a meaningful low-dimensional descriptor space [55, 185]. An "auto-chose" algorithm is used to tailor a chemistry space of typically four to six computed parameters such as polarizability, charge, 2D connectivity or exposed surface area for a given compound set. Mason and coworkers developed "ChemDiverse" using four-point multiple potential 3D pharmacophores as descriptors, allowing the calculation of complementary descriptor profiles for enzymes or receptors of interest [186, 198]. All combinations of atomic descriptors, e.g. H-bond donors/acceptors, acidic/lipophilic centers and their distances including 3D conformational sampling, are combined into a feature key. This key can be used to assess the diversity of individual libraries, to compare the relative diversity of libraries and to design libraries to fill voids in the pharmacophoric space. If a receptor model exists, the pharmacophore profile can be calculated not just for a ligand, but also for enzyme and receptor sites using complementary site points for the "Design-in-Receptor" technique.

A major problem involving the use of 3D descriptors is that they are typically calculated from hypothetical molecular information. There is, of course, no guarantee that such predictions indeed resemble bioactive confirmations. Although 3D descriptors of molecules should contain maximum information achievable and thus be superior to 2D representations, this problem associated with predicting uncertainty often eliminates the principal advantages of 3D methods. Therefore, 2D methods still play a major role in chemoinformatics regarding diversity assessment.

A further step in chemoinformatics regarding diversity is to extract information from the discovered voids in the given chemical space, build a description of these voids in the fingerprint space and then use another algorithm to "reconstruct" new compounds from this description that can fill the diversity gap. This combined approach is currently under investigation in some software companies [199] and will definitely enhance the use of chemoinformatic methods in the library design process.

8.5
Integrated Software Packages

The increasing importance of high-throughput screening and combinatorial chemistry has resulted in an explosion of data, therefore requiring more effective methods to visualize and structure the produced data. Also, as described in Section 8.3, numerous computer algorithms for the analysis of data based on chemical structures like ADMET parameters or as used in virtual screening have been developed. To maximize the potential of the data analysis and expert computer tools by the research scientist, it is desirable to have integrated software packages that combine the most important visualization methods, structure preparation algorithms and exploration tools [162]. Querying pharmacological, pharmacokinetic, physicochemical and chemical data in the databases of companies must be possible by the research scientist using an easy to use query engine. Application of the individual tools should be easy and not require a format change of the existing data. Results should be easy to understand and not require expert knowledge in chemoinformatics. Some companies such as Bayer have introduced an in-house system. The example from Bayer will be described in Section 8.5.2. There are also commercially available packages that have been, in most cases, developed in cooperation with pharmaceutical companies. Representative packages are described in the following section, with no claim to be comprehensive. For further information, the reader is referred to the Internet homepages of the individual developers.

8.5.1
Commercially Available Packages

8.5.1.1 Accelrys: DIVA®

DIVA® [163] allows retrieving and working with large data volumes of chemical structures, assay results and other chemical and biological data and offers easy-to-use tools for data integration, visualization, analysis and reporting (Fig. 8.17). DIVA includes intuitive tools for merging and pivoting of rows and columns, combining data from multiple sources into a single, uniform data set in one convenient spreadsheet that is ready for analysis. 2D or 3D graphs can be built with full interactive user control over the graph display, including subset selection, rotations and transformations. Its histograms allow comparison of how compounds perform in different experiments, e.g. to detect trends in selectivity, while correlation analysis gives a visual display of how overall results for different assays are related. Static data analysis by FIRM (Formal Inference-based Recursive Modeling) is a powerful decision tree analysis tool that automatically finds the mathematical rules that explain structure–activity/property relationships within the data and displays the results in an intuitive graphical form. Because FIRM is automatic and uses robust algorithms, no statistical expert knowledge is needed to derive meaningful results.

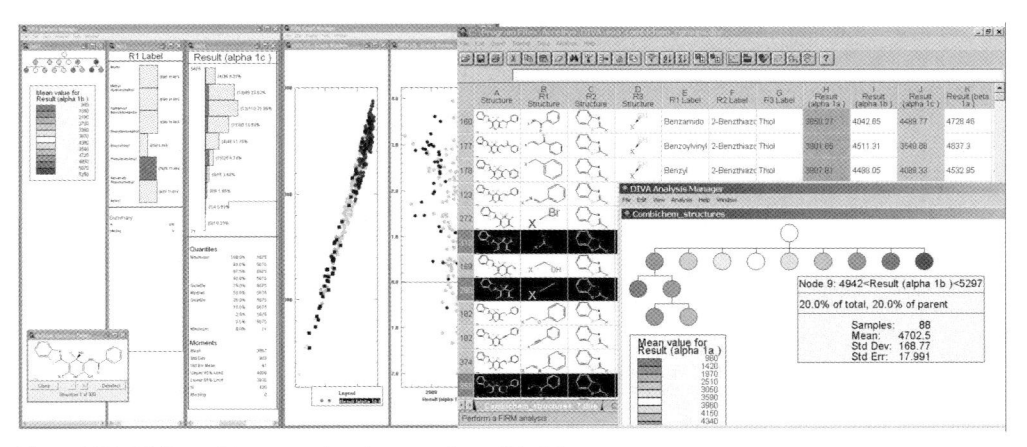

Figure 8.17 DIVA provides a complete, integrated set of decision support tools.

8.5.1.2 BioSolveIT: HTSview

HTSview is a graphical user interface (GUI) [164] for the analysis of up to 500 000 molecules and allows interactive viewing and browsing of structural data (Fig. 8.18). It allows viewing data in different ways, e.g. sorting by identifiers, classes, activities or clusterings. It is especially suited for the drawing of large similarity matrices and comes with several meaningful data plots, e.g. enrichment and

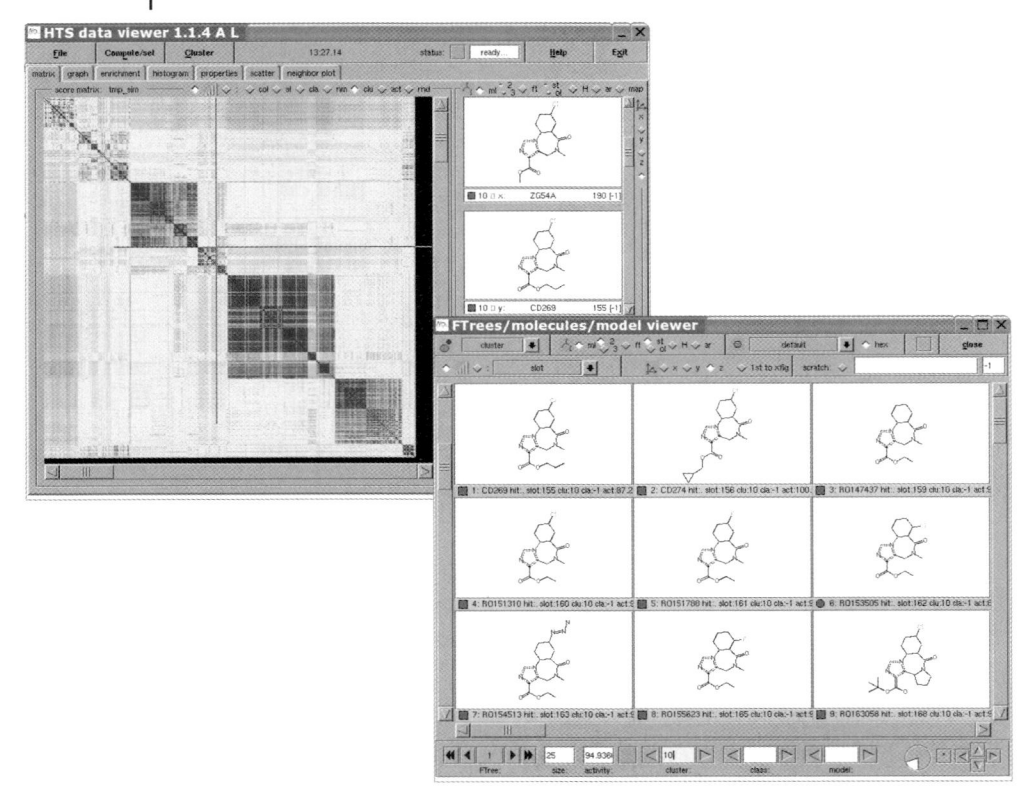

Figure 8.18 HTSview offers facilities to compute similarity matrices, clustering according to similarity and/or activity, allows one to analyze cluster statistics and to browse cluster members.

scatterplots. The GUI is an extension of the FTrees [165]. FTrees performs a descriptor-based fuzzy molecular similarity analysis and allows fast topological alignments of ligands, even for diverse sets of molecules. In addition, a similarity score describing the global and local fit of the structures is provided. The alignments are used to build multiple ligand models or biophore models. HTSview's multiple alignment models can be used to extract preliminary QSAR models; it allows fragment analysis of active and inactive molecules (looking at fragment frequencies). Biophore models can be built which can be used for predicting the activity of new compounds. A visual interpretation of the SAR hypothesis is supported.

8.5.1.3 SciTegic: Pipeline Pilot™

Pipeline Pilot [166] is an easy-to-use graphical interface to set up complex data filters and calculations (Fig. 8.19). *Ad hoc* queries can be asked against flat file and database sources; no centralized database is needed. Data Pipelining is a comple-

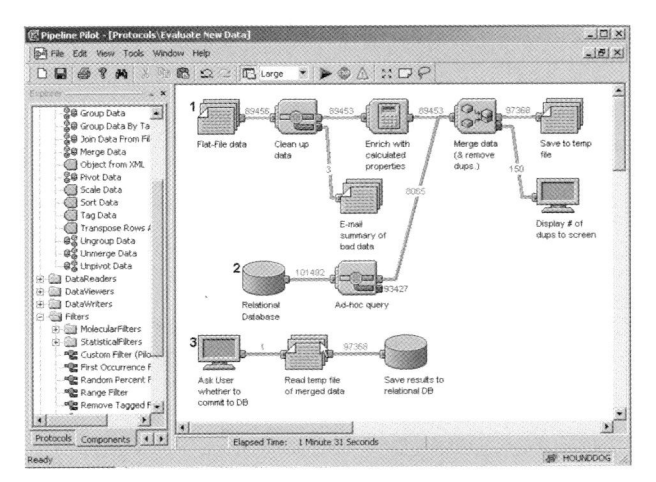

Figure 8.19 Data pipelining is the independent processing of data points through a branching network of computational steps for unprecedented flexibility interacting with data, automating processing and sharing protocols. Data are progressively read into a protocol and analyzed or modified by each component as it flows though the network towards the output end-points.

mentary technology to relational database systems – it is not in itself a data management tool. However, Data Pipelining offers tremendous flexibility advantages for analysis by processing all the data in real time; it is not constrained by precalculated and stored data in a database. Informatics analyses today generally require multiple computational steps and Pipeline Pilot serves as the ideal platform for capturing and executing these procedures. These automated data analysis pipelines can be saved for future use or shared with a broad community of users within the community across the Internet. Newly developed Bayesian statistical characterization methods extend the capabilities by learning from patterns in the data. Predictive categorization models are created and property prediction is available through classical SAR models. Pipeline Pilot provides interfaces to established chemoinformatics tools of other vendors, e.g. MDL, Accelrys and Spotfire.

8.5.1.4 Bioreason: ClassPharmer™ Suite

Bioreason's analysis software [167] (Fig. 8.20) is divided into modules: the suite is made up of a core *Viewer* unit that serves as primary interface with different available functional units. It addresses the needs of lead identification, optimization and predictive toxicology. New unique and proprietary algorithms are applied for manipulation and classification of molecules and associated data. The compound datasets are organized into chemical classes defined by a scaffold. Theses classes are learned from the data via graph-based common scaffold representations of molecular structures. In this non-hierarchical categorization, compounds may appear in as many classes as structural features were defined. The *HTS* module allows the detection of potentially false negatives and false-positive hits via local predictive models, rationalizes library design by considering scaffold diversity, helps to identify different mechanisms of action and finds similar, predicted active compounds from other libraries by computational models for classes leading to activity. The *MedChem* module computes R-group information for classes of compounds and presents the result as an annotated scaffold and R-table and mines structure–activity relationships from compounds within a class based upon R-group properties. Moreover, it is possible to test virtual compounds derived from learned hypotheses.

8.5.1.5 Spotfire Lead Discovery

Spotfire [168] offers an interactive, visual approach to data analysis of chemical structures and any associated data such as pharmacological assay results, ADME or chemical properties (Fig. 8.21). It creates an environment that supports decision making and interactive analysis. Its patented visual analysis technology uses multidimensional display (e.g. scatterplots, profile charts, heat maps, pie charts, histograms) and interactive filters. Filtering and selection change dynamically in all visualizations and empower one to see quickly and easily trends, patterns, on outliers and unanticipated relationships. Spotfire DecisioSite™ for Lead Discovery provides interactive tools for structure searching, activity profiling, clustering on properties and structural descriptors. Results are displayed in heat maps and

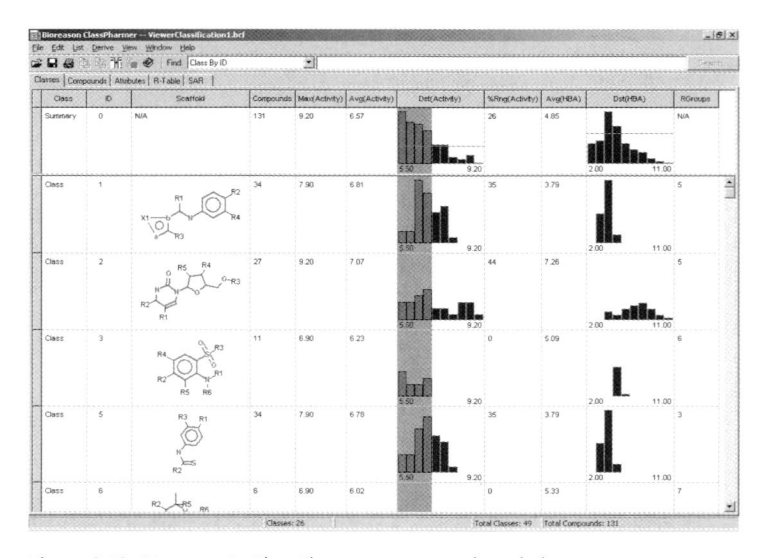

Figure 8.20 Bioreason's ClassPharmer computes knowledge from screening data in a consistent and natural language understandable by the chemist in terms of SAR and pharmacophores.

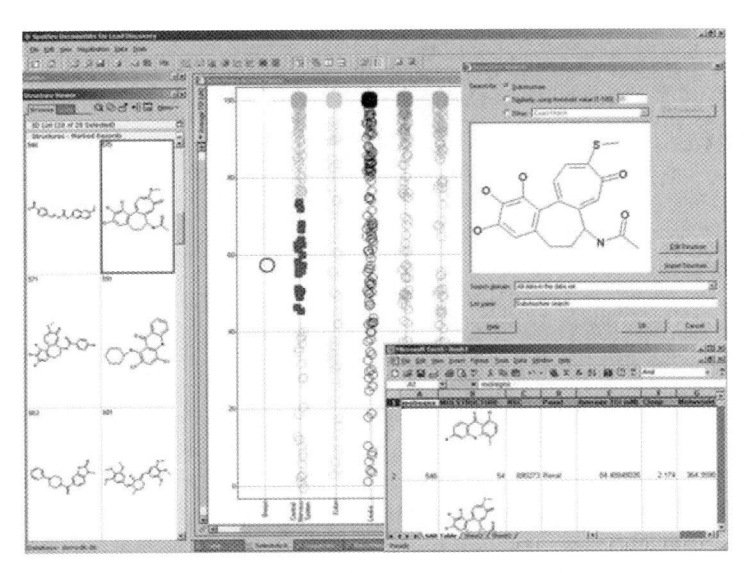

Figure 8.21 Spotfire DecisionSite offers an interactive, visual approach to data analysis that empowers individuals to see quickly and easily trends, patterns, outliers and unanticipated relationships in data.

integrated dendrograms. Principal component analysis (PCA) and interactive statistic tools (e.g. chi-squared statistics) are also included. Incorporation of the latest statistical methods using packages such as S-PLUS, SAS or the gnu-R-Project is also possible. Viewing of chemical structures from an MDL chemistry database is possible, in addition to similarity and substructure searching and capabilities that can be expanded in scope using cartridge technology from Daylight and Accelrys.

Spotfire's shared environment provides user-definable so-called "Guided Scenarios" to implement standard procedures for capture best practice by sharing workflows, analysis and decision points from a centralized repository. Chemists are guided through the various steps for analysis, visualization and reporting of their data by their analysis questions.

8.5.1.6 LeadScope

LeadScope is an interactive computer program for visualizing, browsing and interpreting chemical and biological screening data [169–171] (Fig. 8.22).

The program does not construct possible substructures for a number of given compounds, but organizes chemical structures in a large taxonomy of predefined familiar structural features such as functional groups, aromatics and heterocycles (by far the largest group within the set), each combined with common substituents – the building blocks commonly found in small-molecule drug candidates. Therefore, compounds can be assigned to more than one group depending on the

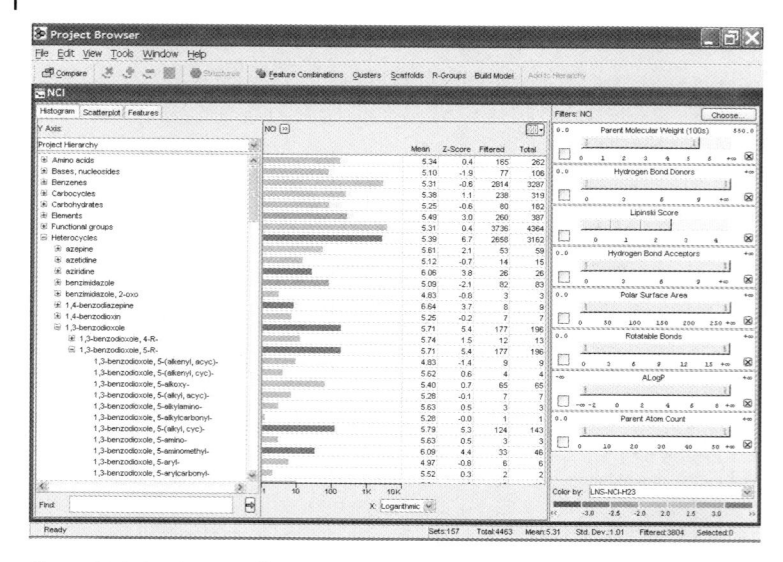

Figure 8.22 LeadScope offers to organize structures in groups with predefined structural features.

structural fragments that they contain. The compound classification can be used to explore SARs in a dataset and to search other databases for related structures. Various powerful pictorial representations including scatterplots, histograms and color coding can be used to summarize the data. Structural classes are highlighted and can be statistically correlated with biological activity. Interactive sliders are used to filter on any imported property with numerical values. Instead of focusing on individual compounds and their activities, the integrated techniques allow users to broaden their view to sets of compounds, correlated structural features and statistics over the full dataset.

8.5.1.7 OmniViz

The OmniViz software [172] consists of a basic module that provides typical spreadsheet functionality and analysis and visualization tools (Fig. 8.23). Filtering options, such as applying Lipinski rules or other custom operations, are accomplished using the Stored Math query option. Several physicochemical, structural and biological descriptors can be calculated. In addition, process data obtained from the synthesis of compounds and biological screening assays can be incorporated into the overall dataset to allow for QC analysis of the experimental data. An extensive list of predefined numeric operations, including ratios, normalizations, arithmetic and many-to-one operations and standard clustering algorithms such as the Jarvis–Patrick, K-means and hierarchical clustering methods are available using several different distance measures, including the extended Tanimoto index. In addition, custom clustering algorithms are supported with a user-configurable

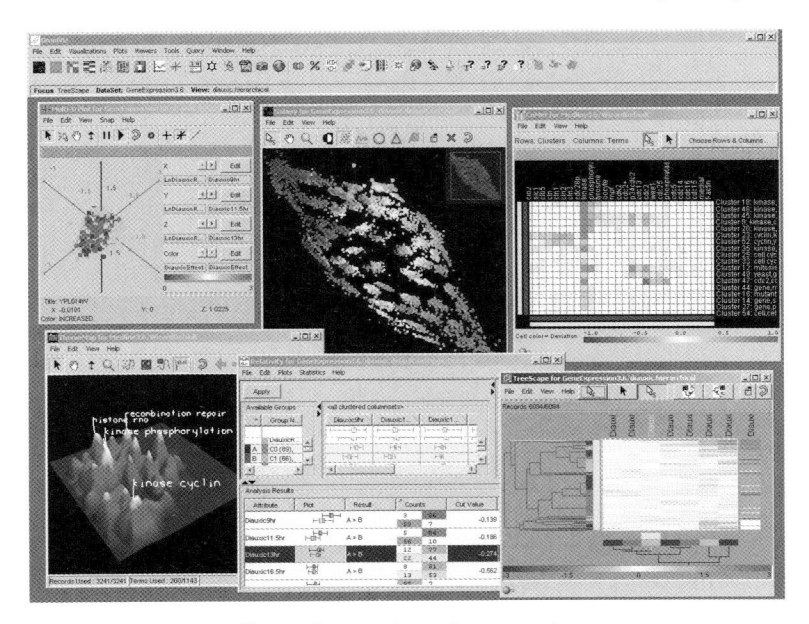

Figure 8.23 OmniViz offers sophisticated visualization tools.

interface. The relation of any record to any other record can be visualized by a proximity map, the Galaxy™ view. Similarity might be based on structural fingerprints, activity or other parameters.

The CorScape™ heatmap provides an interactive overview of patterns in a data set. CoMet™ visualizations can be applied to summarize the correlation between chemical clusters or activity levels and their associated attributes, such as chemical class or screening classifications (active, moderately active, inactive).

The OmniViz Library Sciences module allows the rapid evaluation of compounds of interest in the context of the published literature whereas the Patent Analysis module permits the assessment of the competitive standing.

8.5.1.8 SARNavigator

HTS datasets can be explored in SARNavigator [38], a desktop chemical analysis system, using a spreadsheet, scatterplots, histograms, correlation coefficients and advanced visualization tools such as similarity maps and bull's eye plots (Figure 8.24). SARNavigator consists of three components, SARNavigator Base, HQSAR and HTS.

A wide variety of molecular descriptors can be computed and relationships within chemical-biological datasets investigated. Substructure search capabilities combined with maximum common substructure determination allow users to locate compounds containing a particular scaffold or undesirable molecular fragments. R-group deconvolution can be used to generate SAR tables easily.

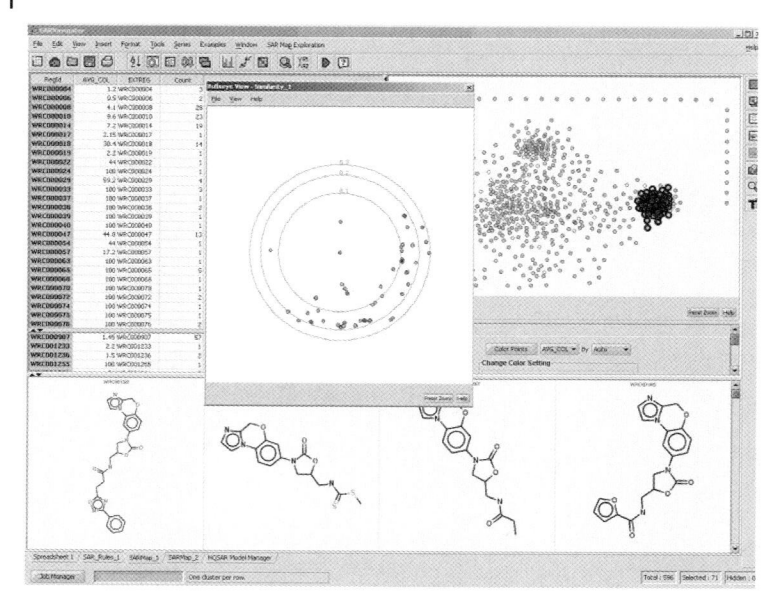

Figure 8.24 SARNavigator provides a high-capacity chemical spreadsheet environment and specialized modules.

The HQSAR module adds the ability to generate QSAR models. Users can predict molecular properties or biological activity and highlight the molecular fragments important for that activity. The resulting models can be shared between users.

The HTS module provides the ability to generate SAR maps based on a novel PCA/NLM procedure, to view and explore the entire SAR landscape from large-scale HTS, uHTS and vHTS (virtual high-throughput screening) experiments, incorporating both hit and similar non-hit compounds. Clustering, drill-down and reprojection tools can be used to explore all areas of chemical information within the screening deck. Included in the HTS module are SAR Rules, that use the same algorithm as described in Section 8.3.8 for Structural Units Analysis.

8.5.2
In-house Packages

Although a variety of software packages are currently available from various vendors, some of the major pharmaceutical companies have initiated their own programs to develop integrated packages. This ensures a highly customized system and also the opportunity for close interaction of software developers with the scientists who will use the software.

Bayer HealthCare has implemented a fully integrated system that covers all informatics processes of drug discovery from screening to the delivery of a candidate to preclinical development [173]. The software package, called "Pharmaco-

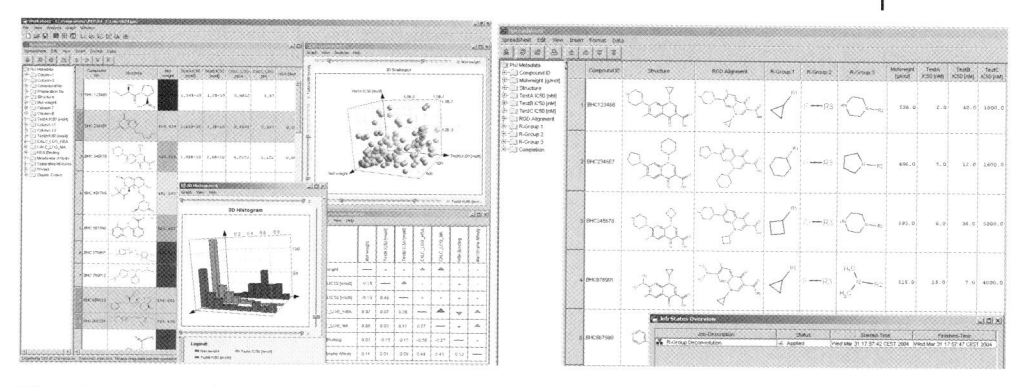

Figure 8.25 Bayer's Pharmacophore Informatics (PIx) platform is an integreted package covering the complete workflow including data retrieval, data visualization and analysis.

phore Informatics" (PIx), includes a chemically intelligent spreadsheet, several data analyzing and visualization methods such as scatterplots, histograms, bar charts, profile plots and grid views that are interactively connected with the spreadsheet (Fig. 8.25). Prediction tools for major physicochemical and structural parameters have been integrated. Many compute scripts are available such as R-group deconvolution, maximum common substructure, clustering methods, similarity grouping, reduced graph algorithms (for an example, Baytree, see 8.3.7), PCA/NLM, structural units analysis for the generation of SAR rules (see SUA, Section 8.3.8) and correlation analysis. Further compute scripts can be easily added to the system.

Along with the development of the software, all databases previously scattered throughout the organization were combined into one large data warehouse. A sophisticated query system enables research scientists to access all pharmacological, pharmacokinetic, physicochemical and structural data from their desktop. Together with predicted data, screening hits can be easily evaluated, e.g. regarding their drug/lead likeness profile. Data mining tools for HTS data are now available to every research scientist, including expert tools that have been made user friendly.

8.6
Conclusions

The initial goal of chemoinformatics was to make drug discovery, including HTS, more efficient. Despite the fact that many success stories describing the potential of chemoinformatic methods in HTS have been reported in the literature, it is difficult to evaluate the impact. Most of these methods have been applied retrospectively after lead and lead series finding for justification. Furthermore, several methods have been developed only recently and publication of data on the resulting compounds in the early stages of clinical development is often restrictive for pharmaceutical companies for competitive and patent reasons. Actually, the question of success stories in supporting screening and the hit to lead process with *in silico* tools can even be tracked back to the question of whether HTS itself has contributed to clinical candidates or not. In an interesting review, Golebiowski et al. [174] cite two references with opposite opinions regarding this topic, specifically "The majority of drugs going into clinical trials will originate from high-throughput chemistry" [175] and "High-throughput screening did *not* have a significant impact on the derivation of recently launched drugs" [176].

As *in silico* methods become more reliable in predicting data and integrated software packages are enabling research scientists to use routinely what were formerly considered to be expert systems, the authors believe that the future will surely bring more success stories in the field of chemoinformatics in high-throughput screening.

Acknowledgments

We would like to thank our colleagues at Bayer HealthCare for many fruitful discussions, especially Dr. Jill E. Wood, Dr. Heinrich Meier and Dr. Jörg Hüser for reviewing the manuscript and the above-mentioned software companies for providing screenshots.

References

1 J. Xu, A. Hagler, *Molecules* **2002**, *7*, 566–600; Chemoinformatics and drug discovery.
2 A. Böcker, G. Schneider, A. Teckentrup, *QSAR Comb. Sci.* **2004**, *23*, 207–213; Status of HTS data mining approaches.
3 W. Guba, O. Roche, in *Chemogenomics in Drug Discovery*, H. Kubinyi, G. Müller (eds.), Wiley-VCH, Weinheim, **2004**, 325–339; Computational filters in lead generation: targeting drug-like chemotypes.
4 F.K. Brown, *Annu. Rep. Med. Chem.* **1998**, *33*, 375–384; Chemoinformatics: what is it and how does it impact drug discovery?
5 T. Olsson, T. A. Oprea, *Curr. Opin. Drug Discov. Dev.* **2001**, *4*, 308–313; Cheminformatics: a tool for decision-makers in drug discovery.
6 M. Hann, R. Green, *Curr. Opin. Chem. Biol.* **1999**, *3*, 379–383; Cheminformatics – a new name for an old problem.
7 B.L. Claus, D. J. Underwood, *Drug Discov. Today* **2002**, *7*, 957–966; Discovery informatics: its evolving role in drug discovery.
8 G.E. Moore, *Electronics* **1965**, *38*, 114–117; Cramming more components onto integrated circuits; http://www.intel.com/research/silicon/mooreslaw.htm.
9 A. Golebiowski, S. R. Klopfenstein, D. E. Portlock, *Curr. Opin. Chem. Biol.* **2001**, *5*, 273–284; Lead compounds discovered from libraries.
10 C.J. Manly, S. Louise-May, J. D. Hammer, *Drug Discov. Today* **2001**, *6*, 1101–1110; The impact of informatics and computational chemistry on synthesis and screening.

11 C.A. Lipinski, A. Hopkins, *Nature* **2004**, *432*, 855–61; Navigating chemical space for biology and medicine.
12 C.A. Lipinski, *J. Pharmacol. Toxicol. Methods* **2000**, *44*, 235–249; Drug-like properties and the causes of poor solubility and permeability.
13 M.F.M. Engels, P. Ventakatarangan, *Curr. Opin. Drug Discov. Dev.* **2001**, *4*, 275–283; Smart screening: approaches to efficient HTS.
14 K.-H. Baringhaus, G. Hessler, *Drug Discov. Today: Technol.* **2004**, *1*, 197–202; Fast Similarity searching and screening hit analysis.
15 S. Varma, L. Fisher, T. Lyons, D. Chen, *Creating a Smart Virtual Screening Protocol, Part II*, http://www.accelrys.com/reference/cases/Studies/smart_virtual_-screening_part2.pdf.
16 A.M. van Rhee, J. Stocker, D. Printzenhoff, C. Creech, P. K. Wagoner, K. L. Spear, *J. Comb. Chem.* **2001**, *3*, 267–277; Retrospective analysis of an experimental high-throughput screening data set by recursive partitioning.
17 D.C. Weaver, *Curr. Opin. Chem. Biol.* **2004**, *8*, 264–270; Applying data mining techniques to library design, lead generation and lead optimization.
18 S.S. Young, R. L. H. Lam, W. J. Welch, *Curr. Opin. Drug Discov. Dev.* **2002**, *5*, 422–427; Initial compound selection for sequential screening.
19 L. Hodes, *J. Chem. Inf. Comput. Sci.* **1989**, *29*, 66–71; Clustering a large number of compounds. 1. Establishing the method on an initial sample.
20 J.Y. Q. Lai, S. Langston, R. Adams, R. E. Beevers, R. Boyce, S. Burckhardt, J. Cobb, Y. Ferguson, E. Figueroa, N. Grimster, A. H. Henry, N. Khan, K. Jenkins, M. W. Jones, R. Judkins, J. Major, A. Masood, J. Nally, H. Payne, L. Payne, G. Raphy, T. Raynham, J. Reader, V. Reader, A. Reid, P. Ruprah, M. Shaw, H. Sore, M. Stirling, A. Talbot, J. Taylor, S. Thompson, H. Wada, D. Walker, *Med. Res. Rev.* **2004**, *24*, 1–21; Preparation of kinase-biased compounds in the search for lead inhibitors of kinase targets.
21 S.L. McGovern, E. Caselli, N. Grigorieff, B. K. Shoichet, *J. Med. Chem.* **2002**, *45*, 1712–1722; A common mechanism underlying promiscuous inhibitors from virtual and high-throughput screening.
22 A. Merckwirth, H. Mauser, T. Schulz-Gasch, O. Roche, M. Stahl, T. Lengauer, *J. Chem. Inf. Comput. Sci.* **2004**, *44*, 1971–1978; Ensemble methods for classification in cheminformatics.
23 O. Roche, P. Schneider, J. Zuegge, W. Guba, M. Kansy, A. Alanine, K. Bleicher, F. Danel, E.-M. Gutknecht, M. Rogers-Evans, W. Neidhart, H. Stalder, M. Dillon, E. Sjörgen, N. Fotouhi, P. Gillespie, R. Goodnow, W. Harris, P. Jones, M. Taniguchi, S. Tsujii, W. v.d. Saal, G. Zimmermann, G. Schneider, *J. Med. Chem.* **2002**, *45*, 137–142; Development of a virtual screening method for identification of "frequent hitters" in compound libraries.
24 A. Alanine, M. Nettekoven, E. Roberts, A. W. Thomas, *Comb. Chem. High Throughput Screen.* **2003**, *6*, 51–66; Lead generation – enhancing the success of drug discovery by investing in the hit to lead process.
25 J.R. Kenseth, S. J. Coldiron, *Curr. Opin. Chem. Biol.* **2004**, *8*, 418–423; High-throughput characterization and quality control of small-molecule combinatorial libraries.
26 G.M. Rishton, *Drug Discov. Technol.* **1997**, *2*, 382–384; Reactive compounds and in vitro false positives in HTS.
27 G.M. Rishton, *Drug Discov. Technol.* **2003**, *8*, 86–96; Nonleadlikeness and leadlikeness in biochemical screening.
28 M. Wagener, V. J. van Geerestein, *J. Chem. Inf. Comput. Sci.* **2000**, *40*, 280–292; Potential drugs and nondrugs: prediction and identification of important structural features.
29 L. Di, E. H. Kerns, *Curr. Opin. Chem. Biol.* **2003**, *7*, 402–408; Profiling drug-like properties in discovery research.
30 M.F. Engels, L. Wouters, R. Verbeeck, G. Vanhoof, *J. Biomol. Screen.* **2002**, *7*, 341–351; Outlier mining in high throughput screening experiments.

31 M. Randic, in *Concepts and Applications of Molecular Similarity*, M. Johnson, G. M. Maggiora (eds.), Wiley, New York, **1990**, 77–146; Design of molecules with desired properties: a molecular similarity approach to property optimization.

32 Y.C. Martin, J. L. Kofron, L. M. Traphagen, *J. Med. Chem.* **2002**, *45*, 4350–4358; Do structurally similar molecules have similar biological activity?

33 Elsevier–MDL, San Leandro, CA, USA, www.litlink.com/company/about/history.jsp.

34 G. Grethe, T. E. Moock, *J. Chem. Inf. Comput. Sci.* **1990**, *30*, 511–520; Similarity searching in REACCS. A new tool for the synthetic chemist.

35 V.J. Gillet, in *Molecular Diversity in Drug Design*, Dean P. M., Lewis, R. A. (eds.), Kluwer, Dordrecht, **1999**, 43–65; Background theory of molecular diversity.

36 B. Debska, B. Guzowska-Swider, *J. Chem. Inf. Comput. Sci.* **2000**, *44*, 4615–4627; Fuzzy definition of molecular fragments in chemical structures.

37 J. Xu, *J. Chem. Inf. Comput. Sci.* **1996**, *36*, 25–34; GMA: a generic match algorithm for structural homomorphism, isomorphism, maximal common substructure match and its applications.

38 Tripos, St. Louis, MO, USA, www.tripos.com.

39 Daylight Chemical Information Systems, Mission Viejo, CA, USA, www.daylight.com.

40 D.M. Bayada, H. Hamersma, V. J. van Geerestein, *J. Chem. Inf. Comput. Sci*, **1999**, *39*, 1–10; Molecular diversity and representativity in chemical databases.

41 X. Chen, A. Rusinko, S. S. Young, *J. Chem. Inf. Comput. Sci*, **1998**, *38*, 1054–1062; Recursive partitioning analysis of a large structure-activity data set using three-dimensional descriptors.

42 R.D. Brown, Y. C. Martin, *J. Chem. Inf. Comput. Sci.* **1996**, *36*, 572–584; Use of structure activity data to compare structure-based clustering methods and descriptors for the use in compound selection.

43 L. Xue, J. Bajorath, *Comb. Chem. High Throughput Screen.* **2000**, *5*, 363–372; Molecular descriptors in chemoinformatics, computational combinatorial chemistry and virtual screening.

44 G. Schneider, W. Neidhart, T. Giller, G. Schmid, *Angew. Chem. Int. Ed.* **1999**, *38*, 2894–2896; "Scaffold-hopping" by topological pharmacophore search: a contribution to virtual screening.

45 J. M. Bernard, G. M. Downs, P. Willett, in *Virtual Screening for Bioactive Molecules*, H.-J. Böhm, G. Schneider (eds.), Wiley-VCH, Weinheim, **2000**, 59–80; Descriptor-based similarity measures for screening chemical databases.

46 J. Hert, P. Willett, D. J. Wilton, P. Acklin, K. Azzaoui, E. Jacoby, A. Schuffenhauer, *Org. Biomol. Chem.* **2004**, *22*, 3256–3266; Comparison of topological descriptors for similarity-based virtual screening using multiple bioactive reference structures.

47 D.R. Flower, *J. Chem. Inf. Comput. Sci.* **1998**, *38*, 379–386; On the properties on bit string-based measures of chemical similarity.

48 R.P. Sheridan, S. K. Kearsley, *Drug. Discov. Today* **2002**, *7*, 903–911; Why do we need so many chemical similarity search methods?.

49 J.D. Holliday, C. Y. Hu, P. Willett, *Comb. Chem. High Throughput Screen.* **2002**, *58*, 155–166; Grouping of coefficients for the calculation of inter-molecular similarity and dissimilarity using 2D fragment bit-strings.

50 Y. Martin, *J. Comb. Chem.* **2001**, *3*, 231–250; Diverse viewpoints on computational aspects of molecular diversity.

51 J.A. Hartigan, *Clustering Algorithms*, Wiley, New York, **1975**.

52 A.K. Jain, M. N. Murty, P. J. Flynn, *ACM Comput. Surv.* **1999**, *31*, 264–323; Data clustering: a review.

53 J.W. Raymond, C. J. Blankley, P. Willett, *J. Mol. Graph. Model.* **2003**, *5*, 421–33; Comparison of chemical clustering methods using graph- and fingerprint-based similarity measures.

54 A. Schnur, *J. Chem. Inf. Comput. Sci*, **1999**, *39*, 36–45; Design and diversity analysis of large combinatorial libraries using cell-based methods.

55 R.S. Pearlman, K. M. Smith, *Perspect. Drug Discov. Des.* **1998**, *9*, 339–353; Novel software tools for chemical diversity.

56 S. Gibson, R. McGuire, D. C. Rees, *J. Med. Chem.* **1996**, *20*, 4065–4072; Principal components describing biological activities and molecular diversity of heterocyclic aromatic ring fragments.

57 L. Xue, J. Bajorath, *J. Chem. Inf. Comput. Sci.* **2000**, *40*, 801–809; Molecular descriptors for effective classification of biologically active compounds based on principal component analysis identified by a genetic algorithm.

58 B. Cuissar, F. Touffet, B. Cremilleux, R. Bureau, S. Rault, *J. Chem. Inf. Comput. Sci.* **2002**, *42*, 1043–1052; The maximum common substructure as a molecular depiction in a supervised classification context: experiments in quantitative structure/biodegradability relationships.

59 G. Cerruela Garcia, I. Luque Ruiz, M. A. Gomez-Nieto, *J. Chem. Inf. Comput. Sci.* **2004**, *1*, 30–41; Step-by-step calculation of all maximum common substructures through a constraint satisfaction based algorithm.

60 Tripos, Inc., *DISTILL*, http://www.tripos.com/data/SYBYL/Distill_072505.pdf.

61 D. Wilton, P. Willett, G. Mullier and K. Lawson, *J. Chem. Inf. Comput. Sci.* **2003**, *43*, 469–474; Comparison of ranking methods for virtual screening in lead-discovery programmers.

62 A. Ormerod, P. Willett and D. Bawden, *Quant. Struct.-Act. Relat.* **1989**, *8*, 115–129; Comparison of fragment weighting schemes for substructural analysis.

63 J. Zupan, J. Gasteiger, *Neural Networks in Chemistry and Drug Design*, 2nd edn., Wiley-VCH, Weinheim, **1999**.

64 J. Gasteiger, A. Teckentrup, L. Terfloth, S. Spycher, *J. Phys. Org. Chem.* **2003**, *16*, 232–245; Neural networks as data mining tools in drug design.

65 H. Bauknecht, A. Zell, H. Bayer, P. Levi, M. Wagener, J. Sadowski, J. Gasteiger, *J. Chem. Inf. Comput. Sci.* **1996**, *36*, 1205–1213; Locating biologically active compounds in medium-sized heterogeneous datasets by topological autocorrelation vectors: dopamine and benzodiazepine agonist.

66 SAS Institute, Cary, NC, USA, http://www.sas.com/, http://support.sas.com/.

67 StatSoft, *Statistica, Classification Trees*, StatSoft, Tulsa, OK, USA, http://www.statsoft.com/textbook/stclatre.html.

68 R Project, *The R Project for Statistical Computing*, http://www.r-project.org/.

69 J.A. Morgan and J. N. Sonquist, *J. Am. Stat. Assoc.* **1963**, *58*, 415–434; Problems in the analysis of survey data and a proposal.

70 Accelrys Inc., *DIVA*, Accelrys, San Diego, CA, USA, http://www.accelrys.com/products/diva/index.html.

71 D.M. Hawkins, S. S. Young, A. Rusinko III, *Quant. Struct.-Act. Relat.* **1997**, *16*, 296–302; Analysis of large structure-activity dataset using recursive partitioning.

72 X. Chen, A. Rusinko III, A. Tropsha, S. S. Young, *J. Chem. Inf. Comput. Sci.* **1999**, *39*, 887–896; Automated pharmacophore identification for large chemical data sets.

73 A.M. van Rhee, J. Stocker, D. Printzenhoff, C. Creech, P. K. Wagoner, K. L. Spear, *J. Comb. Chem.* **2000**, *3*, 267–277; Retrospective analysis of an experimental high-throughput screening data set by recursive partitioning,.

74 A. Rusinko III, S. S. Young, D. H. Drewry, S. W. Gerritz, *Comb. Chem. High Throughput Screen.* **2002**, *2*, 125–133; Optimization of focused chemical libraries using recursive partitioning.

75 J.W. Godden, J. R. Furr, J. Bajorath, *J. Chem. Inf. Comput. Sci.* **2003**, *43*, 182–188; Recursive median partitioning for virtual screening of large databases.

76 V.J. Gillet, P. Willet, J. Bradshaw, *J. Chem. Inf. Comput. Sci.* **2003**, *43*, 338–345; Similarity search using reduced graphs.

77 B. Barker, E. Gardiner, V. J. Gillet, P. Kitts, J. Morris, *J. Chem. Inf. Comput. Sci.* **2003**, *43*, 346–356; Further Development of reduced graphs for identifying bioactive compounds.

78 G.W. Bemis, M. A. Murcko, *J. Med. Chem.* **1996**, *39*, 2887–2893; The properties of known drugs. 1. Molecular frameworks.

79 G.W. Bemis, M. A. Murcko, *J. Med. Chem.* **1999**, *42*, 5095–5099; The properties of known drugs. 2. Side chains.
80 S. Seidler, *PhD Thesis*, LMU Munich, **2004**; BayTree: ein Werkzeug zur gerüstbasierten Visualisierung und Aktivitätsanalyse von Screeningergebnissen chemischer Strukturdatenbanken.
81 *The Comprehensive Medicinal Chemistry (CMC–3D)*, available from MDL and currently holds 8473 molecules, **2005**; http://www.mdli.com/products/knowledge/medicinal_chem/index.jsp.
82 J.L. Jenkins, M. Glick, J. W. Davies, *J. Med. Chem.* **2004**, *47*, 6144–6159; A 3D similarity method for scaffold hopping from known drugs or natural ligands to new chemotypes.
83 S. Reiling, R. Dorfman, S. Burkett, presented at the Third Joint Sheffield Conference on Chemoinformatics, Sheffield, 21 April **2004**; Identification of relevant substructures in screening data.
84 H.-J. Böhm, A. Flohr, M. Stahl, *Drug Discov. Today: Technol.* **2004**, *1*, 217–224; Scaffold hopping.
85 G.A. Patani, E. J. LaVoie, *Chem. Rev.* **1996**, *95*, 3147–3176; Bioisosterism: a rational approach in drug design.
86 P. Ertl, *J. Chem. Inf. Comput. Sci.* **2003**, *43*, 374–380; Chemoinformatics analysis of organic substituents: identification of the most common substituents, calculation of substituent properties and automatic identification of drug-like bioisosteric groups.
87 R.W. Spencer, *J. Biomol. Screen.* **1997**, *2*, 69–70, Diversity analysis in high throughput screening.
88 M. Hann, B. Hudson, X. Lewell, R. Lifely, L. Miller, N. Ramsden, *J. Chem. Inf. Comput. Sci.* **1999**, *39*, 897–902; Strategic pooling of compounds for high-throughput screening.
89 B. Balkenhohl, C. van dem Bussche-Hunnefeld, A. Lansky, C. Zechel, *Angew. Chem. Int. Ed. Engl.* **1996**, *35*, 2288–2337; Combinatorial synthesis of small organic molecules.
90 C.M. Dobson, *Nature* **2004**, *432*, 824–828; Chemical space and biology.
91 A. Schuffenhauer, M. Popov, U. Schopfer, P. Acklin, J. Stanek, E. Jacoby, *Comb. Chem. High Throughput Screen.* **2004**, *7*, 771–781; Molecular diversity management strategies for building and enhancement of diverse and focused lead discovery compound screening collections.
92 W.P. Walters, Ajay, M. A. Murcko, *Curr. Opin. Chem. Biol.* **1999**, *3*, 384–387; Recognizing molecules with drug-like properties.
93 Ajay, *Curr. Top. Med. Chem.* **2002**, *2*, 1273–1286; Predicting drug likeness: why and how?.
94 D.E. Clark, S. D. Pickett, *Drug Discov. Today* **2000**, *5*, 49–58; Computational methods for the prediction of drug likeness.
95 B. Matter, K.-H. Baringhaus, T.Naumann, T. Klabunde, B. Pirard, *Comb. Chem. High Throughput Screen.* **2001**, *4*, 453–475; Computational approaches towards the rational design of drug-like compound libraries.
96 T. Mitchell, G. A. Showell, *Curr. Opin. Drug Discov. Dev.* **2001**, *4*, 314–318; Design strategies for building drug-like chemical libraries.
97 W.J. Egan, W. P. Walters, M. A. Murcko, *Curr. Opin. Drug Discov. Dev.* **2002**, *5*, 540–549; Guiding molecules towards drug likeness.
98 K. Ghose, V. N. Viswanadhan, J. J. Wendoloski, *J. Comb. Chem.* **1999**, *1*, 55–68; A knowledge-based approach in designing combinatorial or medicinal chemistry libraries for drug discovery. 1. A qualitative and quantitative characterization of known drug databases.
99 J. Sadowski, H. Kubinyi, *J. Med. Chem.* **1998**, *41*, 3325–3329; A scoring scheme for discriminating between drugs and nondrugs.
100 J. Xu, J. Stevenson, *J. Chem. Inf. Comput. Sci.* **2000**, *40*, 1177–1187; Drug-like index: a new approach to measure drug-like compounds and their diversity.
101 M.M. Hann, T. I. Oprea, *Curr. Opin. Chem. Biol.* **2004**, *8*, 255–263; Pursuing the leadlikeness concept in pharmaceutical research.
102 C.A. Lipinski, F. Lombardo, B. W. Dominy, P. J. Feeney, *Adv. Drug Deliv. Rev.* **1997**, *23*, 3–25; Experimental and

computational approaches to estimate solubility and permeability in drug discovery and development settings.
103 C.A. Lipinski, A. Hopkins, *Nature* **2004**, *432*, 855–861; Navigating chemical space for biology and medicine.
104 J.R. Proudfoot, *Bioorg. Med. Chem. Lett.* **2005**, *15*, 1087–1090; The evolution of synthetic oral drug properties.
105 T.I. Oprea, A. M. Davis, S. J. Teague, P. D. Leeson, *J. Chem. Inf. Comput. Sci.* **2001**, *41*, 1308–1315; Is there a difference between leads and drugs? A historical perspective.
106 J.-Y. Ortholand, A. Ganesan, *Comb. Chem.* **2004**, *8*, 271–280; Natural products and combinatorial chemistry: back to the future.
107 F.E. Koehn, G. T. Carter, *Nature Rev. Drug Discov.* **2005**, *4*, 206–220; The evolving role of natural products in drug discovery.
108 L.B. Muegge, S. L. Heald, D. Brittelli, *J. Med. Chem.* **2001**, *44*, 1841–1846; Simple selection criteria for drug-like chemical matter.
109 K.-R. Müller, G. Rätsch, S. Sonnenburg, S. Mika, M. Grimm, N. Heinrich, *J. Chem. Inf. Model.* **2005**, *45*, 249–253; Classifying »drug likeness« with kernel-based learning methods.
110 M. Clark, *J. Chem. Inf. Model.* **2005**, *45*, 30–38; Generalized fragment-substructure based property prediction method.
111 B. Yu, A. Adedoyin, *Drug Discov. Today* **2003**, *8*, 852–861; ADME-Tox in drug discovery: integration of experimental and computational technologies.
112 H. van de Waterbeemd, E. Gifford, *Nat. Rev. Drug Discov.* **2003**, *2*, 192–204; ADMET in silico modelling: towards prediction paradise.
113 A.M. Davis, R. J. Riley, *Current. Opin. Chem. Biol.* **2004**, *8*, 378–386; Predictive ADMET studies, the challenges and the opportunities.
114 A.P. Beresford, H. E. Selick, M. H. Tarbit, *Drug Discov. Today* **2002**, *7*, 109–116; The emerging importance of predictive ADME simulation in drug discovery.
115 J.R. Votano, *Curr. Opin. Drug Discov. Develop.* **2005**, *8*, 32–37; Recent uses of toplogical indices in the development of in silico ADMET models.
116 A. Helma, *Curr. Opin. Drug Discov. Develop.* **2005**, *8*, 27–31; *In silico* predictive toxicology: the state-of-the-art and strategies to predict human health effects.
117 D.E. Clark, P. D. J. Grootenhuis, *Curr. Opin. Drug Deliv. Dev.* **2002**, *5*, 382–390; Progress in computational methods for the prediction of ADMET properties.
118 S. Ekins, C. L. Waller, P. W. Swaan, G. Cruciani, S. A. Wrighton, J. H. Wikel, *J. Pharmacol. Toxicol. Methods* **2000**, 251–272; Progress in predicting human ADME parameters *in silico*.
119 M.D. Segall, A. P. Beresford, in *Enabling Technologies: Delivering the Future for Pharmaceutical R&D*, C. Sansom (ed.), PJB Publications, New York **2002**, 93–119; The promise of technology in pre-clinical development.
120 H. Yu, A. Adedoyin, *Drug Discov. Technol.* **2003**, *8*, 852–861; ADME-Tox in drug discovery: integration of experimental and computational technologies.
121 J.S. Delaney, *Drug Discov. Today* **2005**, *10*, 289–295; Predicting aqueous solubility from structure.
122 W.L. Jorgensen, E. M. Duffy, *Adv. Drug Deliv. Rev.* **2002**, *54*, 355–366; Prediction of drug solubility from structure.
123 A.-L.B. Ungell, *Drug Discov. Today: Technol.* **2004**, *1*, 423–430; Caco-2 replace or refine?.
124 F. Jensen, H. H. F. Refsgaard, R. Bro, P. B. Brockhoff, *QSAR Comb. Sci.* **2005**, *24*, 449–457; Classification of membrane permeability of drug candidates: a methodological investigation.
125 A.J. Leo, *Chem. Rev.* **1993**, *93*, 1281–1306; Calculating $\log P_{oct}$ from structures.
126 A. Erös, I. Kövesdi, L. Örfi. Takács-Novák, Gy. Acsády, Gy. Kéri, *Curr. Med. Chem.* **2002**, *9*, 1819–1829; Reliability of $\log P$ predictions based on calculated molecular descriptors: a critical review.
127 B. Palm, K. Luthman, A. Ungell, G. Strandlund, F. Beigi, P. Lundahl, P. Ar-

tursson, *J. Med. Chem.* **1998**, *41*, 5382–5392; Evaluation of dynamic polar molecule surface area as predictor of drug absorption: comparison with other computational and experimental predictors.

128 P. Ertl, B. Rohde, P. Selzer, *J. Med. Chem.* **2000**, *43*, 3714–3717; Fast calculation of molecular polar surface areas a sum of fragment-based contributions and its application to the prediction of drug transport properties.

129 M.D. Hämäläinen, A. Frostell-Karlsson, *Drug Discov. Today: Technol.* **2004**, *1*, 397–405; Predicting the intestinal absorption potential of hits and leads.

130 Y.C. Martin, *J. Med. Chem.* **2005**, *48*, 3164–3170; A bioavailability score.

131 G. Colmenarejo, A. Alvarez-Pedraglio, J.-L. Lavandera, *J. Med. Chem.* **2001**, *44*, 4370–4378; Cheminformatic models to predict binding affinities to human serum albumin.

132 F. Lombardo, J. F. Blake, W. J. Curatolo, *J. Med. Chem.* **1996**, *39*, 4750–4755; Computation of brain–blood partitioning of organic solutes via free energy calculations.

133 P. Crivori, G. Cruciani, P.-A. Carrupt, B. Testa, *J. Med. Chem.* **2000**, *43*, 2204–2216; Predicting blood–brain barrier permeation from three-dimensional molecular structure.

134 U. Norinder, M. Haeberlein, *Adv. Drug. Deliv. Rev.* **2002**, *54*, 291–313; Computational approaches to the prediction of the blood–brain distribution.

135 G.N. Kumar, S. Surapaneni, *Med. Res. Rev.* **2001**, *21*, 397–411; Role of drug metabolism in drug discovery and development.

136 T.N. Thompson, *Med. Res. Rev.* **2001**, *21*, 412–449; Optimization of metabolic stability as a goal of modern drug design.

137 F.P. Guengerich, *Nat. Rev. Drug Discov.* **2002**, *1*, 359–366; Cytochrome P450 enzymes in the generation of commercial products.

138 J.M. Hutzler, D. M. Messing, L. C. Wienkers, *Curr. Opin. Drug Discov. Dev.* **2005**, *8*, 51–58; Predicting drug–drug interactions in drug discovery: where are we now and where are we going?.

139 B. Langowski, A. Long, *Adv. Drug Deliv. Rev.* **2002**, *54*, 407–415; Computer systems for the prediction of xenobiotic metabolism.

140 M.J. de Groot, M. J. Ackland, V. A. Horne, A. A. Alex, B. C. Jones, *J. Med. Chem.* **1999**, *42*, 4062–4070; A novel approach to predicting P450 mediated drug metabolism. CYP2D6 catalyzed N-dealkylation reactions and qualitative metabolite predictions using a combined protein and pharmacophore model for CYP2D6.

141 S. Ekins, G. Bravi, S. Blinkley, J. S. Gillespie, B. J. Ring, J. H. Wikel, S. A. Wrighton, *J. Pharmacol. Exp. Ther.* **1999**, *290*, 429–438; Three- and four-dimensional quantitative structure activity relationship analyses of cytochrome P-450 3A4 inhibitors.

142 J.M. Kriegl, T. Arnold, B. Beck, T. Fox, *QSAR Comb. Sci.* **2005**, *24*, 491–502; Prediction of human cytochrome P450 inhibition using support vector machines.

143 G.M. A. Keseru, *J. Comput.-Aided Mol. Des.* **2001**, *15*, 649–657; Virtual high throughput screen for high affinity cytochrome P450cam substrates. Implication for *in silico* prediction of drug metabolism.

144 S.B. Kirton, C. A. Baxter, M. J. Sutcliffe, *Adv. Drug. Deliv. Rev.* **2002**, *54*, 385–406; Comparative modelling of cytochromes P450.

145 M.J. de Groot, S. Ekins, *Adv. Drug. Deliv. Rev.* **2002**, *54*, 367–383; Pharmacophore modeling of cytochromes P450.

146 P.A. Williams, J. Cosme, A. Ward, H.C. Angove, D. M. Vinković, H. Jhoti, *Nature* **2003**, *424*, 464–468; Crystal structure of human cytochrome P450 2C9 with bound warfarin.

147 A. Williams, J. Cosme, D. M. Vinković, A. Ward, H.C. Angove, P. J. Day, C. Vonrhein, I. J. Tickle, H. Jhoti, *Science* **2004**, *305*, 683–686; Crystal structures of human cytochrome P450 3A4 bound to metyrapone and progesterone; Astex Technologies, GB Patent 2 395 718; Crystal structure of Cytochrome P450 3A4 and its use.

148 G. M. Keserüu, L. Molnár, *J. Chem. Inf. Comput. Sci.* **2002**, *42*, 437–444; META-PRINT: a metabolic fingerprint. application to casette design for high-throughput ADME screening.

149 B. Dickins, *Curr. Top. Med. Chem.* **2004**, *4*, 1745–1766; Induction of cytochromes P450.

150 S. Ekins, *Drug Discov. Today* **2004**, *9*, 276–285; Undesirable drug interactions with promiscuous proteins *in silico*.

151 S. H. Hilal, S. W. Karickhoff, L. A. Carreira, *QSAR Comb. Sci.* **2004**, *23*, 709–720; Prediction of solubility, activity coefficient and liquid/liquid partition coefficient of organic compounds.

152 M. E. Beck, *J. Chem. Inf. Model.* **2005**, *45*, 273–282; Do Fukui function maxima relate to sites of metabolism? A critical case study.

153 S. Shaik, S. Cohen, S. P. de Visser, P. K. Sharma, D. Kumar, S. Kozuch, F. Ogliaro, D. Danovich, *Eur. J. Inorg. Chem.* **2004**, 207–226; The "rebound controversy": an overview and theoretical modeling of the rebound step in C–H hydroxylation by cytochrome P450.

154 B. Manga, J. C. Duffy, P. H. Rowe, M. T. D. Cronin, *QSAR Comb. Sci.* **2003**, *22*, 263–273; A hierarchical QSAR model for urinary excretion of drugs in humans as a predictive tool for biotransformation.

155 R. Benigni, A. Richard, *Methods Enzymol.* **1998**, *14*, 264–276; Quantitative structure-based modeling applied to characterization and prediction of chemical toxicity.

156 B. Greene, *Adv. Drug Deliv. Rev.* **2002**, *54*, 417–431; Computer systems for the prediction of toxicity: an update.

157 A. Fermini, A. A. Fossa, *Nat. Rev. Drug Discov.* **2003**, *2*, 439–447; The impact of drug-induced QT interval prolongation on drug discovery and development.

158 L. Muzikant, R. C. Penland, *Curr. Opin. Drug Discov. Dev.* **2002**, *5*, 127–135; Models for profiling the potential QT prolongation risk of drugs.

159 B. Roche, G. Trube, J. Zuegge, P. Pflimlin, A. Alanine, G. Schneider, *ChemBioChem* **2002**, *3*, 455–459; A virtual screening method for prediction of the hERG potassium channel liability of compound libraries.

160 M. Keserü, *Bioorg. Med. Chem. Lett.* **2003**, *13*, 2773–2775; Prediction of hERG potassium channel affinity by traditional and hologram QSAR methods.

161 R. Nilakantan, F. Immermann, K. Haraki, *Comb. Chem. High Throughput Screen.* **2002**, *5*, 105–110; A novel approach to combinatorial library design.

162 B. Gedeck, P. Willet, *Curr. Opin. Chem. Biol.* **2001**, *5*, 389–395; Visual and computational analysis of structure-activity relationships in high-throughput screening data.

163 Accelrys Inc., San Diego, CA, USA, www.accelrys.com.

164 BioSolveIT, Sankt Augustin, Germany, www.biosolveit.com.

165 M. Rarey, J. S. Dixon., *J Comput.-Aided Mol. Des.* **1998**, *5*, 471–490; Feature trees: a new molecular similarity measure based on tree matching.

166 Scitegic Inc., San Diego, CA, USA. www.scitegic.com.

167 Bioreason, Santa Fe, NM, USA, www.bioreason.com.

168 Spotfire, Göteborg, Sweden, www.spotfire.com.

169 Leadscope, Columbus, OH, USA, www.leadscope.com.

170 G. Roberts, G. J. Myatt, W. P. Johnson, K. P. Cross, P. E. Blower, Jr., *J. Chem. Inf. Comput. Sci.* **2000**, *40*, 1302–1314; LeadScope: software for exploring large sets of screening data.

171 P. E. Blower, Jr., K. P. Cross, M. A. Fligner, G. J. Myatt, J. S. Verducci, C. Yang, *Curr. Opin. Discov. Technol.* **2004**, *1*, 37–47; Systematic analysis of large screening sets in drug discovery.

172 Omniviz, Maynard, MA, USA, www.omniviz.com.

173 P. G. Nell, W. J. Scott, ACS Presentations CINF 97/98, 229th ACS Annual Meeting, San Diego, CA, **2005**; Integrated approaches to informatics – Bayer HealthCare pharmacophore informatics platform.

174 A. Golebiowski, S. R. Klopfenstein, D. E. Portlock, *Curr. Opin. Chem. Biol.* **2003**, *7*, 308–325; Lead compounds discovered from libraries: Part 2.

175 S. Borman, *Chem. Eng. News* **2001**, *8*, 49–58; Combinatorial chemistry: novel strategies for drugs and materials.

176 J.R. Proudfoot, *Bioorg. Med. Chem. Lett.* **2002**, *12*, 1647–1650; Drugs, leads and drug likeness: an analysis of some recently launched drugs.

177 A. Brazma, J. Vilo, *FEBS Lett.* **2000**, *480*, 17–24; Gene expression data analysis.

178 W. Shannon, R. Culverhouse, J. Duncan, *Pharmacogenomics* **2003**, *4*, 41–52; Analyzing microarray data using cluster analysis.

179 S.J. Haggarty, P. A. Clemons, S. L. Schreiber, *J. Am. Chem. Soc.* **2003**, *35*, 10543–10545; Chemical genomic profiling of biological networks using graph theory and combinations of small molecule perturbations.

180 R.A. Butcher, S. L. Schreiber, *Curr. Opin. Chem. Biol.* **2005**, *1*, 25–30; Using genome-wide transcriptional profiling to elucidate small-molecule mechanism.

181 A. Jensen, S. Seidler (Bayer Aktiengesellschaft), Patent WO 2002/074035 A2; Method for generating a hierarchical topological tree of 2D or 3D-structural formulas of chemical compounds for property optimization of chemical compounds.

182 A.M. Davis, D. J. Keeling, J. Steele, N. P. Tomkinson, A. C. Tinker, *Curr. Top. Med. Chem.* **2005**, *5*, 421–439; Components of successful lead generation.

183 H. Matter, *J. Med. Chem.* **1997**, *40*, 1219–1229; Selecting optimally diverse compounds from structure databases: a validation study of two-dimensional and three-dimensional molecular descriptors.

184 T. Potter, H. Matter, *J. Med. Chem.* **1998**, *41*, 478–488; Random or rational design? Evaluation of diverse compound subsets from chemical structure databases.

185 R.S. Pearlman, http://www.netsci.org/Science/Combichem/feature08.html; Novel software tools for addressing chemical diversity.

186 J.S. Mason, M. A. Hermsmeier, *Curr. Opin. Chem. Biol.* **1999**, *3*, 342–349; Diversity assessment.

187 R.D. Clark, J. Kar, L. Akella, F. Soltanshahi, *J. Chem. Inf. Comput. Sci.* **2003**, *43*, 829–836; OptDesign: extending optimizable k-dissimilarity selection to combinatorial library design.

188 F. Brown, *Curr. Opin. Drug Discov. Dev.* **2005**, *8*, 298–302; Editorial opinion: chemoinformatics – a ten year update.

189 A.M. Davis, D. J. Keeling, J. Steele, N. P. Tomkinson, A. C. Tinker, *Curr. Top. Med. Chem.* **2005**, *5*, 421–439; Components of successful lead generation.

190 S.J. Haggarty, *Curr. Opin. Chem. Biol.* **2005**, *9*, 296–303; The principle of complementarity: chemical versus biological space.

191 H. Sun, *Curr. Comput.-Aided Drug Des.* **2005**, *1*, 179–193; Predicting ADMET properties by projecting onto chemical space – benefits and pitfalls.

192 H.-J. Roth, *Curr. Opin. Chem. Biol.* **2005**, *9*, 293–295; There is no such thing as "diversity"!

193 A. Reayi, P. Arya, *Curr. Opin. Chem. Biol.* **2005**, *9*, 240–247; Natural product-like chemical space: search for chemical dissectors of macromolecular interactions.

194 D.V. S. Green, S. D. Picket, *Mini-Rev. Med Chem.* **2004**, *4*, 1067–1076; Methods for library design and optimization.

195 E.A. Jamois, *Curr. Opin. Chem. Biol.* **2003**, *7*, 326–330; Reagent-based and product-based computational approaches in library design.

196 T.R. Webb, *Curr. Opin. Drug Discov. Dev.* **2005**, *8*, 303–308; Current directions in the evolution of compound libraries.

197 D.B. Kitchen, F. L. Stahura, J. Bajorath, *Mini-Rev. Med. Chem.* **2004**, *4*, 1029–1039; Computational techniques for diversity analysis and compound classification.

198 S.D. Pickett, J. S. Mason, I. M. McLay, *J. Chem. Inf. Comput. Sci*, **1996**, *36*, 1214–1223; Diversity profiling and design using 3D pharmocophores – pharmacophore-derived queries (PDQ).

199 R. Dorfman, R. D. Clark, Tripos, St. Louis, MO, USA, personal communication, www.tripos.com.

200 J.N. Weinstein, T. G. Myers, P. M. O'Connor, S. H. Friend, A. J. Fornace, Jr., K. W. Kohn, T. Fojo, S. E. Bates, L. V. Rubinstein, N. L. Anderson, J. K. Buolamwini, W. W. van Osdol, A. P. Monks, D. A. Scudiero, E. A. Sausville, D. W. Zaharevitz, B. Bunow, V. N. Viswanadhan, G. S. Johnson, R. E. Wittes and K. D. Paull, *Science* **1997**, *275*, 343–349; An information-intensive approach to the molecular pharmacology of cancer.

201 Z.E. Perlman, M. D. Slack, Y. Feng, T. J. Mitchison, L. F. Wu, S. J. Altschuler, *Science*, **2004**, *306*, 1194–1198; Multidimensional drug profiling by automated microscopy.

202 P.R.N. Wolohan, L. B. Akella, R. J. Dorfman, P. G. Nell, S. M. Mundt, R. D. Clark, *J. Chem. Inf. Model.*, **2006**, *46*, 1188–1183; *Structural Units Analysis Identifies Lead Series and Facilitates Scaffold Hopping in Combinatorial Chemistry.*

9
Combinatorial Chemistry and High-throughput Screening
Roger A. Smith and Nils Griebenow

9.1
Introduction

Since the early 1990 s, "combinatorial chemistry" has emerged and developed as a powerful technology for the generation of large collections of compounds or "compound libraries", primarily to facilitate the identification of new actives for drug targets via high-throughput screening techniques. One of the first articles to describe the synthesis of combinatorial libraries to support biological screening was published in 1991 by researchers at Affymax Research Institute [1]. In this work, a combination of solid-phase synthesis, photolabile protecting groups and photolithography was applied to achieve spatially addressable parallel chemical synthesis and an array of 1024 peptides was thereby generated for screening. In 1992, Brenner and Lerner described a process for preparing compound libraries in which each compound was encoded with a unique nucleotide sequence or genetic tag [2] and the methodology to apply this process to prepare a peptide library was subsequently reported [3]. A combinatorial library of 117 649 peptides, generated by solid-phase synthesis and encoded with inert chemical tags, was reported by Ohlmeyer et al. in 1993 [4]. In 1992, Ellman's group reported the solid-phase synthesis of some nonpolymeric organic compounds, namely 1,4-benzodiazepine derivatives [5], and subsequently described a 192-member combinatorial library prepared by parallel solid-phase synthesis on arrays of "Geysen pins" [6]. In a related effort, a parallel reaction apparatus designed for solid-phase organic synthesis was utilized by DeWitt et al. to prepare libraries of hydantoins and benzodiazepines [7].

These and other key achievements formed the groundwork for combinatorial chemistry to be further investigated, developed and applied as an efficient technology to generate large numbers of diverse chemical structures. A historical account of some of the earlier combinatorial synthesis methods has been published [8], and also commentaries by the authors of several "classical" papers in combinatorial chemistry [9]. Excellent reviews from this early period of combinatorial chemistry were published by Gordon and colleagues [10, 11].

High-Throughput Screening in Drug Discovery. Edited by Jörg Hüser
Copyright © 2006 WILEY-VCH Verlag GmbH & Co. KGaA, Weinheim
ISBN: 3-527-31283-8

A flurry of articles in the 1990 s hailed combinatorial chemistry as a technology that would significantly accelerate the pace of drug development and save millions of dollars in R&D costs and change the whole notion of drug discovery and development (e.g. [12–15]). As the technology has evolved during the last decade, a wide variety of approaches and techniques have been explored and refined. Certain combinatorial chemistry methodologies have, for the most part, been rejected by most pharmaceutical researchers and some recent articles have discussed, questioned and critiqued the technology and its impact on drug discovery [16–18]. Nevertheless, combinatorial chemistry is generally recognized as a valuable and powerful drug discovery tool, when appropriately applied to lead identification and optimization [19, 20].

9.2
Categories of Compound Libraries for High-throughput Screening

Compound libraries used for high-throughput screening can be categorized by a wide variety of criteria, such as the synthesis phase (solid phase, solution phase), synthesis process concept (e.g. parallel, pool/split), synthesis chemistry (e.g. Ugi, olefin metathesis), chemical structure class (e.g. benzodiazepines), sample type (discrete compounds or mixtures of compounds), research objective (e.g. lead identification, lead optimization), the source (e.g. vendor, natural products, custom synthesis) and so forth. In an attempt to clarify the terminology used in the field of combinatorial chemistry, the International Union of Pure and Applied Chemistry (IUPAC) prepared a technical report with a glossary of almost 150 terms [21]. More recently, a perspective on the structural representation of combinatorial chemistry was published [22]. In this chapter, the terminology and structural representations are intended to conform to these two reports.

Fundamentally, the term "combinatorial chemistry" refers to a process in which sets of compounds are prepared from combinations of building blocks (chemical reagents). A set of compounds obtained by this process, called a "combinatorial library", will generally fall into one of the following two broad categories:

1. *Unbiased library.* Also known as a "random library", this library type is generally designed and obtained by synthesis-driven concepts, without bias towards any molecular target. The library is intended to consist of a diverse collection of chemical structures, for the identification of "screening hits" for any number of targets. Normally, the screening hit results are further investigated, with the goal of generating validated new "lead" structures. It follows that these libraries are also commonly referred to as "lead identification", "lead discovery" or "lead finding" libraries. Typically, an unbiased library will consist of structures having a common chemical core or scaffold (or template) and in some cases the compound scaffold is described as a "privileged structure". This term, originally introduced by Evans et al. [23] in describing benzodiazepine-based CCK-A antagonists, usually refers to a chemical scaffold that is semi-rigid, allowing substituent groups to be

positioned in specific orientations. Such "privileged structures" often confer favorable drug-like properties on the compounds incorporating the scaffold structure and they are often capable of providing ligands for more than one molecular target. Reviews have been published that describe this concept and list a variety of privileged structures [24, 25]. Several examples of combinatorial libraries based on privileged structures have been described in the literature [26–29].

2. *Directed library.* Also known as a "biased" or "focused library", this library type is prepared from a limited number of building blocks and/or a particular scaffold and is designed to contain compounds that are active against a particular target or family of targets. That is, many of the compounds in a directed library will contain structural features that define the pharmacophore for the target, based on pre-existing information or hypotheses. Such libraries may be designed, prepared and screened to generate new leads that are related to known actives (e.g. competitor compounds) or to improve on the properties of an existing lead series. This type of library may also be known as a "lead optimization" library or referred to as "target directed", "project directed" or "thematic". A directed library may be designed on the basis of a "privileged structure", but it should be appreciated that a library based on privileged structures may well contain actives against multiple targets [30, 31].

In addition to the combinatorial libraries that can be categorized as unbiased libraries or directed libraries, there are other compound collections used in high-throughput screening. Research facilities engaged in drug discovery efforts will typically have a corporate compound collection, sometimes referred to as a "compound library" or "screening bank", consisting of historical compounds prepared in-house and compound collections purchased from external vendors. In addition, "natural product libraries" consisting of compounds isolated from various sources have been used in conjunction with high-throughput screening to identify new leads for drug discovery [32, 33]. In general, the corporate synthetic compound collections and natural product libraries can be classified as unbiased libraries, used to identify new leads for any molecular target.

In Section 9.5, various examples of lead structures identified from unbiased and directed libraries will be reviewed, and also the impact of combinatorial libraries on drug discovery and development.

9.3
Synthesis Techniques and Library Formats

Two decades of exceptional growth in combinatorial chemistry and the continuing development in this area have led to a wide array of associated synthesis and automation techniques. These techniques include the "classical" solid-phase and solution-phase methods, introduced in the early days of combinatorial chemistry,

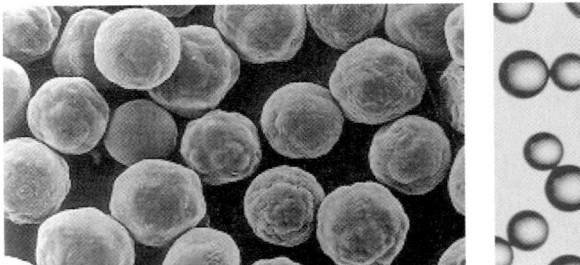

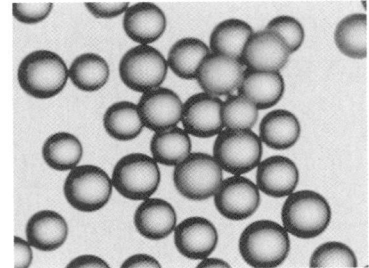

Figure 9.1 Polystyrene beads imaged with an electron microscope (left) and a light microscope (right). Reprinted with the permission of Rapp Polymere GmbH, Tübingen [38].

in addition to recently emerging techniques, such as microwave-assisted chemistry, dynamic combinatorial chemistry and fluorous-phase methods. Some excellent reviews introduce these topics and describe best practices for their application in the drug discovery process [34–37]. In this chapter, the key concepts of combinatorial synthesis techniques and some feature applications of these methods will be discussed, with a particular focus on issues regarding library production for high-throughput screening.

9.3.1
Solid-phase Synthesis

The term solid-phase synthesis describes the stepwise assembling of substrates (combinatorial building blocks) to prepare compounds on an insoluble, functionalized, polymeric material (Fig. 9.1). Suitable supports for this purpose include various sorts of polystyrene beads (crosslinked with divinylbenzene) and different grafted plastic pins.

Immobilization of the synthesized compounds on the solid support (usually via a linker fragment or functionality) offers simple purification via filtration. In addition, excesses of reagents and reactants can be used to drive reactions to completion, since they can also be easily removed by filtration. Further chemical elaboration of the product can then be performed by additional reaction/filtration cycles until the desired product has been assembled. In a subsequent cleavage step, the product is released from the support, leaving a clean compound (Fig. 9.2).

The concept of solid-phase synthesis dates back to 1963, when Merrifield developed a method of synthesizing peptides on solid support [39]. In subsequent work, solid-phase synthesis found great utility in polypeptide and oligonucleotide chemistry, especially with the advent of automated methods. The first combinatorial solid-phase syntheses of peptide libraries were presented by Geysen et al. [40] and Houghten [41] in 1984 and 1985, respectively. Geysen et al. used grafted plastic pins for the production of peptide arrays, whereas Houghten established the so-called "tea-bag" method for multiple peptide synthesis. However, at that time, solid-phase synthesis suffered from limited application, as only a few organic reactions had been transferred to solid supports. Some early examples were reported by Leznoff and Wong [42, 43] and Crowley and Rapoport [44] in 1973 and 1976, respectively, but their work remained somewhat unappreciated until solid-phase organic chemistry was applied to combinatorial library synthesis. Some of the earliest examples of solid-phase synthesis towards libraries of non-peptidic, drug-like compounds were published by Bunin and Ellman [5] and DeWitt et al. [7]. Stimulated by these reports, the expansion of the scope of solid-phase synthesis became an extremely active research area. Thus, today's repertoire of solid-phase transformations comprises a wide array of classical heterocyclic chemistry, functional group transformations and sophisticated organometallic reactions [45].

Another key aspect for the success of solid-phase synthesis is the appropriate choice of functionality for attachment to the resin. Typically, tethering to a solid support is accomplished by "linkers" that function as polymeric protecting groups. The challenge is to identify a linker that is stable under the reaction conditions during synthesis but allows for the facile cleavage of the product into solution without decomposition. Developments in this area are characterized by a steady evolution in the range of available linkers. Frequently used linkers include acid-labile linkers, nucleophile-labile handles and photolabile linkers [46, 47]. However, an intrinsic characteristic of solid-phase synthesis is that attachment to the linker requires a functional group, which is typically revealed after cleavage from the resin. Thus, every member of a combinatorial library bears the same functional group, which in some cases may be less than desirable. To overcome this inherent problem,

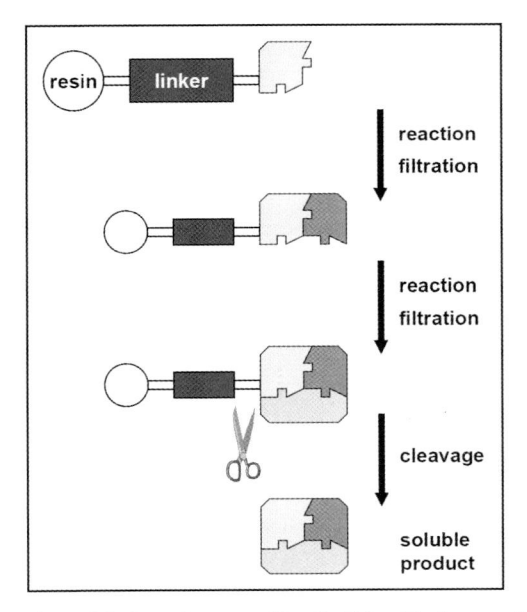

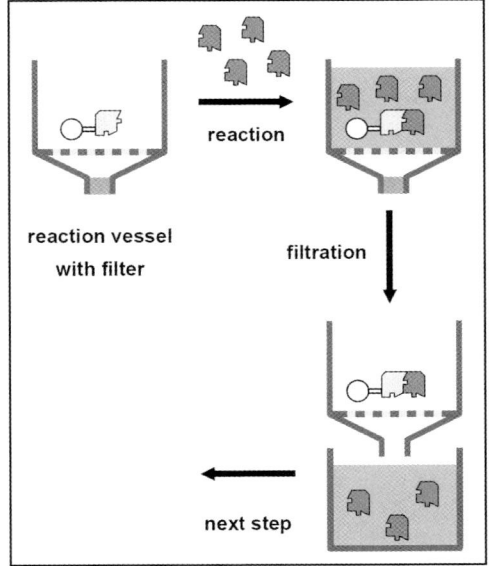

Figure 9.2 Stepwise assembly of building blocks on solid support (left). Process for performing a solid-phase reaction step in a reaction vessel (e.g. syringe) with filter frit (right).

Figure 9.3 Example of a highly acid-labile silicon linker [49].

specific linkers have been developed that do not leave their "trace" on the final compound. Frequently, cleavage in this case results in an imperceptible C–H bond, which gives no information about the history of immobilization. Today, a variety of these so-called "traceless" linkers exist, including selenium-, germanium- and silicon-based linkers and triazine and decarboxylative traceless linkers (Fig. 9.3) [48].

Apart from all the benefits, however, the use of solid-phase synthesis has its limitations. On the one hand, it can be difficult to adapt solution-phase chemistry to solid-phase synthesis. For example, hydrogenation on a solid support requires tailor-made reduction conditions since the normal conditions using heterogeneous catalysts are problematic. Instead, diimide has been used for this purpose [50]. On the other hand, monitoring of solid-phase reactions tends to be elaborate, because typical methods of analysis such as solution-phase NMR or LC–MS cannot be applied directly, owing to the insolubility of the material. In addition, the quantity

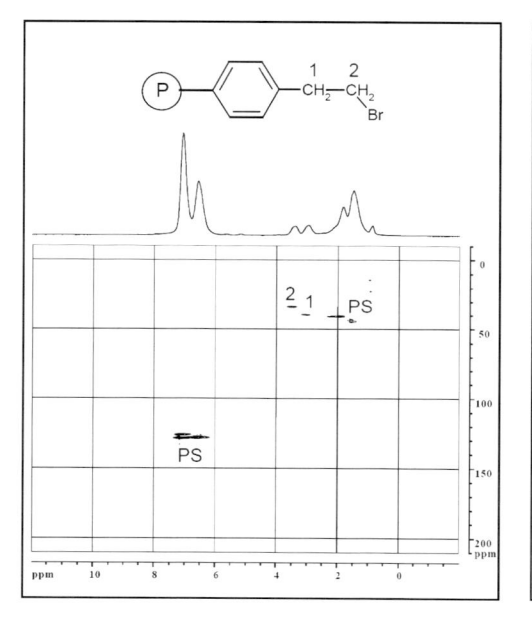

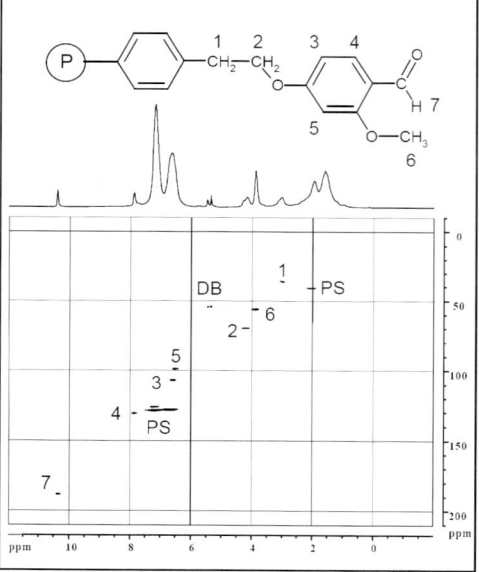

Figure 9.4 Sample analysis on solid support with HR-MAS-NMR. 300 MHz HSQC spectrum of 2-(bromoethyl)polystyrene (left) and aldehyde-loaded resin (right). PS, polystyrene; DB, signal of the double bond of a side product, resulting from elimination.

of the resin-bound compounds is small, compared with the polymer backbone. Hence sophisticated methods such as gel-phase HR-MAS-NMR [51–54] (Fig. 9.4), FTIR [55–57], on-resin MS [58–61] and electrochemical impedance spectroscopy [62] are required for compound analysis or it is necessary to cleave the product from an aliquot of the resin prior to analysis.

To overcome these limitations, special resins such as dendrimer resins [63] and polyethylene glycol polymers have been developed [64]. The dendrimer resins have a significantly higher loading, providing more material for analysis, whereas the polyethylene glycol polymers are soluble in certain solvents, allowing for routine solution-phase analysis. Further, the polyethylene glycol polymer can be precipitated out of solution by crystallization for purification purposes after synthesis.

The techniques of solid-phase synthesis are mainly divided into two categories, the parallel synthesis of discrete compounds and the "pool/split" procedure used predominantly for mixture synthesis.

9.3.1.1 Parallel Solid-phase Synthesis Techniques and Tools

In the parallel synthesis approach, different chemical structures are prepared separately in arrays of discrete reaction vessels. From hundreds to thousands of reaction vessels are used to perform the reactions and the reagents are delivered either by laboratory robots or by manual tools such as multi-channel pipettes. By these methods, the synthesized compounds can be easily identified (by their spatial location) and sufficient material of several milligrams can be prepared for characterization, screening and archiving. Examples of devices for parallel solid-phase synthesis are shown in Fig. 9.5.

9.3.1.2 Pool/Split Techniques with Encoding

An important synthesis approach pioneered by Furka et al. [66] is the "pool/split" process, also known by the terms "split/pool", "split-and-mix", "divide–couple–

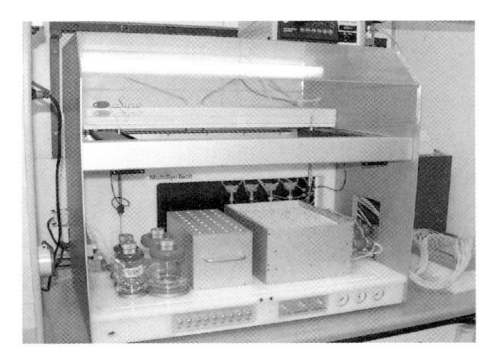

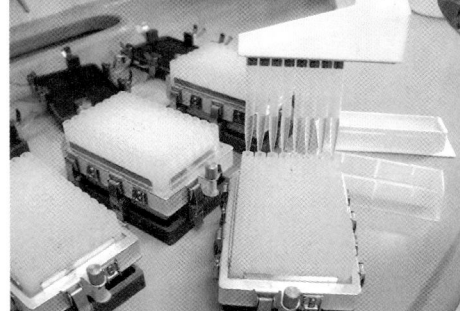

Figure 9.5 Fully automated synthesizer with 96 reaction vessels (left) [65]. Reagent delivery into a manual reaction block with a 96-well microtiter plate format (right).

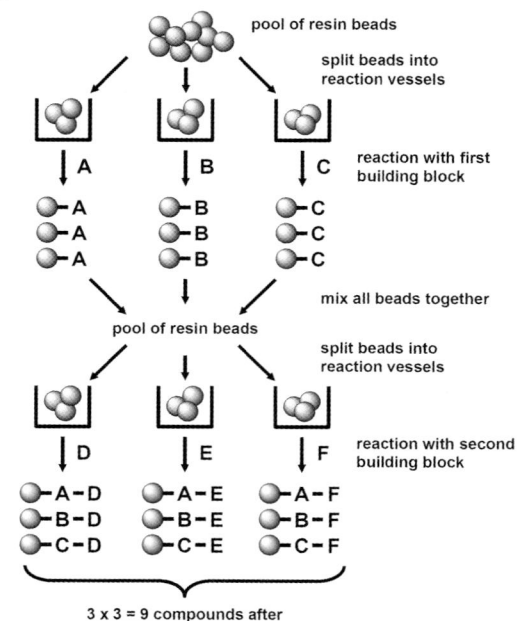

Figure 9.6 Principle of the "pool/split" procedure [36].

recombine", "portioning–mixing" and "split-and-combine". In this method, resin beads or micro-particles are used. For each step, a batch of beads is divided into equal portions in individual reaction vessels and each portion is reacted with a different building block. After work-up, the beads are pooled and mixed, then divided again into individual reaction vessels. The separated aliquots of beads are reacted with the next set of building blocks and the process is continued until the desired library has been assembled. The size of a library is governed by the number of building blocks and reaction steps. In the example shown in Fig. 9.6, two combinatorial cycles using three building blocks in each step delivers nine compounds ($3 \times 3 = 9$). Hence only six reaction steps are needed for the synthesis of nine compounds. With an increase in the number of building blocks and/or combinatorial reaction steps employed for library synthesis, the effect of "saving" reaction steps is considerably higher.

As an example, the preparation of a 1000-compound library via a three-step pool/split synthesis using 10 building blocks in each step needs only 30 reactions. Compared with parallel synthesis, the same library would require a minimum of 1110 reaction steps. However, the pool/split synthesis has two drawbacks when compared with parallel synthesis. Although each bead contains a discrete compound (one-bead, one-compound libraries), the quantity of the released compounds is low (depending on the resin loading capacity, each bead carries typically 0.1–1 nmol of compound) and the structure is unknown. Therefore, if a compound proves to be biologically active in screening, the identification of the active library member must be carried out. For this purpose, several strategies have been established. Among these, iterative and orthogonal deconvolution, positional scanning, analysis of the released compound by mass spectrometry and chemical tagging are frequently used techniques [34, 67, 68]. Chemical tagging is based on chemical modification of the resin bead to give coded information regarding the structure of the assembled compound [69]. In detail, an individual marker is reacted with the resin bead whenever a building block is coupled, thereby tracking the chemical history of each bead. Because the tags are added during the synthesis process, this kind of labeling is frequently called in-process tagging. Finally, cleavage and analysis of the tagging molecules reveal a code for the structural identity of the compound formed. As an example, Ohlmeyer et al. in 1993 introduced a powerful technique for chemical tagging by indexing the beads using a defined set of haloaromatic carbenes [4].

Radiofrequency (RF) tagging is another intriguing approach to exploit the productivity of the pool/split procedure [70–72]. This technology employs glass-encased transponder tags that are placed together with a portion of synthesis resin into semi-porous containers (micro-reactors) (Fig. 9.7). Since the reactors have a loading capacity of 30–300 mg of resin, the method allows for the synthesis of discrete compounds on a multi-milligram scale. With the aid of the transponder code, each reactor is individually guided through the synthesis sequence. In detail, the tag information is handed over to a synthesis database that allows control of the entire combinatorial process. Because the reactors are labeled prior to synthesis, this approach is usually termed "pre-encoded tagging".

Synthesis is carried out by allowing the reaction solvents and reagents to pass through the outer mesh walls of the micro-reactors and can be performed using standard laboratory glassware (Fig. 9.8). During each combinatorial synthesis cycle, the micro-reactors are sorted into the reaction vessels (directed sorting), either manually or by automation, then reacted with the respective building block and pooled for the washing procedure. The process is repeated until a discrete compound has been assembled in each individual micro-reactor. Finally, the micro-reactors are arrayed into cleavage blocks where the individual products are released from the resin. Representative steps of the workflow are illustrated in Fig. 9.8. In a similar manner, radio-frequency encoding has been developed for library production on pins (crowns, lanterns) [73].

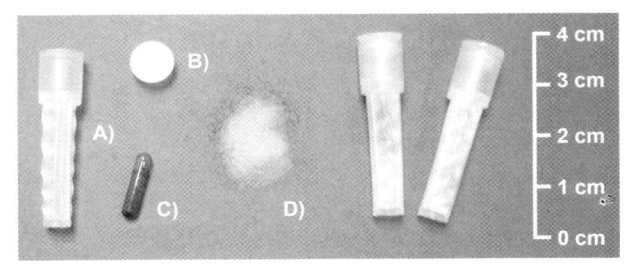

Figure 9.7 (A) Semi porous micro-reactor; (B) reactor cap; (C) glass-encased RF transponder; (D) polystyrene resin.

Figure 9.8 Automated directed sorting of micro-reactors (top left). Sets of micro-reactors are reacted with their respective building block in round-bottle flask (top right). Washing of the pooled reactors (bottom left). Micro-reactors arrayed into a cleavage block (bottom right).

An exciting recent advance in the directed sorting technology is the use of so-called nano-reactors, which are labeled with machine-readable 2D-barcodes [28]. This optical encoding technology allows for the synthesis of libraries with up to 100 000 members, each in milligram quantities.

9.3.2
Solution-phase Synthesis

Whereas solid-phase synthesis initially dominated combinatorial chemistry, parallel solution-phase techniques have received increasing attention over recent years, as indicated by the growing number of publications. Advantages that characterize solution-phase combinatorial chemistry are faster validation times, assessed chemical transformations and the ease of monitoring reactions by standard analytical protocols. However, solution-phase combinatorial chemistry suffers from one principal disadvantage. Few organic reactions proceed to give the desired compounds in quantitative yields and with acceptable levels of impurities from precursors, by-products or reagents. Therefore, work-up and purification are the main bottleneck. This is especially the case for high-throughput library synthesis. To help in overcoming these difficulties, a variety of purification methodologies such as liquid–liquid extraction, polymer-assisted solution-phase synthesis and automated high-throughput chromatography have been established.

9.3.2.1 Polymer-supported Reagents and Scavengers

Common approaches within the field of polymer-assisted solution-phase techniques are the application of "scavenger" resins and the use of polymer-supported reagents (Fig. 9.9) [74]. Scavenger resins contain reactive functional groups that bind selectively to unwanted reagents and by-products. They are added to the reaction mixture upon completion and simple filtration allows for removal of the resin-bound impurities, leaving the purified product in solution.

Suitable scavenger resins contain functional groups that are complementary in reactivity to the non-product species requiring sequestration. Today, tailor-made resins for a wide variety of different chemical transformations are available.

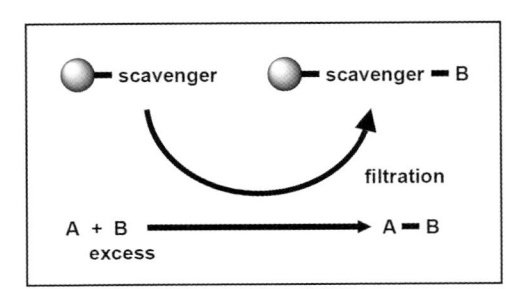

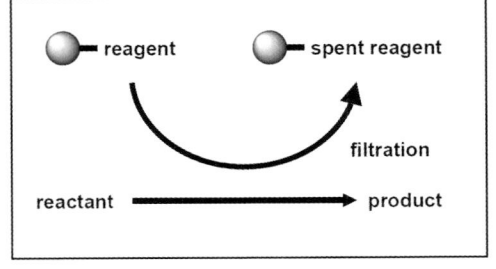

Figure 9.9 The use of scavenger resin techniques (left) and the application of solid-supported reagents (right) [34].

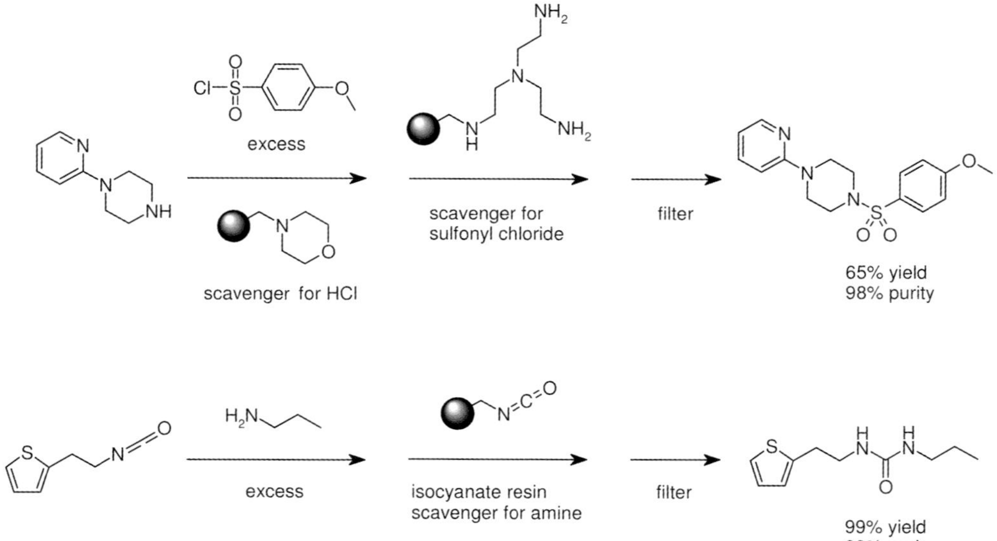

Figure 9.10 Current Applications of scavenger techniques [75, 86].

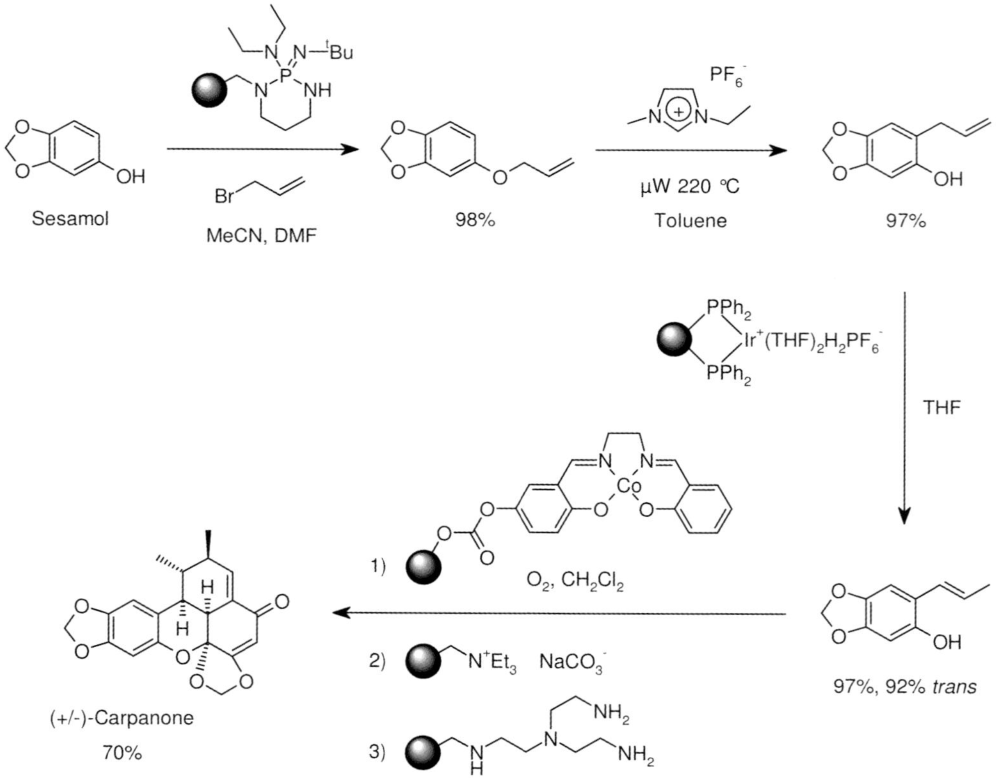

Figure 9.11 Polymer-assisted solution-phase synthesis applied to the synthesis of the natural product carpanone [88].

Frequently used scavenger resins contain nucleophilic groups such as amines [75, 85], hydrazides [76–78] and pyridines [79, 80] or various electrophilic functionalities such as isocyanates [75, 81, 82, 85], aldehydes [83] and isatoic anhydride [84]. Furthermore, ion-exchange resins have been broadly applied for sequestering. One of the first examples of the use of scavenger resins for library production was reported by Kaldor et al. in 1996 [85]. In this work, both solid-supported nucleophiles and electrophiles were employed to facilitate the work-up and purification procedure of different amine alkylations and acylations. Examples of some early applications of scavenger resins are shown in Fig. 9.10 [75, 86].

The use of solid-supported reagents is another intriguing area of polymer-assisted solution-phase synthesis that is becoming increasingly important [87]. With this approach, the reagents employed are bound to the polymer and, again, the work-up and purification procedure is reduced to a simple filtration. An impressive example of the multiple use of polymer-supported reagents and scavengers was given by Baxendale et al. [88], whereby the natural product carpanone was synthesized without the need for conventional purification techniques (Fig. 9.11).

9.3.2.2 Extraction Techniques for Purification

Liquid–liquid extraction is a well-known purification method and is carried out during the work-up procedure after almost every traditional organic chemical reaction. The principle behind this method is to partition substances between two liquid phases that are non-miscible. Usually, water and an organic solvent are used. Depending on their individual partition coefficients, the substances are enriched in either of the two phases and, therefore, they can be isolated via phase separation. Boger and coworkers introduced this purification principle for the generation of combinatorial libraries [89–91]. In this work, an acid–base extraction was applied for the isolation of amines. Today, a wide variety of liquid–liquid separation techniques have found application in solution-phase combinatorial synthesis. Examples include liquid-handling robots with and without liquid interface-detecting capabilities. Typically, these robots are used both for work-up and in the synthesis procedure (Fig. 9.12). Hence the entire process can be accomplished completely unattended. An interesting alternative for automated phase separation is the so-called "lollipop" technique [92]. The technique takes advantage of rapid freezing of the aqueous phase. Utilizing some PEEK (polyether ether ketone) pins in a microtiter plate format, the aqueous phase freezes to the pins, allowing for removal from the organic phase (Fig. 9.12).

However, liquid–liquid extraction fails when emulsions are formed or when impurities have the same solubility properties as the library members. Another approach in liquid–liquid extraction is the fluorous phase technique, pioneered by Horvath and Rabai [94]. This technique relies on the finding that perfluorohydrocarbons such as perfluorohexane are insoluble in most organic solvents and in water. The basic strategy is to attach the compounds or the reagents to a fluorous tag that consists of a large perfluorinated moiety. This makes the molecules soluble

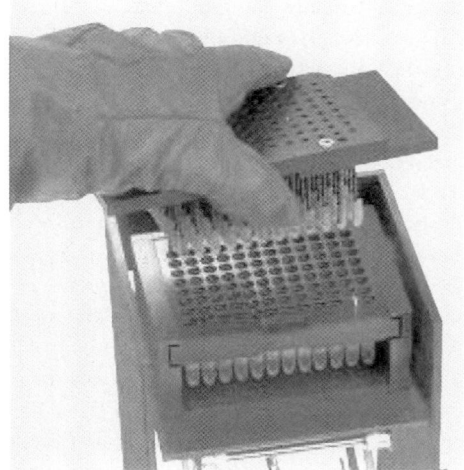

Figure 9.12 Liquid handling robot used for synthesis and work-up (left). Lollipop phase separation in a 96-well microtiter plate format (right). Reprinted with the permission of Zinsser Analytic GmbH, Frankfurt [93].

in fluorous solvents, such that purification can then be performed by a fluorous–organic or fluorous–aqueous extraction, whereby only the compounds containing a fluorous tag remain in the fluorous phase. Some recent reviews describe the applications of fluorous technologies in combinatorial synthesis [95].

Solid-phase extraction is a further purification method that is frequently used in combinatorial solution-phase synthesis. By this approach, the compounds to be extracted are partitioned between a solid and a liquid phase. Suitable adsorbents include various sorts of silica- and alumina-based gels, fluorinated gels and ion-exchange resins. In general, solid-phase extraction is performed in a stepwise

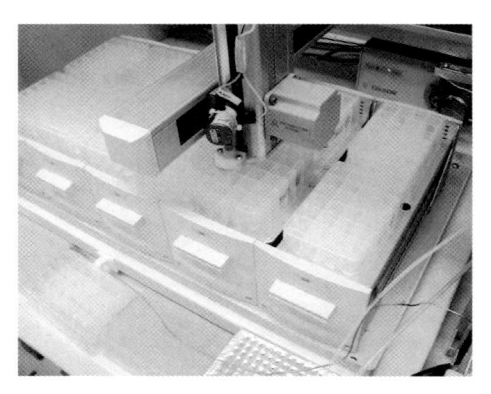

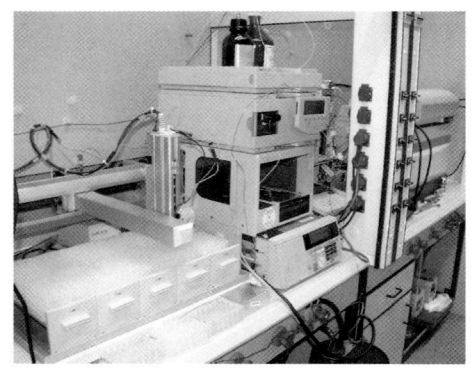

Figure 9.13 Example of an MS-directed automated preparative HPLC system.

manner. Initially, the crude reaction mixture is loaded on to the solid adsorbent. Then, undesired components are washed away with a solvent in which only the impurities are soluble, leaving the desired product adsorbed. Finally, the purified compound is eluted into a collection vessel by using another solvent. When applied to parallel synthesis, solid-phase extraction is usually performed using arrays of extraction cartridges or extraction plates with a microtiter plate footprint [96].

9.3.2.3 Purification by Chromatography

Recent years have seen a clear trend towards the use of high-quality libraries for screening purposes and much more emphasis has been given to purity assessment and the development of high-throughput chromatographic purification methods. These efforts have led to the rapid evolution of automated preparative liquid chromatography, including dramatic improvements in throughput and analytical capabilities. Amongst the plethora of available chromatographic purification techniques, MS-directed purification with fully automated preparative HPLC has emerged as the method of choice for library purification. Several reports describe the set-up and application of typical purification systems as well as their impact on the purity of combinatorial libraries (Fig. 9.13) [97–99].

9.3.3
Library Formats

Depending on their origin, combinatorial libraries differ significantly in terms of size (number of compounds), amount, purity and structural complexity and can be categorized into the following three basic groups: (1) one-bead one-compound libraries resulting from the pool/split protocol on loose resin beads, (2) pre-encoded libraries prepared by RF or optical tagging methods and (3) spatially addressable libraries obtained from parallel solution- or solid-phase synthesis approaches (Table 9.1) [100].

Table 9.1 Library formats resulting form the different synthesis approaches.

Preparation method	Library size	Amount of sample	Comments
(1) One-bead, one-compound libraries			
Pool/split solid-phase synthesis with loose resin beads	10 000–2 000 000	0.1–1 nmol (40–400 ng)	Best for very large libraries, deconvolution necessary, purity is assessed only for selected members and depends strongly on reliability of chemistry, usually libraries are arranged as mixtures

Preparation method	Library size	Amount of sample	Comments
(2) Pre-encoded libraries			
Pool/split solid-phase synthesis with labeled nano- or micro-reactors	1000–100 000	5–100 µmol (2–40 mg)	Most productive for discrete compounds in mg quantities, RF or optical labeling of different sized reactors, flexible chemistry because reactions are performed in standard laboratory glassware, purity of smaller libraries assessed prior to screening[a]
(3) Spatially addressable libraries			
Parallel solid-phase synthesis in arrayed reaction vessels	100–20 000	10–100 µmol (4–40 mg)	Variety of reaction devices range from microtiter plate-based reaction blocks to fully automated workstations, purity assessed prior to screening[a]
Parallel solution-phase synthesis in arrayed reaction vessels	100–5000	10–250 µmol (4–100 mg)	Reaction devices range from microtiter plates to fully automated workstations with sophisticated glassware reaction vessels, purity assessed prior to screening, samples often purified by chromatography[b]

a Typically, samples with a purity lower than 60–80 % are flagged or discarded.
b Samples separated by high-throughput HPLC typically have a purity higher than 90 %.

9.3.3.1 One-bead One-compound Libraries

The pool/split (split-and-mix) approach on loose resin beads is still the best method for the preparation of very large libraries. Some impressive examples include a natural product-like library of polycyclic compounds with approximately 2.18 million members reported by Schreiber and coworkers [101, 102], a library of 27 000 vancomycin analogs [103] and a triazine library consisting of more than 46 600 compounds [104]. Typically, the one-bead, one-compound libraries are plated and screened as mixtures. This is mainly due to the nature of the pool/split process and some practical considerations. To ensure that statistically each individual library member is formed during the pool/split process, the number of beads used for synthesis needs to be significantly higher than the intended number of library members to be synthesized. With a threefold redundancy of beads per library member, about 95 % of the theoretical compounds are present in the library [105]; typically, redundancies in the range 200–300 are used for library production to ensure that all members are synthesized [106]. However, plating of each individual bead into one well to release discrete compounds is often impractical, especially in the case of very large libraries, because 200–300 copies of the library have to be plated. Therefore, the libraries are frequently arranged as mixtures (typically each resin aliquot from the last reaction step is portioned into one individual well in which the mixture is released from the resin). However, plating and screening of single compounds of libraries with a lower redundancy (e.g. of three) have been reported [107]. Although the pool/split process is the most productive approach for the synthesis of very large libraries, it has several drawbacks. One is the lack of analytical control during synthesis, restricting the chemistry to highly reliable chemical transformations. Another limitation is the inherent need for structure elucidation of active compounds after biological screening. This additional deconvolution step is time consuming, i.e. when mixtures are screened and the rate of false positives is high. Finally, a further restriction is the small quantity of the compounds (0.1–1 nmol). This allows for only a few experiments per sample and invariably requires the resynthesis of every active compound in larger amounts for follow-up screening and further analytical and biological profiling. Overall, these drawbacks frequently outbalance the advantages of the pool/split productivity. Hence this approach is mostly used for specific screening applications, e.g. on-bead affinity screening [104, 108], and is rarely applied in an industrial high-throughput screening environment.

9.3.3.2 Pre-encoded Libraries

With the advent of pre-encoding technologies such as RF labeling and optical encoding, as described above, major drawbacks of the pool/split (mix-and-split) procedure have been solved while maintaining its productivity. Primary advantages include the higher quantity of the products (up to several milligrams) and the fact that discrete compounds with known identities are obtained from the process. A further improvement is that the statistical nature of the pool/split process is replaced by directed sorting of the reactors. Hence the method allows one to work with a redundancy of one, avoiding the synthesis of additional library replicates. In addition, quality control during the synthesis can be performed by guiding through some extra reactors for analytical purposes. When RF labeling is applied in combination with automated sorting, libraries of 10 000 members are easily created, providing discrete compounds in the range 25–100 µmol (ca. 25–40 mg) [109]. The optical encoding further extends the pre-encoded method to libraries of up to 100 000 compounds in quantities of about 5 µmol (ca. 2 mg) [28]. Typically, in the case of smaller libraries (below 10 000 members), the purity of the compounds is assessed prior to screening, e.g. by LC–MS, and compounds with a purity below a certain cutoff (60–80%) are flagged or discarded. In summary, the pre-encoding technology is highly productive, affording very large libraries of discrete compounds in sufficient amounts for characterization, screening and archiving. The quality of the libraries in terms of purity again depends strongly on the reliability of the chemistry protocols; however, it can be improved when purity assessment is applied. Hence this method is increasingly used to expand the compound collections for high-throughput screening purposes. However, the technology suffers to some extent from the cost of automation devices and consumables and the fact that it is limited to solid-phase synthesis.

9.3.3.3 Spatially Addressable Libraries

This category covers all the libraries that stem from parallel synthesis approaches performed in arrays of test-tubes, microtiter plates or reactors either on a solid support or in solution phase. Because of the wide variety of reaction devices and associated synthesis methods used for the production of spatially addressable libraries, their properties such as size, quantity, purity and complexity can differ widely. However, a certain classification can be made regarding the size of the library that governs its properties. The production of larger libraries, for example with several thousand members, requires extensive parallelization and miniaturization of the reaction devices [110]. Hence the quantities of the desired products are limited and sacrifices must often be made regarding the chemical reactions (e.g. restrictions in temperature control, insufficient prevention of moisture or absence of an inert gas atmosphere). This can adversely affect the chemical complexity and purity of the library. However, various miniaturized reaction devices have been successfully used for library production, predominantly applied as reaction blocks in a 96-well microtiter plate footprint [111–115]. When solid-phase synthesis is performed, these reaction blocks allow for the synthesis of libraries of up to 20 000 members in amounts of 10–100 µmol [110]. In the case of solution phase, the libraries resulting from this approach are frequently smaller with a maximum of 5000 members. This is because solution-phase combinatorial synthesis typically relies on one to two diversification steps, whereas solid-phase synthesis approaches are usually based on three to four point variations. Hence the combinatorial matrix is smaller for solution-phase chemistry. The quantities of such solution-phase libraries range between 10 and 250 µmol. Usually, the purity of both solid-phase and solution-phase libraries is assessed and compounds below a certain cutoff are flagged or discarded. Alternatively, in the case of solution-phase

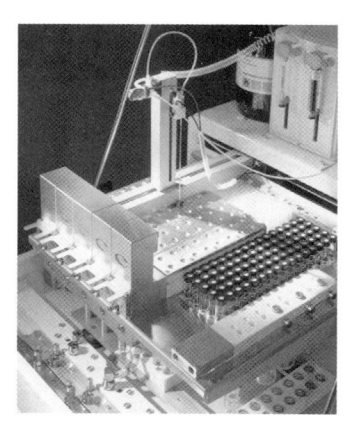

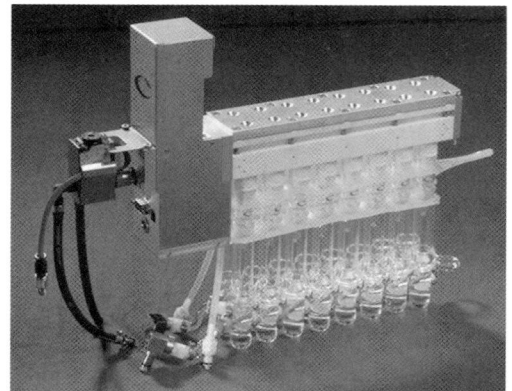

Figure 9.14 Fully automated workstation with glass reactors allowing the operation of parallel reactions in a temperature range from −70 to 150 °C under an inert gas atmosphere. Reprinted with the permission of Chemspeed Technologies, Augst, Switzerland [118].

synthesis, the compounds are purified by high-throughput chromatography to provide products with a common purity higher than 90% [98]. In general, these types of spatially addressable libraries are commonly used in drug discovery campaigns and form the major part of the compound collections used in high-throughput screening. When more sophisticated reaction devices that allow for more chemical flexibility are used for library production, such as fully automated workstations (Fig. 9.14) [116, 117], the libraries tend to be smaller and quantities of the compounds are usually higher. Hence the libraries are predominantly used for lead optimization purposes and further biological profiling.

9.4
Library Design and Profiling Approaches

Over the past two decades, there has been a continuous increase in the number of drug candidates originating from the high-throughput screening of combinatorial libraries [119]. However, the effectiveness in advancing hit compounds from combinatorial libraries into drug candidates is still not as high as predicted in the early 1990 s [120–123]. Detailed retrospective analyses of high-throughput screening campaigns have revealed that the large diversity-driven libraries from the early days of combinatorial chemistry did not deliver progressable hit structures to the extent expected because of their limited drug-likeness and crucial physicochemical values such as molecular weight and lipophilicity [124, 125]. Hence a wide array of guiding principles and computational methods for the design of combinatorial libraries have been developed to improve success rates. Library design guided by the biological target structure has become a smarter, more efficient way to obtain drug-like libraries with an increased output of progressable hit compounds [126, 127]. Additionally, consideration of drug-like physicochemical and ADMET properties early in library design are further concepts to produce good initial lead compounds [128–130]. Excellent overviews of the wide array of computational tools for library design and profiling are provided in Chapter 8 and in several review articles [131–136].

9.5
Impact of Combinatorial Libraries on Drug Discovery

During the period 1991–2003, almost 2500 combinatorial libraries were described in the literature, as compiled in several excellent reviews published by Dolle [137–144]. Based on the literature citations in these comprehensive compilations, it is apparent that the number of libraries being published each year shows no sign of decline (Fig. 9.15). In order to assess the impact of combinatorial libraries on drug discovery, however, one must consider the various libraries that have been described in conjunction with biological data, which is in excess of 600 libraries [137–144]. Certainly, a comprehensive review of these libraries is beyond the scope of

9 Combinatorial Chemistry and High-throughput Screening

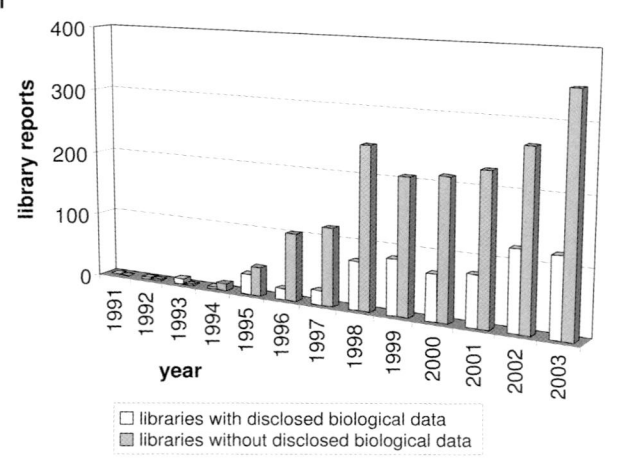

Figure 9.15 Published descriptions of combinatorial libraries, based on Dolle's review articles [137–144].

this chapter. In addition to Dolle's reviews, several articles have been published in recent years that discuss the utility of combinatorial libraries for drug discovery [145–153]. In this chapter, we will present several representative examples where combinatorial libraries facilitated lead identification and optimization and also cases where combinatorial libraries have had a significant impact on the generation of clinical candidates.

9.5.1
Lead Identification

As mentioned in Section 9.2, new lead identification efforts usually involve high-throughput screening of corporate compound collections, which may be considered as "unbiased libraries". Such compound collections can often consist of 250 000–500 000 compounds or more and with the advent of combinatorial chemistry some corporate collections have now reached 1–2 million compounds [145, 146, 154]. It has been pointed out that journal articles in recent years routinely describe the identification of a lead compound from high-throughput screening of a corporate compound collection [145, 146]. Combinatorial libraries designed for lead identification, commonly incorporating a "privileged structure" or based on molecular modeling, may be screened as a distinct effort or as part of the corporate collection screen. An excellent example of lead identification from an unbiased combinatorial library consisting of compound mixtures was described by Willoughby et al. at Merck [155]. In this work, a library of 128 000 2-arylindoles (**1**) were prepared by pool/split solid-phase synthesis, isolated as 320 pools (compound mixtures) consisting of 400 compounds per pool. Screening against 16 G-protein coupled receptor targets revealed that some pools were both active and selective for certain receptors. Deconvolution by resynthesis and testing of sub-mixtures and

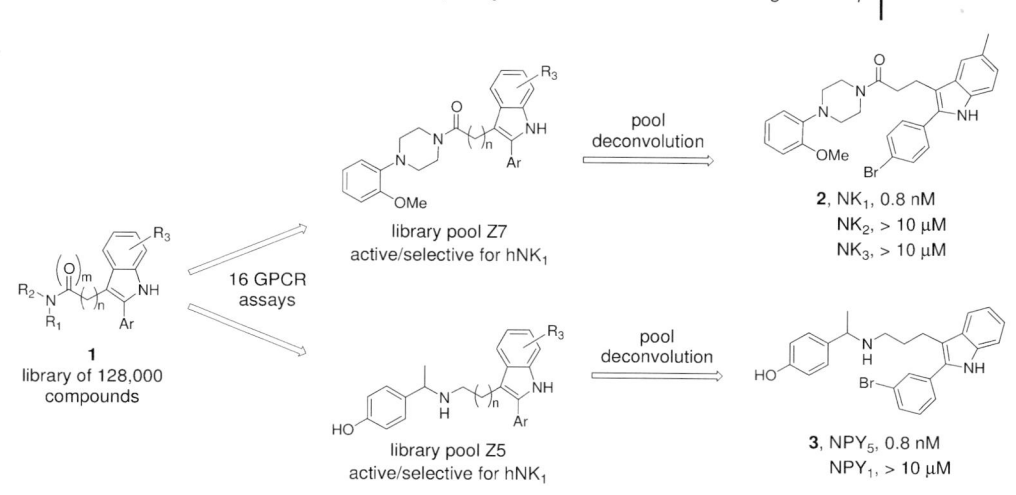

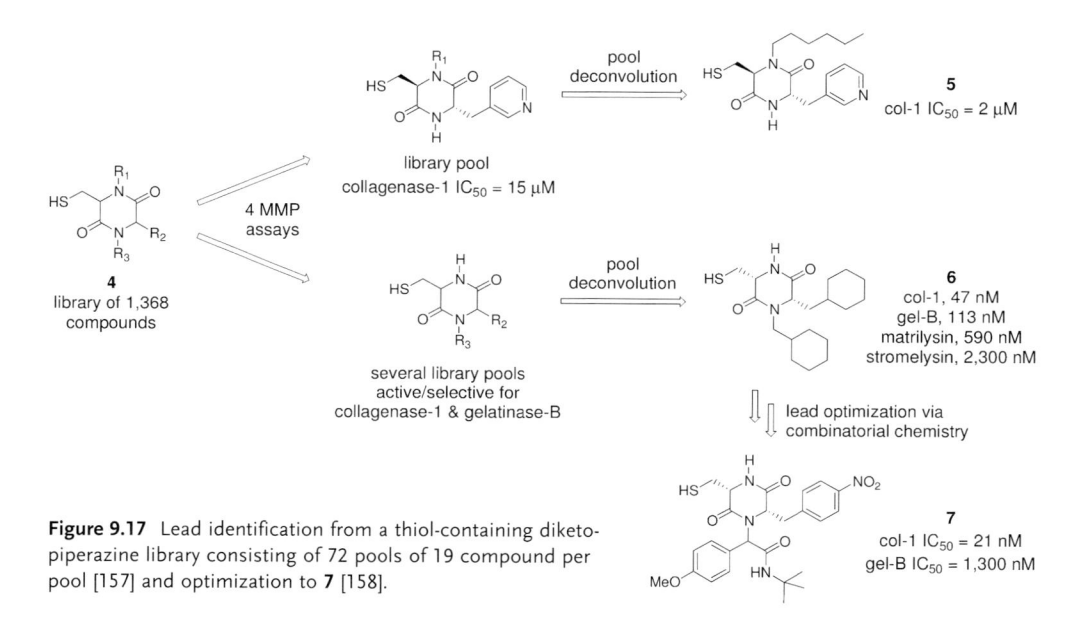

Figure 9.16 Lead identification from a 2-arylindole library consisting of 320 pools of 400 compounds per pool [155].

individual compounds led to specific potent compounds, such as **2** and **3** (Fig. 9.16). Compound **2**, which showed excellent potency for hNK$_1$ and high selectivity against other NK receptors, has served as a viable lead for a medicinal chemistry program [156].

From modeling studies with various crystal structures of matrix metalloproteases (MMPs), Szardenings and colleagues at the Affymax Research Institute considered that a library based on a 2,5-diketopiperazine (DKP) scaffold and

Figure 9.17 Lead identification from a thiol-containing diketopiperazine library consisting of 72 pools of 19 compound per pool [157] and optimization to **7** [158].

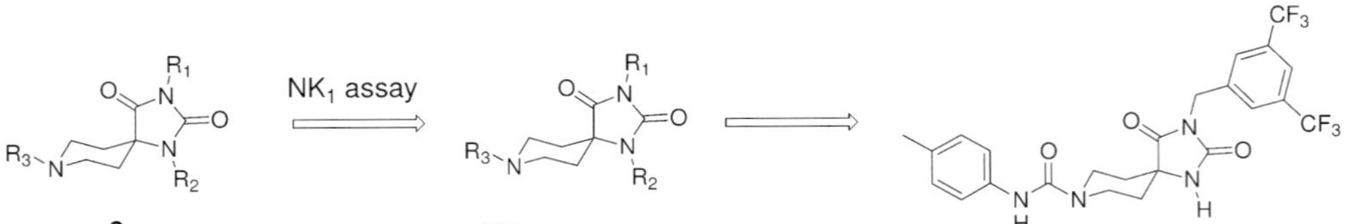

Figure 9.18 Lead identification from a spirohydantoin library consisting of 136 discrete compounds [161, 162].

incorporating a cysteine group as the MMP zinc ligand might afford new MMP inhibitor leads [157]. Two libraries were prepared by pool/split solid-phase synthesis, each consisting of 684 members as 36 pools of 19 compounds per pool (Fig. 9.17). Deconvolution of the most active pools afforded potent inhibitors (IC_{50} < 100 nM) of collagenase-1 and gelatinase-B, e.g. **6**, showing IC_{50} = 47 nM against collagenase-1. In further lead optimization efforts, additional libraries were investigated to generate diketopiperazines such as **7**, having 60-fold selectivity for collagenase-1 over gelatinase-B [158]. A review of the use of combinatorial libraries to identify inhibitors of various proteinases and other enzymes has been published by Batra et al. [159].

Although the pool/split solid-phase combinatorial synthesis technique has the power to produce very large compound libraries and has afforded numerous lead discoveries, the quality control of such libraries has been a concern for many years. To address this issue, for example, scientists at Pharmacopeia developed a statistical sampling protocol to assess the quality of large encoded libraries [160]. For a library of 25 500 statin amides, analysis of 1900 randomly retrieved resin beads revealed an overall 85 % positive confirmation, suggesting that only about 21 400 of the library compounds were actually synthesized. The trend in the major pharmaceutical industry during the past 10 years has been a move in favor of smaller libraries produced by parallel synthesis. In this case, the discrete compounds produced are obtained in milligram amounts and characterized by traditional analytical methods such as LC–MS and NMR. As an example, parallel solution-phase and solid-phase synthesis methodologies were applied at Hoffmann-La Roche to generate an array of compounds containing both the privileged structure

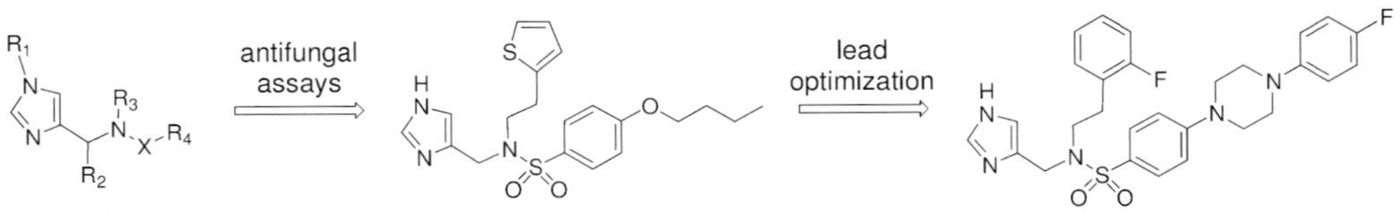

Figure 9.19 Lead identification and optimization from imidazole libraries to identify novel antifungal agents [163, 164].

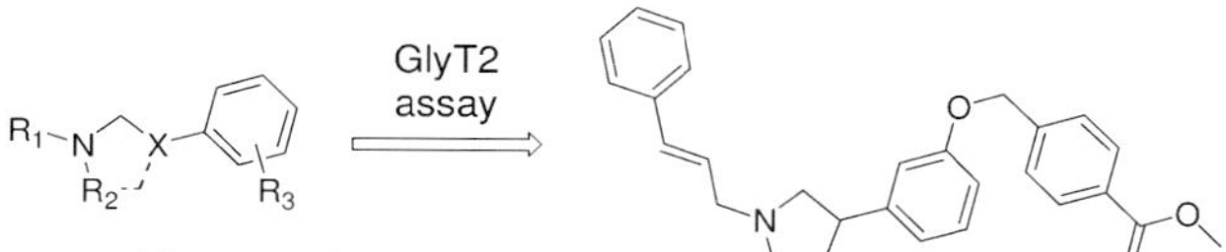

13
library of 3,042
compounds

14, IC_{50} = 1.3 μM

Figure 9.20 Lead identification from a library designed for CNS bioavailability [166].

spirohydantoin as a template and the 3,5-bis(trifluoromethyl)phenyl motif as a neurokinin receptor "needle" [161, 162]. From a library of 136 such compounds, 97 were found to have NK_1 K_i values of <10 μM, e.g. compound **9** (Fig. 9.18).

Saha and colleagues at Janssen Research prepared a library of about 1000 4-substitued imidazoles, designed on the basis of the azole phenethylamine pharmacophore found in many antifungal agents [163, 164]. The discovery library **10** was prepared by solid-phase parallel synthesis and potent leads such as **11** were identified (Fig. 9.19). Further optimization, involving the preparation of an additional combinatorial library, afforded the compound **12**, which was active against a range of fungi. Recently, a review of the identification of antifungal agents using combinatorial chemistry approaches has been published by Gupta et al. [165].

A discovery library designed with a bias for compounds having physiochemical properties favoring central nervous system (CNS) permeability was generated at Organon Laboratories [166]. The synthesis was conducted by solid-phase methods using Syro II reactors or IRORI MicroKans, to produce 3042 tertiary amines as discrete compounds. Screening against the glycine transporter-2 (GlyT2) provided the lead arylamine **14** (Fig. 9.20).

Based on molecular modeling studies with the X-ray crystal structure of raloxifene bound to estrogen receptor-α and -β (ER-α and -β), researchers at Bayer designed a 2-aroylbenzofuran library for the identification of new estrogen receptor modulators [167]. Parallel solid-phase synthesis afforded a library of 320 compounds **15** (Fig. 9.21). Screening against ER-α and -β provided compound **16**,

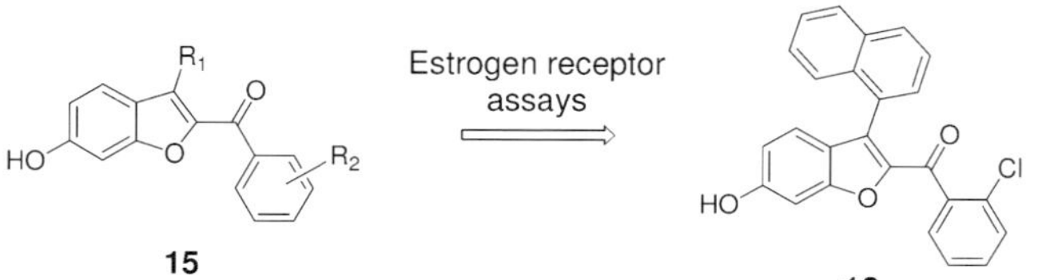

15
library of 320
compounds

16
ER-α IC_{50} = 30 nM
ER-β IC_{50} = 63 nM
bone pit IC_{50} = 530 nM

Figure 9.21 Lead identification from a library designed for estrogen receptor ligands [167].

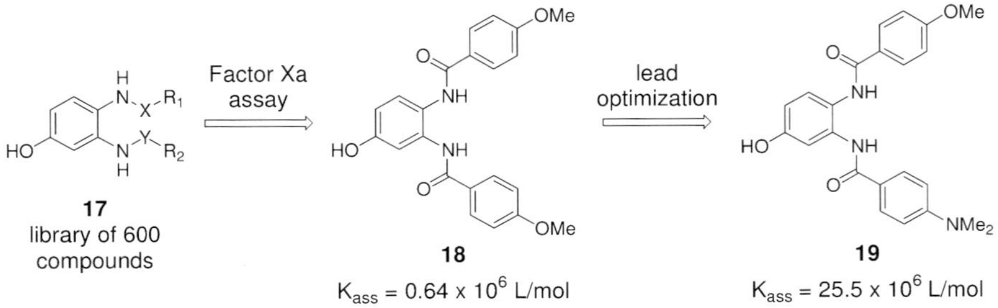

Figure 9.22 Lead identification from a library of phenolic compounds [168, 169].

which exhibited IC_{50} values of <100 nM for both receptors. This compound and related leads were also active in a bone pit assay, which serves as a model for bone resorption.

From the screening of a library of 600 phenolic compounds **17** prepared by parallel solid-phase synthesis at Sphinx/Eli Lilly [168], the simple 1,2-dibenzamidobenzene **18** was found as a lead for human factor Xa (fXa), with apparent K_{ass} = 0.64 × 10^6 L mol^{-1} [169] (Fig. 9.22). This compound was validated as a lead, as further optimization efforts provided compound **19**, exhibiting a 40-fold increase in fXa activity and 100-fold selectivity against human thrombin [169].

Two combinatorial libraries incorporating a biphenyl privileged structure core were prepared by solid-phase synthesis methods at Bayer [170, 171] and both of these libraries provided lead structures as vitronectin receptor (v3) antagonists (Fig. 9.23). The first library (**20**) afforded mostly inactive compounds, but a set of urea derivatives such as **21** and **22** proved to be fairly potent [170]. A second library (**23**) generated from *p*-bromophenylalanine also afforded potent vitronectin antagonists, which could be optimized further with the aid of molecular modeling to afford **26**.

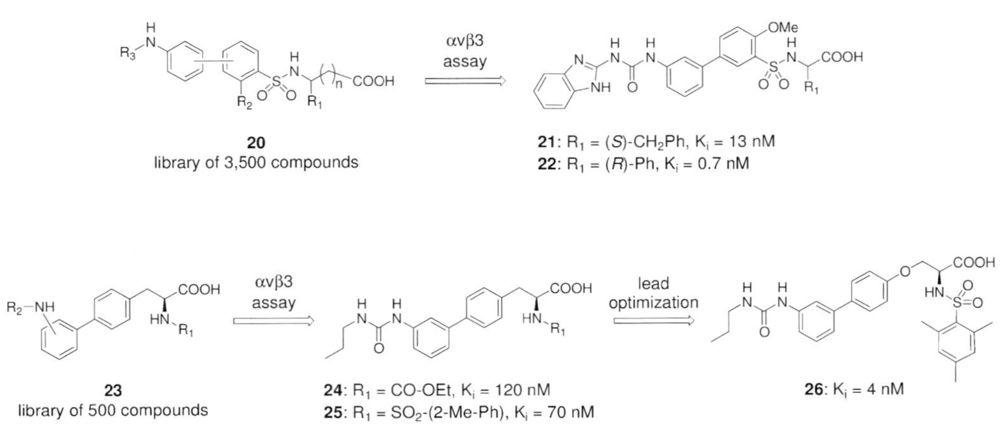

Figure 9.23 Lead identification from two combinatorial biphenyl libraries [170, 171].

The examples described above clearly illustrate that compound libraries, both large and small, can effectively provide new leads for a variety of molecular targets. For numerous additional examples, the reader is referred to the compilations published by Dolle [137–142, 144].

9.5.2
Lead Optimization

As described in the examples above, lead identification libraries can range in size and consistency from hundreds of structures existing as discrete compounds to several tens of thousands of structures existing as compound pools. In contrast, combinatorial libraries designed for lead optimization are generally prepared as discrete compounds, with compound numbers ranging from just tens to 1 000–2 000. Indeed, during the past decade, the preparation of lead optimization libraries has increasingly been carried out by "traditional" medicinal chemists, by both solution-phase and solid-phase methods, rather than by specialized combinatorial chemists.

As a first example of successful lead optimization via focused combinatorial libraries, researchers at Aventis carried out parallel solution-phase synthesis based on a bis-arylimidazole lead **27** that had been identified to inhibit TNFα release from human monocytes (Fig. 9.24). This effort generated seven directed libraries and ultimately produced the "development candidate" RPR-200765 [172] as a potent p38 MAP kinase inhibitor. This compound displayed good oral anti-arthritic efficacy in a rat model. Further optimization was carried out by investigation of an algorithmically guided combinatorial library prepared by solid-phase synthesis, affording the very potent pyrimidine analog AVE-8677 (RPR-238677) [173]. This compound was described as a "potential drug candidate" and may be in clinical studies for the treatment of rheumatoid arthritis [174, 175].

Researchers at Bayer carried out lead optimization in an erythropoietin (EPO) program by using various combinatorial synthesis techniques (automated solid-phase synthesizer, RF tagging) [176]. High-throughput screening of ~650 000 compounds provided the initial lead **28**, which displayed $EC_{50} = 230$ nM in a

27 (RPR132331)
lead compound
TNFα EC_{50} = 800 nM

combinatorial chemistry lead optimization
7 focused libraries, 190 compounds

RPR-200765
TNFα EC_{50} < 200 nM
p38 IC_{50} = 50-110 nM
"development candidate"
(rheumatoid arthritis)

combinatorial chemistry lead optimization
>570 compounds

AVE-8677 (RPR-238677)
p38 IC_{50} = 4 nM
"potential drug candidate"
(rheumatoid arthritis)

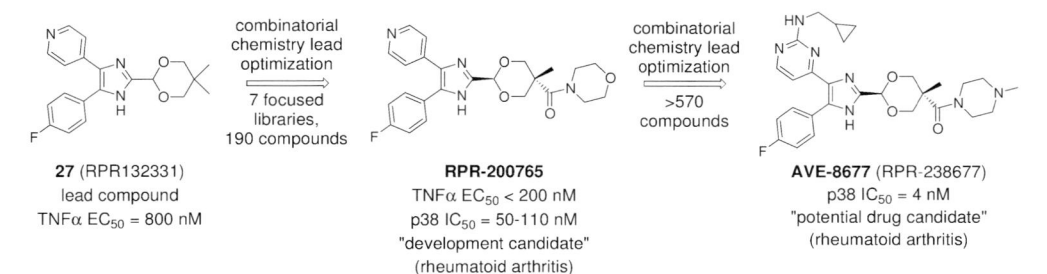

Figure 9.24 Lead identification, optimization and generation of a TNFα/p38 inhibitor for the treatment of rheumatoid arthritis through the aid of combinatorial chemistry [172–175].

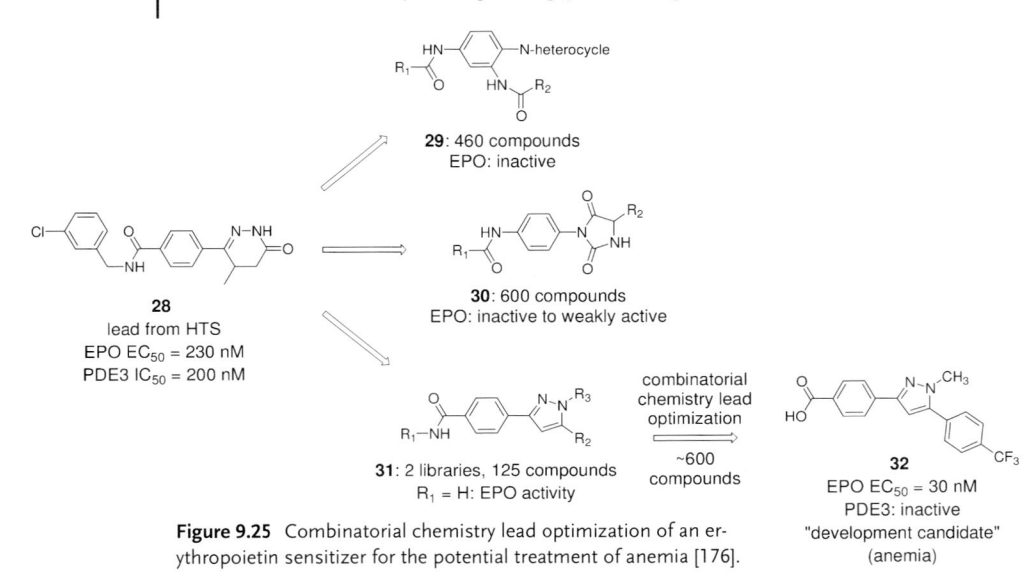

Figure 9.25 Combinatorial chemistry lead optimization of an erythropoietin sensitizer for the potential treatment of anemia [176].

human EPO assay, but also showed comparable phosphodiesterase-3 activity (PDE3, IC_{50} = 200 nM) (Fig. 9.25). An initial library effort designed to screen a broad variety of heterocyclic head groups provided no active compounds, possibly

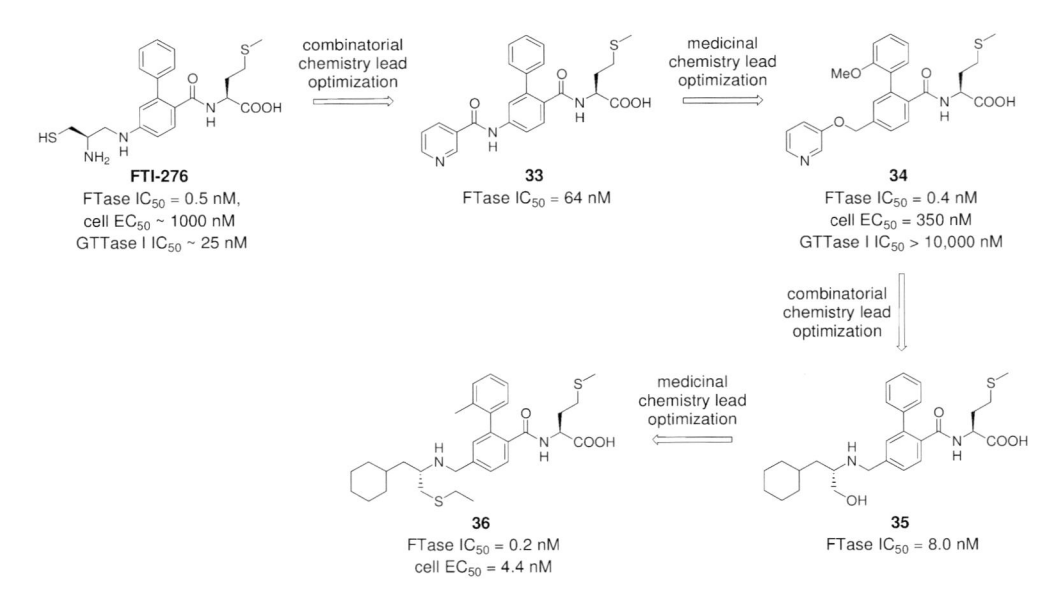

Figure 9.26 Optimization of a farnesyltransferase inhibitor lead by combinatorial and medicinal chemistry approaches, towards a potential treatment for cancer [177, 178].

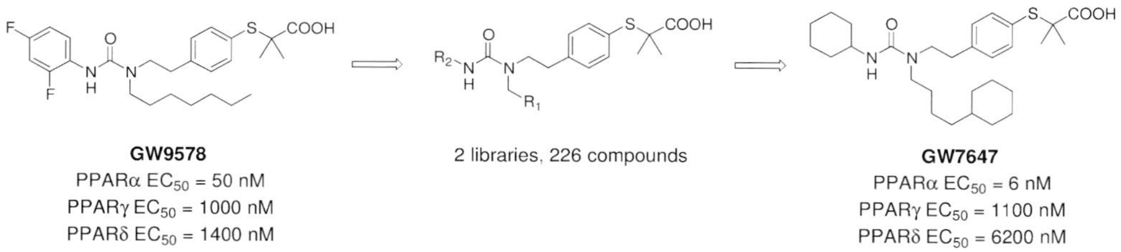

GW9578
PPARα EC$_{50}$ = 50 nM
PPARγ EC$_{50}$ = 1000 nM
PPARδ EC$_{50}$ = 1400 nM

2 libraries, 226 compounds

GW7647
PPARα EC$_{50}$ = 6 nM
PPARγ EC$_{50}$ = 1100 nM
PPARδ EC$_{50}$ = 6200 nM

Figure 9.27 Optimization of a PPARδ agonist by combinatorial chemistry, towards a potential treatment for dyslipidemia [179].

owing to the presence of an additional substituent on the central core (**29**). Subsequent libraries focused on specific phenyl–heterocycle scaffolds and whereas a phenyl–hydantoin library (**30**) gave some weak actives, a phenyl–pyrazole library (**31**) provided an important breakthrough in the project. Optimization by several follow-up libraries ultimately afforded the potent and selective compound **32**, which was selected for preclinical development as an EPO sensitizer for the potential treatment of anemia [176].

At Abbott Laboratories, the very potent peptidomimetic protein farnesyltransferase (FTase) inhibitor FTI-276 was optimized via combinatorial solid-phase synthesis to the non-cysteine-containing compound **33** (Fig. 9.26) [177]. Then, more traditional medicinal chemistry efforts were applied to afford the very potent compound **34**, which was also highly selective against GGTase I [177]. Subsequent optimization efforts by combinatorial solid-phase synthesis gave rise to **35**, which was improved further by medicinal chemistry work to provide **36**. This analog exhibits good potency in both enzyme and cellular assays, and also in inhibits tumor growth in a MiaPaCa-2 tumor xenograft mouse model [178].

In order to identify a PPARδ agonist with improved selectivity and physical properties, scientists at GlaxoSmithKline prepared two libraries totaling 226 compounds by parallel solid-phase synthesis (Fig. 9.27) [179]. From this effort, compound GW7647 was identified as a potent human PPARδ agonist with ~200-fold selectivity over the other receptor subtypes, which also exhibited *in vivo* lipid-lowering activity [179].

In an effort to identify human immunodeficiency virus protease (HIVP) inhibitors with superior properties to the AIDS drug indinavir, researchers at Merck prepared two libraries by a pool/split solid-phase synthesis approach (Fig. 9.28) [180, 181]. In the first example, the two terminal groups were varied to produce a 902-member library as 41 pools each containing 22 compounds. By resin archive and iterative deconvolution, potent compounds such as **37** were identified [180]. In a second library, the two end groups and a central fragment were varied, to afford a 300-member library. Testing followed by deconvolution provided compound **38**, which was substantially more potent than indinavir against both wild-type and A-44 mutant HIV protease [181].

The results described above represent just a few of the vast number of examples reported in the literature, whereby combinatorial synthesis was applied in con-

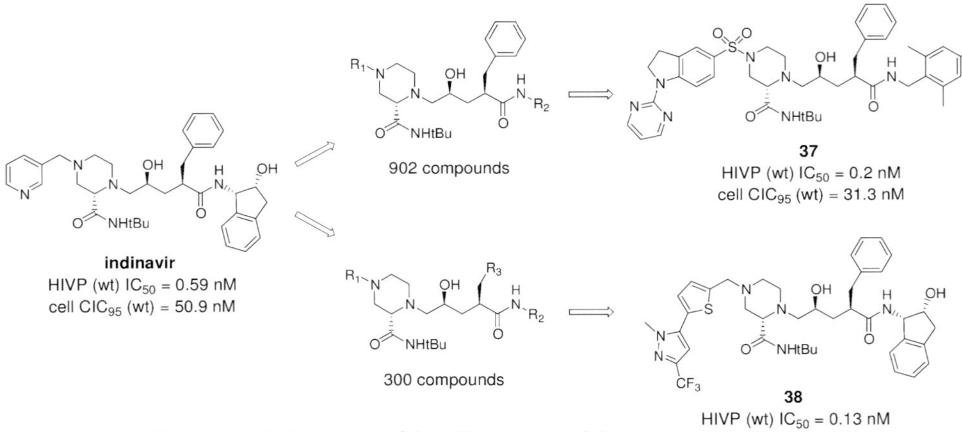

Figure 9.28 Optimization of the HIV protease inhibitor indinavir pool/split combinatorial chemistry, towards an improved treatment for AIDS [180, 181].

junction with high-throughput screening to expedite lead optimization. For numerous additional examples, the reader is referred to other review articles [137–142, 144–146].

9.5.3
Clinical Drug Candidates

As described above, the literature provides clear evidence that combinatorial chemistry approaches have been successfully applied to provide novel leads for a variety of drug targets, and also to permit efficient and effective optimization of lead structures. However, the impact of combinatorial chemistry on the identification of new drugs is less clear, although it is apparent that the early predications of the technology [12–15] were considerably overstated. Certain articles have cited several examples where combinatorial chemistry has played a role in the development of clinical candidates [145–147, 153], but it is difficult to substantiate many of

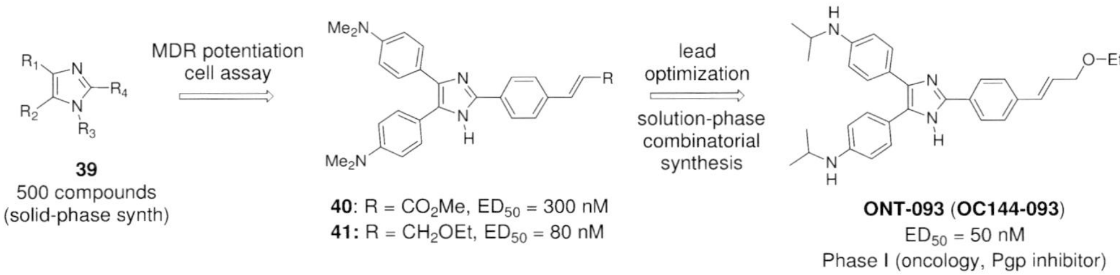

Figure 9.29 Lead identification, optimization and generation of a Pgp-modulator clinical candidate via combinatorial libraries of substituted imidazoles [182–185].

these examples with information in the public domain. Nevertheless, the cases that follow have clear documentation demonstrating that combinatorial libraries can contribute significantly to the development of clinical drug candidates.

Zhang et al. at Ontogen generated a lead discovery library of 500 substituted imidazoles **39** by solid-phase synthesis, with building blocks selected on the basis of the structural features of known P-glycoprotein (Pgp) modulators (Fig. 9.29) [182]. Several lead structures such as **40** and **41** were identified as modulators of Pgp in a multi-drug resistance (MDR) potentiation assay in the CEM/VLB1000 cell line. Follow-up solution-phase combinatorial libraries and traditional medicinal chemistry efforts resulted in the identification of ONT-093 (OC144–093) [183], which is currently in Phase I as a Pgp modulator for the potential reversal of MDR in patients undergoing cancer chemotherapy [184, 185].

In a collaboration of Bayer and Onyx Pharmaceuticals, high-throughput screening of the corporate compound collection identified the commercially-available 3-thienylurea **42** as a modest inhibitor of raf-1 kinase (IC_{50} = 17 μM) (Fig. 9.30) [186]. Extensive analoging provided **43**, having 10-fold improved potency; however, the series of compounds could not be improved further by traditional medicinal chemistry approaches. A combinatorial library of ca. 1000 ureas was therefore prepared, using a total of 75 isocyanates and ca. 300 amines, including some custom-prepared building blocks. Screening of this library provided the second-generation lead **44**, which was just slightly more potent than **43**, but was partic-

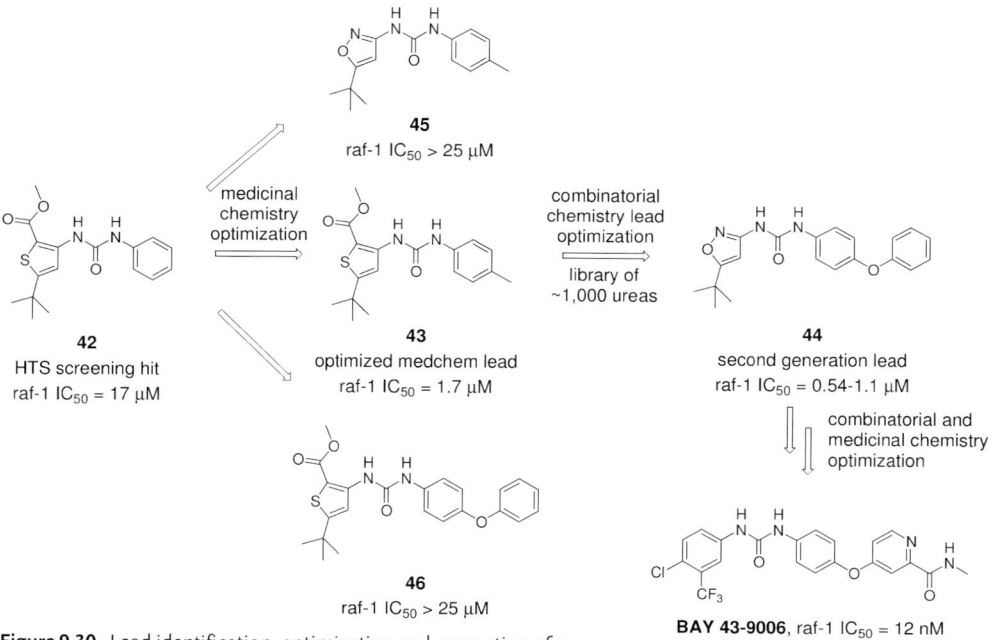

Figure 9.30 Lead identification, optimization and generation of a raf kinase clinical candidate impacted by combinatorial chemistry [186–189].

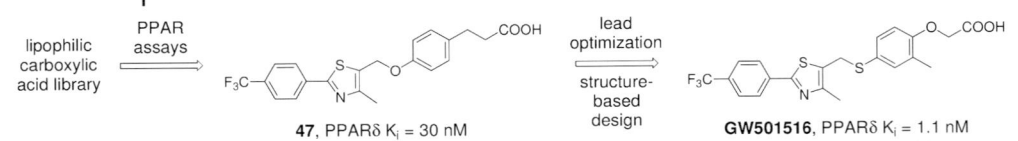

Figure 9.31 Lead identification, optimization and generation of a selective PPARδ agonist for the treatment of dyslipidemia through the aid of combinatorial chemistry [190–192].

ularly striking as it deviated from the SAR established through single-point modifications of **42** or **43** (cf. **45**, **46**). Hence the less biased "multiple-point modifications" involved in the combinatorial chemistry approach allowed the identification of the second-generation lead **44**, which could be optimized further by complementary medicinal and combinatorial chemistry efforts to ultimate produce BAY 43–9006 as a potent raf-1 kinase inhibitor [187]. BAY 43-9006 was subsequently characterized as a multi-kinase inhibitor, exhibiting activities against additional protein kinases such as VEGFR-2 and PDGFR [188]. BAY 43-9006 (sorafenib) is currently in Phase III trials for renal cell carcinoma and Phase II trials for various other cancer types [189].

Scientists at GlaxoSmithKline identified a peroxisome proliferator activating receptor-δ (PPARδ)-selective agonist with the aid of combinatorial chemistry (Fig. 9.31) [190]. Screening of combinatorial libraries of lipophilic carboxylic acids provided **47** (PPARδ K_i = 30 nM) and this lead was subsequently optimized with the aid of structure-based design to the highly potent and selective agonist GW501516 [191], which is currently in Phase II trials for the treatment of dyslipidemia [192].

Solid-phase combinatorial chemistry was utilized by scientists at Novartis to identify a dipeptidyl peptidase (DPP)-IV inhibitor lead **48**, toward the development of a treatment for Type-2 diabetes (Fig. 9.32) [193]. Optimization of the lead DDP-IV inhibitor led to DPP-728, a potent inhibitor of DPP-IV, which was also demonstrated to reduce plasma glucose levels in nonhuman primates after oral admin-

Figure 9.32 Lead identification, optimization and generation of a DPP-IV inhibitor for the treatment of Type-2 diabetes through the aid of combinatorial chemistry [193–197].

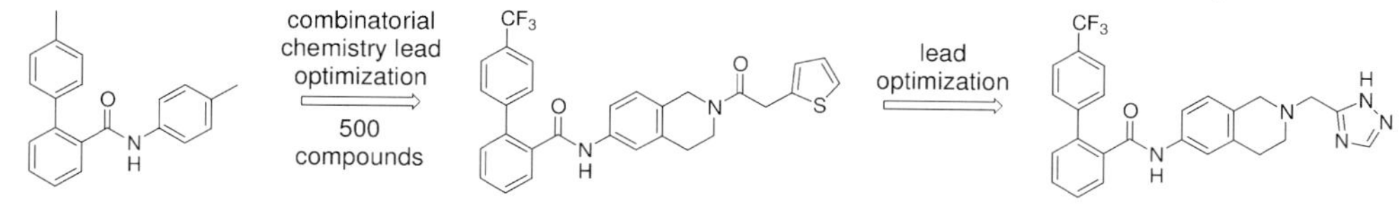

9.5 Impact of Combinatorial Libraries on Drug Discovery | 289

Figure 9.33 Lead optimization and identification of an MTP inhibitor for the treatment of dyslipidemia through the aid of combinatorial chemistry [198–200].

49 screening hit rat MTP IC_{50} = 250 nM

50 rat MTP IC_{50} = 7 nM

CP-346086 rat MTP IC_{50} = 2 nM Phase I (dyslipidemia)

istration. DPP-728 progressed to Phase II clinical trials [194], but has been discontinued in favor of the more potent and chemically stable analog LAF-237 (vildagliptin) [195], which is currently in Phase III clinical studies for Type-2 diabetes [196, 197].

Researchers at Pfizer generated 500 analogs of a lead microsomal triglyceride transfer protein (MTP) inhibitor **49** by robotics-assisted parallel synthesis to afford the more potent compound **50** (Fig. 9.33) [198]. Further optimization to improve oral exposure provided CP-346086 [199], which is believed to be in Phase I clinical trials as a potential therapeutic for dyslipidemia [200].

Also in the MTP inhibitor research area, scientists at Bristol-Myers Squibb optimized the screening hit **51** to produce the more potent analog **52** (Fig. 9.34) [198]. This lead was then further optimized with the aid of robotics-assisted parallel synthesis [201] to afford BMS-201038 as a potential treatment for dyslipidemia [202]. BMS-201038 progressed in clinical trials to Phase II, but has been discontinued due to the observation of adverse events in liver function [203].

Through structure-based design, scientists at R. W. Johnson generated **53** (RWL-50042) as a lead antagonist for the fibrinogen receptor (GPIIb/IIIa) (Fig. 9.35). To permit efficient lead optimization, combinatorial libraries were prepared by parallel solid-phase synthesis and the analog RWJ-53308 was ultimately iden-

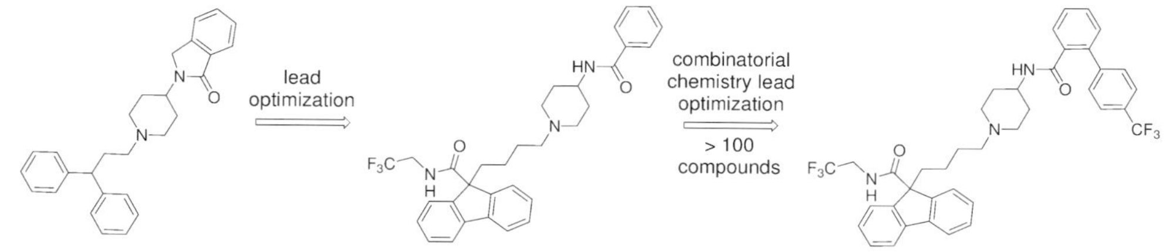

51 screening hit human MTP IC_{50} = 2,200 nM

52 human MTP IC_{50} = 23 nM

BMS-201038 human MTP IC_{50} = 0.5 nM Phase II - discontinued (dyslipidemia)

Figure 9.34 Lead optimization and identification of an MTP inhibitor for the treatment of dyslipidemia through the aid of combinatorial chemistry [198, 201–203].

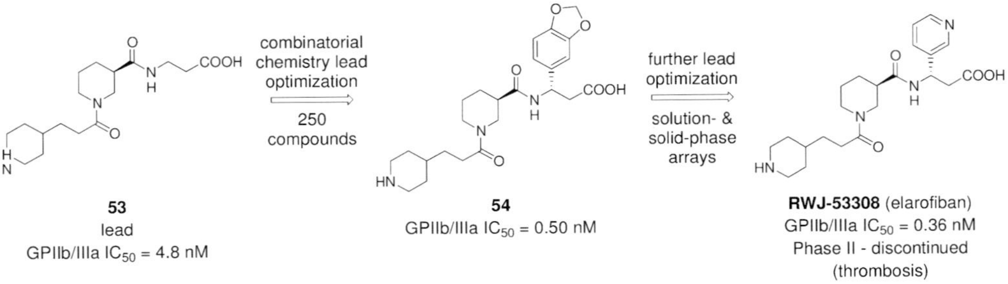

Figure 9.35 Lead optimization and identification of a GPIIb/IIIa antagonist for the treatment of thrombosis through the aid of combinatorial chemistry [204–206].

tified, with significantly improved potency, oral absorption and duration of action [204, 205]. RWJ-53308 progressed to Phase II clinical trials as a platelet aggregation inhibitor for the potential treatment of thrombosis; however, further development was suspended, as related GPIIb/IIIa antagonists in the clinic were discontinued owing to lack of efficacy and/or safety concerns [206].

9.6
Conclusion

Since the early 1990 s, a wide variety of combinatorial chemistry approaches have been investigated and applied, in conjunction with high-throughput screening techniques, to enhance drug development efforts. For new lead identification and lead optimization, the technology has achieved considerable success. Indeed, the technology has evolved and been refined to the extent that many "traditional" medicinal chemists now routinely employ combinatorial/parallel synthesis methods in their work. With respect to producing drug development candidates, the impact of combinatorial chemistry to date has been limited. However, with increased attention to biopharmaceutical properties and advances in high-throughput assays to address pharmacokinetics and toxicology, for example, the contributions of combinatorial chemistry to drug development are expected to increase.

References

1 Fodor, S. P. A., Read, J. L., Pirrung, M. C., Stryer, L., Lu, A. T., Solas, D. *Science* **1991**, *251* (4995), 767–773.
2 Brenner, S., Lerner, R. A. *Proc. Natl. Acad. Sci. USA* **1992**, *89*, 5381–5383.
3 Nielsen, J., Brenner, S., Janda, K. D. *J. Am. Chem. Soc.* **1993**, *115*, 9812–9813.
4 Ohlmeyer, M. H. J., Swanson, R. N., Dillard, L., Reader, J. C., Asouline, G., Kobayashi, R., Wigler, M., Still, W. C. *Proc. Natl. Acad. Sci. USA* **1993**, *90*, 10922–10926.
5 Bunin, B. A., Ellman, J. A. *J. Am. Chem. Soc.* **1992**, *114*, 10997–10998.

6 Bunin, B. A., Plunkett, M. J., Ellman, J. A. *Proc. Natl. Acad. Sci. USA* **1994**, *91*, 4708–4712.

7 DeWitt, S. H., Kiely, J. S., Stankovic, C. J., Schroeder, M. C., Cody, D. M. R., Pavia, M. R. *Proc. Natl. Acad. Sci. USA* **1993**, *90*, 6909–6913.

8 Furka, A. *Drug Dev. Res.* **1995**, *36*, 1–12.

9 Lebl, M. *J. Comb. Chem.* **1999**, *1*, 3–24.

10 Gallop, M. A., Barrett, R. W., Dower, W. J., Fodor, S. P. A., Gordon, E. M. *J. Med. Chem.* **1994**, *37*, 1233–1251.

11 Gordon, E. M., Barrett, R. W., Dower, W. J., Fodor, S. P. A., Gallop, M. A. *J. Med. Chem.* **1994**, *37*, 1385–1401.

12 Alper, J. *Science* **1994**, *264*, 1399–1401.

13 Service, R. F. *Science* **1996**, *272*, 1266–1268.

14 Rotman, D. *Chem. Week* **1995**, June 28, 16–17.

15 McNamee, D. *Lancet* **1995**, *345*, 1167–1168.

16 Wildonger, R. A., Deegan, T. L., Lee, J. W. *J. Autom. Methods Manage. Chem.* **2003**, *25*, 57–61.

17 Geysen, H. M., Schoenen, F., Wagner, D., Wagner, R. *Nat. Rev. Drug Discov.* **2003**, *2*, 222–230.

18 Landers, P. *Wall Street J.* **2004**, February 24, A1.

19 Lombardino, J. G., Lowe, J. A. *Nat. Rev. Drug Discov.* **2004**, *3*, 853–862.

20 Wess, G., Urmann, M., Sickenberger, B. *Angew. Chem. Int. Ed.* **2001**, *40*, 3341–3350.

21 Maclean, D., Baldwin, J. J., Itanov, V. T., Kato, Y., Shaw, A., Schneider, P., Gordon, E. M. *Pure Appl. Chem.* **1999**, *71*, 2349–2365.

22 Maclean, D., Martin, E. J. *J. Comb. Chem.* **2004**, *6*, 1–11.

23 Evans, B. E., Rittle, K. E., Bock, M. G., DiPardo, R. M., Freidinger, R. M., Whitter, W. L., Lundell, G. F., Veber, D. F., Anderson, P. S., Chang R. S. L., Lotti, V. J., Cerino, D. J., Chen, T. B., Kling, P. J., Kunkel, K. A., Springer, J. P., Hirshfieldt, J., *J. Med. Chem.* **1988**, *31*, 2235–2246.

24 Patchett, A. A., Nargund, R. P. *Annu. Rep. Med. Chem.* **2000**, *35*, 289–298.

25 DeSimone, R. W., Currie, K. S., Mitchell, S. A., Darrow, J. W., Pippin, D. A. *Comb. Chem. High Throughput Screen.* **2004**, *7*, 473–493.

26 Horton, D. A., Bourne, G. T., Smythe, M. L. *Chem. Rev.* **2003**, *103*, 893–930.

27 Horton, D. A., Bourne, G. T., Smythe, M. L. *Mol. Diversity* **2002**, *5*, 289–304.

28 Nicolaou, K. C., Pfefferkorn, J. A., Mitchell, H. J., Roecker, A. J., Barluenga, S., Cao, G.-Q., Affleck, R. L., Lillig, J. E. *J. Am. Chem. Soc.* **2000**, *122*, 9954–9967.

29 Guo, T., Hobbs, D. W. *Assay Drug Dev. Technol.* **2003**, *1*, 579–592.

30 Bondensgaard, K., Ankersen, M., Thogersen, H., Hansen, B. S., Wulff, B. S., Bywater, R. P. *J. Med. Chem.* **2004**, *47*, 888–899.

31 Mueller, G. *Drug Discov. Today* **2003**, *8*, 681–691.

32 Jia, Qi. *Stud. Nat. Prod. Chem.* **2003**, *29* (Part J), 643–718.

33 Ortholand, J.-Y., Ganesan, A. *Curr. Opin. Chem. Biol.* **2004**, *8*, 271–280.

34 Sanchez-Martin, R. M., Mittoo, S., Bradley, M. *Curr. Top. Med. Chem.* **2004**, *4*, 653–669.

35 Bondy, S. S. *Curr. Opin. Drug. Discov. Dev.* **1998**, *1*, 116–119.

36 Easson, M. A. M., Rees, D. C. In *Medicinal Chemistry: Principles and Practice*, King, F. D. (ed.), Royal Society of Chemistry, Cambridge, **2002**, Ch. 16, 359–381.

37 Reader, J. C. *Curr. Top. Med. Chem.* **2004**, *4*, 671–686.

38 Rapp Polymere, Tübingen, Germany, www.rapp-polymere.com.

39 Merrifield, R. B. *J. Am. Chem. Soc.* **1963**, *85*, 2149–2154.

40 Geysen, H. M., Meloen, R. H., Barteling, S. J. *Proc. Natl. Acad. Sci. USA* **1984**, *81*, 3998–4002.

41 Houghten, R. A. *Proc. Natl. Acad. Sci. USA* **1985**, *82*, 5131–5135.

42 Wong, J. Y., Leznoff, C. C. *Can. J. Chem.* **1973**, *51*, 2452–2456.

43 Leznoff, C. C., Wong, J. Y. *Can. J. Chem.* **1973**, *51*, 3756–3764.

44 Crowley, J. R., Rapoport, H. *Acc. Chem. Res.* **1976**, *9*, 135–144.

45 Bräse, S., Köbberling, J., Griebenow, N. In *Handbook of Organopalladium Chemistry for Organic Synthesis*, Negishi, E. (ed.), Wiley, New York, **2002**, Vol. 2, Ch. X.
46 Gordon, K., Balasubramanian, S. *J. Chem. Technol. Biotechnol.* **1999**, *74*, 835–851.
47 Morphy, J. R. *Curr. Opin. Drug Discov. Dev.* **1998**, *1*, 59–65.
48 Reitz, B. *Curr. Opin. Drug Discov. Dev.* **1999**, *2*, 358–364.
49 Hone, N. D., Davies, S. G., Devereux, N. J., Taylor, S. L., Baxter, A. D. *Tetrahedron Lett.* **1998**, *39*, 897–900.
50 Lacombe, P., Castagner, B., Gareau, Y., Ruel, R. *Tetrahedron* **1998**, *39*, 6785–6786.
51 Keifer, P. A., Baltusis, L., Rice D. M., Tymiac, A. A., Shoolery, J. N. *J. Magn. Reson. A* **1996**, *119*, 65–75.
52 Sarkar, S. K., Garigipati, R. S., Adams, J. L., Keifer, P. A. *J. Am. Chem. Soc.* **1996**, *118*, 2035–2036.
53 Fitch, W. L., Detre, G., Holmes, C. P., Shoolery, J. N., Keifer, P. A. *J. Org. Chem.* **1994**, *59*, 7955–7956.
54 Rousselot-Pailley, P., Ede, N. J., Lippens, G. *J. Comb. Chem.* **2001**, *3*, 559–563.
55 Yan, B. *Acc. Chem. Res.* **1998**, *31*, 621–630.
56 Yan, B., Kumaravel, G., Anjaria, H., Wu, A., Petter, R., Jewell, C. F., Jr., Wareing, J. R. *J. Org. Chem.* **1995**, *60*, 5736–5738.
57 Yan, B., Fell, J. B., Kumaravel, G. *J. Org. Chem.* **1996**, *61*, 7467–7472.
58 Metzger, J. W., Kempter, C., Wiesmuller K. H., Jung G. *Anal. Biochem.* **1994**, *219*, 261–277.
59 Youngquist, R. S., Fuentes, G. R., Lacey M. P., Keough, T. *J. Am. Chem. Soc.* **1995**, *117*, 3900–3906.
60 Hoffmann, C., Blechschmidt, D., Kruger, R., Karas M., Griesinger, C. *J. Comb. Chem.* **2002**, *4*, 79–86.
61 Enjalbal, C., Maux, D., Martinez, J., Combarieu R., Aubagnac J. L. *Comb. Chem. High Throughput Screen.* **2001**, *4*, 363–373.
62 Hutton, R. S., Adams, J. P., Trivedi, H. S. *Analyst* **2003**, *28*, 103–108.
63 Swali, V., Wells, N. J., Langley, G. J., Bradley, M. *J. Org. Chem.* **1997**, *62*, 4902–4903.
64 Han, H., Wolfe, M. M., Brenner, S., Janda, K. D. *Proc. Natl. Acad. Sci. USA* **1995**, *92*, 6419–6423.
65 MultiSynTech, Witten, Germany, http://www.multisyntech.com.
66 Furka, Á., Sebestyén, F., Asgedom, M., Dibó, G. *Int. J. Pept. Protein Res.* **1991**, *37*, 487–493.
67 Pinilla, C., Appel, J. R., Blanc, P., Houghten, R. A. *Biotechniques* **1992**, *13*, 901–905.
68 Lam, K. S., Lebl, M., Krchňák, V. *Chem. Rev.* **1997**, *97*, 411–448.
69 Chabala, J. C. *Curr. Opin. Biotechnol.* **1995**, *6*, 632–639.
70 Nicolaou, K. C., Xiao, X.-Y., Parandoosh, Z., Senyei, A., Nova, M. P. *Angew. Chem. Int. Ed. Engl.* **1995**, *34*, 2289–2291.
71 Xiao, X.-Y., Parandoosh, Z., Nova, M. P. *J. Org. Chem.* **1997**, *62*, 6029–6033.
72 Discovery Partners International, San Diego, CA, www.discoverypartners.com.
73 Rasoul, F., Ercole, F., Pham, Y., Bui, C. T., Wu, Z., James, S. N., Trainor, R. W., Wickham, G., Maeji, N. J. *Pept. Sci.* **2000**, *55*, 207–216.
74 Flynn, D. L., Devraj, R. V., Parlow, J. J. *Curr. Opin. Drug. Discov. Dev.* **1998**, *1*, 41–50.
75 Booth, R. J., Hodges, J. C. *J. Am. Chem. Soc.* **1997**, *119*, 4882–4886.
76 Emerson, D. W., Emerson, R. R., Joshi, S. C., Sorensen, E. M., Nrek, J. *J. Org. Chem.* **1979**, *44*, 4634–4640.
77 Kamogawa, H., Kanzawa, A., Kodoya, M., Naito, T., Nanasawa, M. *Bull. Chem. Soc. Jpn.* **1983**, *56*, 762–765.
78 Galioglu, O., Akar, A. *Eur. Polym. J.* **1989**, *25*, 313–316.
79 Tomoi, M., Akada, Y., Kakiuchi, H. *Makromol. Chem. Rapid Commun.* **1982**, *3*, 537–542.
80 Shai, Y., Jacobson, K. A., Patchornik, A. *J. Am. Chem. Soc.* **1985**, *107*, 4249–4252.
81 Rebek, J., Brown, D., Zimmerman, S. *J. Am. Chem. Soc.* **1975**, *97*, 4407–4408.
82 Creswell, M. W., Bolton, G. L., Hodges, J. C., Meppea, M. *Tetrahedron* **1998**, *54*, 3983–3998.
83 Frechet, J. M., Schuerch, C. *J. Am. Chem. Soc.* **1971**, *93*, 492–496.
84 Coppola, G. M. *Tetrahedron Lett.* **1998**, *39*, 8233–8236.
85 Kaldor, S. W., Siegel, M. G., Fritz, J. E., Dressman, B. A., Hahn, P. J. *Tetrahedron Lett.* **1996**, *37*, 7193–7196.

86 Booth, J. R., Hodges, J. C. *Acc. Chem. Res.* **1999**, *32*, 18–26.
87 Bhalay, G., Dunstan, A., Glen, A. *Synlett* **2000**, *12*, 1846–1859.
88 Baxendale, I. R., Lee, A.-L., Ley, S. V. *J. Chem. Soc., Perkin Trans. 1* **2002**, *16*, 1850–1857.
89 Cheng, S., Tarby, C. M., Comer, D. D., Williams, J. P., Caporale, L. H., Myers. P. L, Boger, D. L. *Bioorg. Med. Chem. Lett.* **1996**, *4*, 727–737.
90 Cheng, S., Comer, D. D., Williams, J. P., Myers, P. L., Boger, D. L. *J. Am. Chem. Soc.* **1996**, *118*, 2567–2573.
91 Boger, D. L., Tarby, C. M., Myers, P. L., Caporale, L. H. *J. Am. Chem. Soc.* **1996**, *118*, 2109–2110.
92 Radleys Discovery Technologies, Essex, UK, http://www.radleys.com.
93 Zinsser Analytic GmbH, Frankfurt, Germany, http://www.zinsser-analytic.com.
94 Horvath, I. T., Rabai, J. *Science* **1994**, *266*, 72–75.
95 Zhang, W. *Tetrahedron* **2003**, *59*, 4475–4489.
96 Argonaut Technologies, Redwood City, CA, http://www.argotech.com.
97 Zeng, L., Burton, L., Yung, K. Shushan, B. Kassel, D. B. *J. Chromatogr. A* **1998**, *794*, 3–13.
98 Bauser, M. *J. Chromatogr. Sci.* **2002**, *40*, 292–296.
99 Isbell, J., Xu, R., Cai, Z., Kassel, D. B. *J. Comb. Chem.* **2002**, *4*, 600–611.
100 Affleck, R. L. *Curr. Opin. Chem. Biol.* **2001**, *5*, 257–263.
101 Tan, D. S., Foley, M. A., Shair, M. D., Schreiber, S. L. *J. Am. Chem. Soc.* **1998**, *120*, 8565–8566.
102 Tan, D. S., Foley, M. A., Stockwell, B. R., Shair, M. D., Schreiber, S. L. *J. Am. Chem. Soc.* **1999**, *121*, 9073–9087.
103 Ahrendt, K. A., Olsen, J. A., Wakao, M., Trias, J., Ellman, J. A. *Bioorg. Med. Chem. Lett.* **2003**, *68*, 2143–2150.
104 Silen, J. L., Lu, A. T., Solas, D. W., Gore, M. A., MacLean, D., Shah, N. H., Coffin, J. M., Bhinderwala, N. S., Wang, Y., Tsutsui, K. T., Look, G. C., Campbell, D. A., Hale, R. L., Navre, M., DeLuca-Flaherty, C. R. *Antimicrob. Agents Chemother.* **1998**, *42*, 1447–1453.

105 Butin, J. A., Fauchère, A. L. *Trends Pharm. Sci.* **1996**, *17*, 8–12.
106 Dolle, R. E., Guo, J., Li, W., Zhao, N., Connelly, J. A. *Mol. Diversity* **2001**, *5*, 35–49.
107 Kwon, O., Park, S. B., Schreiber, S. L. *J. Am. Chem. Soc.* **2002**, *124*, 13402–13404.
108 Leon, S., Quarrell, R., Lowe G. *Bioorg. Med. Chem. Lett.* **1998**, *8*, 2997–3002.
109 Reactors with a higher resin capacity (up to 300 mg) are usually used for the preparation of smaller libraries.
110 Very large spatially addressable libraries are synthesized in repetitive synthesis runs.
111 SciGene, Sunnyvale, CA, http://www.scigene.com.
112 Mettler-Toledo, Columbus, OH, http://www.bohdan.com.
113 Charybdis Technologies, Carlsbad, CA, http://www.charybdis.com.
114 Barnstead International, Dubuque, IA, http://www.barnsteadthermolyne.com.
115 Zinsser, W. *GIT Labor-Fachzeitschrift* **2001**, *45*, 66–68.
116 http://www.combinatorial.com/instrumentsmaster.html.
117 Braendli, C., Maiwald, P., Schroeer, J. *Chimia* **2003**, *57*, 284–289.
118 Chemspeed Technologies, Augst, Switzerland, http://www.chemspeed.com.
119 Mullin, R. *Chem. Eng. News* **2004**, *82*, 23–32.
120 Merritt, A. T., Gerritz, S. W. *Curr. Opin. Chem. Biol.* **2003**, *7*, 305–307.
121 Lahana, R. *Drug Discov. Today* **1999**, *4*, 447–448.
122 Ramesha, C. S. *Drug Discov. Today* **2000**, *5*, 43–44.
123 Kubinyi, H. *Nat. Rev. Drug Discov.* **2003**, *2*, 665–668.
124 Lipinski, C. A., Lombardo, F., Dominy, B. W., Feeney, P. J. *Adv. Drug Deliv. Rev.* **1997**, *23*, 3–25.
125 Lipinski, C. A. *J. Pharmacol. Toxicol. Methods* **2000**, *44*, 235–249.
126 Böhm, H.-J., Stahl, M. *Curr. Opin. Chem. Biol.* **2000**, *4*, 283–286.
127 Blake, J. F. *Curr. Opin. Chem. Biol.* **2004**, *8*, 407–411.
128 Darvas, F., Dorman, G. *Chim. Oggi* **1999**, *17*, 10–13.

129 Boobis, A., Gundert-Remy, U., Kremers, P., Macheras, P., Pelkonen, O. *Eur. J. Pharm. Sci.* **2002**, *17*, 183–193.

130 Clark, R. D., Wolohan, P. R. N., Hodgkin, E. E., Kelly, J. H., Sussman, N. L. *J. Mol. Graph. Model.* **2004**, *22*, 487–497.

131 Olsson, T., Oprea, T. I. *Curr. Opin. Drug Discov. Dev.* **2001**, *4*, 308–313.

132 Valler, M. J., Green, D. *Drug Discov. Today* **2000**, *5*, 286–293.

133 Beresford, A. P., Selick, H. E., Tarbit, M. H. *Drug Discov. Today* **2002**, *7*, 109–116.

134 Rose, S., Stevens, A. *Curr. Opin. Chem. Biol.* **2003**, *7*, 331–339.

135 Yu, H., Adedoyin, A. *Drug Discov. Today* **2003**, *8*, 852–861.

136 Teague, S. J., Davis, A. M., Leeson, P. D., Oprea, T. I. *Angew. Chem. Int. Ed.* **1999**, *38*, 3743–3748.

137 Dolle, R. E. *J. Comb. Chem.* **2004**, *6*, 623–679.

138 Dolle, R. E. *J. Comb. Chem.* **2003**, *5*, 693–753.

139 Dolle, R. E. *J. Comb. Chem.* **2002**, *4*, 369–418.

140 Dolle, R. E. *J. Comb. Chem.* **2001**, *3*, 477–517.

141 Dolle, R. E. *J. Comb. Chem.* **2000**, *2*, 383–433.

142 Dolle, R. E. *J. Comb. Chem.* **1999**, *1*, 235–282.

143 Dolle, R. E. *Mol. Diversity* **1998**, *4*, 233–256.

144 Dolle, R. E. *Mol. Diversity* **1997–1998**, *3*, 199–233.

145 Golebiowski, A., Klopfenstein, S. R., Portlock, D. E. *Curr. Opin. Chem. Biol.* **2003**, *7*, 308–325.

146 Golebiowski, A., Klopfenstein, S. R., Portlock, D. E. *Curr. Opin. Chem. Biol.* **2001**, *5*, 273–284.

147 Adang, A. E. P., Hermkens, P. H. H. *Curr. Med. Chem.* **2001**, *8*, 985–998.

148 Lee, A., Breitenbucher, J. G. *Curr. Opin. Drug Discov. Dev.* **2003**, *6*, 494–508.

149 Gupta, P., Singh, S. K., Srinivasan, T., Kundu, B. *Curr. Med. Chem.: Anti-Infect. Agents* **2003**, *2*, 113–133.

150 Batra, S., Srinivasan, T., Rastogi, S. K., Kundu, B. *Curr. Med. Chem.* **2002**, *9*, 307–319.

151 Nettekoven, M., Thomas, A. W. *Curr. Med. Chem.* **2002**, *9*, 2179–2190.

152 Ortholand, J.-Y. *Actual. Chim.* **2003**, 103–107.

153 Hermkens, P. H. H., Mueller, G. *E. Schering Res. Found. Workshop* **2003**, *42*, 201–220.

154 Spencer, R. W. *Biotech. Bioeng.* **1998**, *61*, 61–67.

155 Willoughby, C. A., Hutchins, S. M., Rosauer, K. G., Dhar, M. J., Chapman, K. T., Chicchi, G. G., Sadowski, S., Weinberg, D. H., Patel, S., Malkowitz, L., Di Salvo, J., Pacholok, S. G., Cheng, K. *Bioorg. Med. Chem. Lett.* **2002**, *12*, 93–96.

156 Dinnell, K., Chicchi, G. G., Dhar, M. J., Elliott, J. M., Hollingworth, G. J., Kurtz, M. M., Ridgill, M. P., Rycroft, W., Tsao, K.-L., Williams, A. R., Swain, C. J. *Bioorg. Med. Chem. Lett.* **2001**, *11*, 1237–1240.

157 Szardenings, A. Katrin, Harris, D., Lam, S., Shi, L., Tien, D., Wang, Y., Patel, D. V., Navre, M., Campbell, D. A. *J. Med. Chem.* **1998**, *41*, 2194–2200.

158 Szardenings, A. K., Antonenko, V., Campbell, D. A., DeFrancisco, N., Ida, S., Shi, L., Sharkov, N., Tien, D., Wang, Y., Navre, M. *J. Med. Chem.* **1999**, *42*, 1348–1357.

159 Batra, S., Srinivasan, T., Rastogi, S. K., Kundu, B. *Curr. Med. Chem.* **2002**, *9*, 307–319.

160 Dolle, R. E., Guo, J., O'Brien, L., Jin, Y., Piznik, M., Bowman, K. J., Li, W., Egan, W. J., Cavallaro, C. L., Roughton, A. L., Zhao, Q., Reader, J. C., Orlowski, M., Jacob-Samuel, B., Carroll, C. D. *J. Comb. Chem.* **2000**, *2*, 716–731.

161 Bleicher, K. H., Nettekoven, M., Peters, J.-U., Wyler, R. *Chimia* **2004**, *58*, 588–600.

162 Bleicher, K. H., Wuethrich, Y., Adam, G., Hoffmann, T., Sleight, A. J. *Bioorg. Med. Chem. Lett.* **2002**, *12*, 3073–3076.

163 Saha, A. K., Liu, L., Marichal, P., Odds, F. *Bioorg. Med. Chem. Lett.* **2000**, *10*, 2735–2739.

164 Saha, A. K., Liu, L., Simoneaux, R. L., Kukla, M. J., Marichal, P., Odds, F. *Bioorg. Med. Chem. Lett.* **2000**, *10*, 2175–2178.

165 Gupta, P., Singh, S. K., Srinivasan, T., Kundu, B. *Curr. Med. Chem.: Anti-Infect. Agents* **2003**, *2*, 113–133.

166 Barn, D., Caulfield, W., Cowley, P., Dickins, R., Bakker, W. I., McGuire, R., Morphy, J. R., Rankovic, Z., Thorn, M. *J. Comb. Chem.* **2001**, *3*, 534–541.

167 Smith, R. A., Chen, J., Mader, M. M., Muegge, I., Moehler, U., Katti, S., Marrero, D., Stirtan, W. G., Weaver, D. R., Xiao, H., Carley, W. *Bioorg. Med. Chem. Lett.* **2002**, *12*, 2875–2878.

168 Meyers, H. V., Dilley, G. J., Durgin, T. L., Powers, T. S., Winssinger, N. A., Zhu, H., Pavia, M. R. *Mol. Diversity* **1995**, *1*, 13–20.

169 Herron, D. K., Goodson, T., Wiley, M. R., Weir, L. C., Kyle, J. A., Yee, Y. K., Tebbe, A. L., Tinsley, J. M., Mendel, D., Masters, J. J., Franciskovich, J. B., Sawyer, J. S., Beight, D. W., Ratz, A. M., Milot, G., Hall, S. E., Klimkowski, V. J., Wikel, J. H., Eastwood, B. J., Towner, R. D., Gifford-Moore, D. S., Craft, T. J., Smith, G. F. *J. Med. Chem.* **2000**, *43*, 859–872.

170 Urbahns, K., Härter, M., Vaupel, A., Albers, M., Schmidt, D., Brüggemeier, U., Stelte-Ludwig, B., Gerdes, C., Tsujishita, H. *Bioorg. Med. Chem. Lett.* **2003**, *13*, 1071–1074.

171 Urbahns, K., Harter, M., Albers, M., Schmidt, D., Stelte-Ludwig, B., Brüggemeier, U., Vaupel, A., Gerdes, C. *Bioorg. Med. Chem. Lett.* **2002**, *12*, 205–208.

172 McLay, I. M., Halley, F., Souness, J. E., McKenna, J., Benning, V., Birrell, M., Burton, B., Belvisi, M., Collis, A., Constan, A., Foster, M., Hele, D., Jayyosi, Z., Kelley, M., Maslen, C., Miller, G., Ouldelkhim, M.-C., Page, K., Phipps, S., Pollock, K., Porter, B., Ratcliffe, A. J., Redford, E. J., Webber, J., Slater, B., Thybaud, V., Wilsher, N. *Bioorg. Med. Chem.* **2001**, *11*, 537–554.

173 McKenna, J. M., Halley, F., Souness, J. E., McLay, I. M., Pickett, S. D., Collis, A. J., Page, K., Ahmed, I. *J. Med. Chem.* **2002**, *45*, 2173–2184.

174 Foster, M. L., Halley, F., Souness, J. E. *Drug News Perspect.* **2000**, *13*, 488–497.

175 IDdb Drug Report for AVE-8677, *The Investigational Drugs Database* (update as of 31 January 2005), Current Drugs Ltd.

176 Hinzen, B., Braeunlich, G., Gerdes, C., Kraemer, T., Lustig, K., Nielsch, U., Sperzel, M., Pernerstorfer, J. *Handb. Comb. Chem.* **2002**, *2*, 784–805.

177 Augeri, D. J., O'Connor, S. J., Janowick, D., Szczepankiewicz, B., Sullivan, G., Larsen, J., Kalvin, D., Cohen, J., Devine, E., Zhang, H., Cherian, S., Saeed, B., Ng, S.-C, Rosenberg, S. *J. Med. Chem.* **1998**, *41*, 4288–4300.

178 Henry, K. J., Wasicak, J., Tasker, A. S., Cohen, J., Ewing, P., Mitten, M., Larsen, J. J., Kalvin, D. M., Swenson, R., Ng, S.-C., Saeed, B., Cherian, S., Sham, H., Rosenberg, S. H. *J. Med. Chem.* **1999**, *42*, 4844–4852.

179 Brown, P. J., Stuart, L. W., Hurley, K. P., Lewis, M. C., Winegar, D. A., Wilson, J. G., Wilkison, W. O., Ittoop, O. R., Willson, T. M. *Bioorg. Med. Chem. Lett.* **2001**, *11*, 1225–1227.

180 Cheng, Y., Rano, T. A., Huening, T. T., Zhang, F., Lu, Z., Schleif, W. A., Gabryelski, L., Olsen, D. B., Stahlhut, M., Kuo, L. C., Lin, J. H., Xu, X., Jin, L., Olah, T. V., McLoughlin, D. A., King, R. C., Chapman, K. T., Tata, J. R. *Bioorg. Med. Chem. Lett.* **2002**, *12*, 529–532.

181 Raghavan, S., Yang, Z., Mosley, R. T., Schleif, W. A., Gabryelski, L., Olsen, D. B., Stahlhut, M., Kuo, L. C., Emini, E. A., Chapman, K. T., Tata, J. R. *Bioorg. Med. Chem. Lett.* **2002**, *12*, 2855–2858.

182 Zhang, C., Sarshar, S., Moran, E. J., Krane, S., Rodarte, J. C., Benbatoul, K. D., Dixon, R., Mjalli, A. M. M. *Bioorg. Med. Chem. Lett.* **2000**, *10*, 2603–2605.

183 Sarshar, S., Zhang, C., Moran, E. J., Krane, S., Rodarte, J. C., Benbatoul, K. D., Dixon, R., Mjalli, A. M. M. *Bioorg. Med. Chem. Lett.* **2000**, *10*, 2599–2601.

184 Mistry, P., Folkes, A. *Curr. Opin. Invest. Drugs* **2002**, *3*, 1666–1671.

185 IDdb Drug Report for ONT-093, *The Investigational Drugs Database* (update as of 31 January 2005), Current Drugs Ltd., Thomson Scientific, London, UK.

186 Smith, R. A., Barbosa, J., Blum, C. L., Bobko, M. A., Caringal, Y. V., Dally, R., Johnson, J. S., Katz, M. E., Kennure, N., Kingery-Wood, J., Lee, W., Lowinger, T. B., Lyons, J., Marsh, V., Rogers, D. H., Swartz, S., Walling, T., Wild, H. *Bioorg. Med. Chem. Lett.* **2001**, *11*, 2775–2778.

187 Lowinger, T. B., Riedl, B., Dumas, J., Smith, R. A. *Curr. Pharm. Des.* **2002**, *8*, 2269–2278.

188 Wilhelm, S. M., Carter, C., Tang, L., Wilkie, D., McNabola, A., Rong, H., Chen, C., Zhang, X., Vincent, P., McHugh, M., Cao, Y., Shujath, J., Gawlak, S., Eveleigh, D., Rowley, B., Liu, L., Adnane, L., Lynch, M., Auclair, D., Taylor, I., Gedrich, R., Voznesensky, A., Riedl, B., Post, L. E., Bollag, G., Trail, P. A. *Cancer Res.* **2004**, *64*, 7099–7109.

189 IDdb Drug Report for Sorafenib, *The Investigational Drugs Database* (update as of 31 January 2005), Current Drugs Ltd., Thomson Scientific, London, UK.

190 Sznaidman, M. L., Haffner, C. D., Maloney, P. R., Fivush, A., Chao, E., Goreham, D., Sierra, M. L., LeGrumelec, C., Xu, H. E., Montana, V. G., Lambert, M. H., Willson, T. M., Oliver, W. R., Sternbach, D. D. *Bioorg. Med. Chem. Lett.* **2003**, *13*, 1517–1521.

191 Oliver, W. R., Shenk, J. L, Snaith, M. R., Russell, C. S., Plunket, K. D., Bodkin, N. L., Lewis, M. C., Winegar, D. A., Sznaidman, M. L., Lambert, M. H., Xu, H. E., Sternbach, D. D., Kliewer, S. A., Hansen, B. C., Willson, T. M. *Proc. Natl. Acad. Sci. USA* **2001**, *98*, 5306–5311.

192 IDdb Drug Report for GW-501516, *The Investigational Drugs Database* (update as of 31 January 2005), Current Drugs Ltd., Thomson Scientific, London, UK.

193 Villhauer, E. B., Brinkman, J. A., Naderi, G. B, Dunning, B. E., Mangold, B. L., Mone, M. D., Russell, M. E., Weldon, S. C., Hughes, T. E. *J. Med. Chem.* **2002**, *45*, 2362–2365.

194 IDdb Drug Report for DPP-728, *The Investigational Drugs Database* (update as of 31 January 2005), Current Drugs Ltd., Thomson Scientific, London, UK.

195 Villhauer, E. B., Brinkman, J. A., Naderi, G. B., Burkey, B. F., Dunning, B. E., Prasad, K., Mangold, B. L., Russell, M. E., Hughes, T. E. *J. Med. Chem.* **2003**, *46*, 2774–2789.

196 Barlocco, D. *Curr. Opin. Invest. Drugs* **2004**, *5*, 1094–1100.

197 IDdb Drug Report for Vildagliptin, *The Investigational Drugs Database* (update as of 31 January 2005), Current Drugs Ltd., Thomson Scientific, London, UK.

198 Chang, G., Ruggeri, R. B., Harwood, H. J. *Curr. Opin. Drug Discov. Dev.* **2002**, *5*, 562–570.

199 Chandler, C. E., Wilder, D. E., Pettini, J. L., Savoy, Y. E., Petras, S. F., Chang, G., Vincent, J., Harwood, H. J. *J. Lipid Res.* **2003**, *44*, 1887–1901.

200 IDdb Drug Report for CP-346086, *The Investigational Drugs Database* (update as of 31 January 2005), Current Drugs Ltd., Thomson Scientific, London, UK.

201 Wetterau, J. R. II, Sharp, D. Y., Gregg, R. E., Biller, S. A., Dickson, J. A., Lawrence, R. M., Magnin, D. R., Poss, M. A., Robl, J. A., Sulsky, R. B., Tino, J. A., Lawson, J. E., Holava, H. M., Partyka, R. A. *PCT Int. Appl.*, WO-9626205, **1996**.

202 Wetterau, J. R., Gregg, R. E., Harrity, T. W., Arbeeny, C., Cap, M., Connolly, F., Chu, C.-H., George, R. J., Gordon, D. A., Jamil, H., Jolibois, K. G., Kunselman, L. K., Lan, S.-J., Maccagnan, T. J., Ricci, B., Yan, M., Young, D., Chen, Y., Fryszman, O. M., Logan, J. V. H., Musial, C. L., Poss, M. A., Robl, J. A., Simpkins, L. M., Slusarchyk, W. A., Sulsky, R., Taunk, P., Magnin, D. R., Tino, J. A., Lawrence, R. M., Dickson, J. K., Biller, S. A. *Science* **1998**, *282*, 751–754.

203 IDdb Drug Report for BMS-201038, *The Investigational Drugs Database* (update as of 31 January 2005), Current Drugs Ltd., Thomson Scientific, London, UK.

204 Hoekstra, W. J., Maryanoff, B. E., Andrade-Gordon, P., Cohen, J. H., Costanzo, M. J., Damiano, B. P., Haertlein, B., Harris, B. D., Kauffman, J. A., Keane, P., McComsey, D. F., Mitchell, J. A., Scott, L., Shah, R. D., Yabut, S. C. *Bioorg. Med. Chem. Lett.* **1996**, *6*, 2371–2376.

205 Hoekstra, W. J., Maryanoff, B. E., Damiano, B. P., Andrade-Gordon, P., Cohen, J. H., Costanzo, M. J., Haertlein, B. J., Hecker, L. R., Hulshizer, B. L., Kauffman, J. A., Keane, P., McComsey, D. F., Mitchell, J. A., Scott, L., Shah, R. D., Yabut, S. C. *J. Med. Chem.* **1999**, *42*, 5254–5265.

206 IDdb Drug Report for Elarofiban, *The Investigational Drugs Database* (update as of 31 January 2005), Current Drugs Ltd., Thomson Scientific, London, UK.

10
High-throughput Screening and Data Analysis
Jeremy S. Caldwell and Jeff Janes

10.1
Introduction

The advent of the published rough draft of the human genome sequence represents a fundamental shift in the way in which biology is studied. Having access to the full transcript of human genes indicated that the genome was simultaneously vast but more importantly finite, such that ascertaining the function of all genes in the genome became a conceivable exercise. The surge of thought as to how one might go about this comprehensive definition of gene function has given rise to the realization that there is no one perfect application of the human genome sequence, but hundreds of applications for thousands of scientists. As a result, a series of post-genomic tools have been developed to begin functionalizing the human genome, ranging through functional cDNA, siRNA, focused small molecule library screens, gene expression profiling, proteomic applications and more. In parallel, advances in high-throughput screening (HTS), automation, cell biology, imaging, combinatorial libraries and computation have allowed researchers once again to ask fundamental biological questions in new and exciting ways.

An outgrowth of the genomics revolution has also rejuvenated an age-old approach to drug discovery: cell-based screening. The ability to survey small-molecule space (and now genomic libraries) in disease-relevant cellular pathway screens affords the opportunity to identify rapidly bioactive compounds, which, when coupled with advances in genomic, proteomic and affinity-based methods, has imparted the ability to deconvolute the relevant biological (DNA, RNA, protein or other) target of the compound with increasing success. Perhaps more interestingly, computational models and *in silico* deconvolution tools borrowed from precepts in statistical science have allowed one to make sound predictions about the target of a small molecule. HTS databases cataloguing small-molecule activity data across panels of cellular assays ranging from inflammation, metabolic disease, infectious disease, oncology and other fundamental biology including toxicity, transcription and transport can now be mined for correlations between compounds

High-Throughput Screening in Drug Discovery. Edited by Jörg Hüser
Copyright © 2006 WILEY-VCH Verlag GmbH & Co. KGaA, Weinheim
ISBN: 3-527-31283-8

with known targets/mechanisms of action and new, uncharacterized compounds to assign function, biological context if not also the specific target.

This emerging enterprise represents a new era in *in silico* biology and hypothesis generation; however, its practice can only be realized in full with "good" data from HTS. Toward this goal, the automation of high-throughput cellular screening coupled with robust and consistent assay performance is beginning to realize datasets worthy of these meta-data mining goals. Poor data quality is the possible bane of all HTS-based drug discovery campaigns. In this chapter, we initially focus on data analysis methods for quality control of high-throughput cellular screening data, identifying and removing noise and artifacts, an obvious important activity for realizing mineable datasets, which require some features distinct from biochemical assay data handling. Within this section, we explore methods to enrich for screen hits relevant to a pathway -based screen, for more optimal lead and target discovery. Then we handle issues related to the generation and analysis of meta-data from multiple screens initially focused on the removal of compounds and library sets with nonspecific biological effects ranging from assay-dependent activity (e.g. inhibition of luciferase) to general pathway-independent biology (toxicity, transcriptional regulation, etc.). In the second section, on massively parallel cellular screening, we delve into methods to generate and analyze large-scale meta-data sets to forge connections between compounds and their mechanism of action (MOA). Furthermore, we discuss the utility of such multidimensional screening paradigms to guide lead optimization and predict positive and negative attributes of a compound earlier in the drug discovery process. Ultimately the utility of these methods and the latest automation/informatics couplings for advanced lead finding, opportunistic drug discovery and toxicity/side-effect profiling using predictive cellular assays will, once borne out in the academic and pharmaceutical industry, expedite the drug discovery process and realize the promise of the human genome.

10.2
Analysis of Cellular Screening Data

10.2.1
Quality Control and Analysis

The United States National Cancer Institute (NCI) conducts an anti-cancer drug discovery program in which over 100 000 distinct chemical entities have been screened and 10 000 new entities are screened annually in a panel of human cancer cell lines from different tissues. The resulting database represents the records of the screens and describe a comprehensive set of cellular experiments which, if appropriately curated, can be mined in clever ways to extract knowledge regarding compound mechanism of action and testable hypotheses in cancer research. The importance of appropriate quality control and analysis of the NCI dataset is underscored by the fact it is being used to identify the next generation of new chemical entities (NCEs) for oncology therapy. Therefore, special consideration of the methods by which the NCI dataset and other cell-based datasets generated by those pursuing similar screening strategies is mandatory for efficient follow-up data-mining exercises.

In the widely mined NCI60 dataset (see below), all of the cellular assays are antiproliferative or toxicity assays performed in dose–response under nearly identical conditions, differing in little other than the cell line used. In contrast, mining of datasets built up as a result of the accumulation of data from ultra-high-throughput screening (uHTS) assays primarily designed for lead discovery purposes must contend with a wide variety of assay formats, readouts and assay conditions, in addition to the generally lower quality given by uHTS assays. In addition, the larger size and diversity of the screened compound collection leads to a greater variety of artifacts which are reproducible but are not acting via the mechanism which the assay was designed to probe.

The majority of cell-based assays used in drug discovery fall into two broad categories, proliferation/toxicity assays common in oncology and immunology fields and reporter gene assays (RGAs). Proliferation and toxicity assays generally detect either the active metabolic machinery of cells using redox-sensitive colorimetric indicators or high-energy molecules released from the lysis of live cells. The reducible dyes Alamar Blue [1] and MTT [2] are examples of the former. They may suffer from interference from highly colored compounds. As an example of the latter, Promega's CellTiter-Glo® Luminescent Cell Viability Assay [3] makes use of the ATP dependence of firefly luciferase to generate a luminescence signal proportional to the ATP content of the cells. This method is faced with many of the same artifacts as RGA which utilize luciferase as the reporter gene.

Reporter gene assays use cells which have been engineered to place the expression of a protein which is readily detectable through luminescent substrates (luciferase, β-galactosidase and secreted alkaline phosphatase) or direct fluorescent detection (green fluorescent protein and its derivatives) under the direct or indirect control of the biological process of interest. That biological process can be an entire pathway (forward chemical genetics) or it can be focused on a particular target. Target-based screens are run in cells because the biological activity cannot be stably reconstituted in purified form, since the purification of sufficient amounts of enzyme for uHTS is not economically feasible or because there is no reliable nonbiological detection method for the assay end-point. The difference between pathway-based and target-based cellular assays lies not so much in the design of the experimental system as in the intended use of the hits obtained. When mining historical data to aid in the interpretation of a cell-based screen, the methods used may differ based on the goal of the reference screen. However, once the assay is entered into the historical dataset for use in cross-assay data mining, the distinction between target-based and pathway-based is unimportant.

A general treatment of data normalization, quality control (QC) and visualization is handled elsewhere in this volume (see Chapter 7). However, some of these issues as they pertain to the particulars of cell-based screening and cross-assay data mining must be addressed here.

Cellular screening at ultra-high levels of throughput poses many challenges which are more severe than with the corresponding biochemical assays. Cell-based assays commonly have prolonged incubation or proliferation stages during which minute imperfections in plate lid seals can lead to evaporation, which creates dramatic "edge effects" in some of the peripheral wells. Dispenser tips delivering nonhomogeneous solutions, such as cells, clog more frequently, resulting in fractions of plates receiving inconsistent or no volumes of cells. While most if not all of these data are unsalvageable, even under the most sophisticated statistical correction algorithms, the areas affected by evaporation or mis-dispensing may constitute less than 5% of the data in an affected micro-well plate. Traditional QC methods would usually call for the entire plate to be censored and repeated, which is often not economically justifiable when a large number of plates are each affected over a small region. Even worse, automated QC methods based on statistical analysis of high and low controls [4, 5] may overlook problematic plates altogether if the problem does not encroach into the control regions. In addition, such irreparable wells cause derangement of the statistical efforts to remove systematic error from other, salvageable data. One solution to overcome these difficulties, applied at the Genomics Institute of the Novartis Research Foundation (GNF), has been to implement a suite of tools, based on Web applications and SpotFire DecisionSite, to facilitate both manual and automated detection and flagging of irreparable wells. Once the obviously poor data have been eliminated, the remainder are normalized on a plate-by-plate basis to accommodate signal drift over time and extensive heat-map images are generated. Similarly to the products SLIMS [6], StatServer HTS® [7] and GeneData [8], the heat maps reflect not only the plate maps, but also various other arrangements of the data which are designed to bring out variations that occur within a dispense tip over time and systematic variations between dispense tips, rows and columns. Another heat map highlights any carryover effect caused by compounds sticking to pins in the compound transfer pintool. These visualizations often reveal additional irreparable data which would produce false positives if not removed by QC. They also give insight into the data adjustment methods which may be most appropriate and provide feedback to the screen operators for future improvements.

Once data which contain no meaningful signal have been removed, the remaining data can be subjected to various statistical adjustment methods to remove "noise" variation, which is a result of systematic variations in robotics and other equipment operations. For performing these adjustments, the variations are frequently broken down into row and column effects. Many of the physical sources of variation, such as row-based dispensers or moving-head plate readers, are strongly correlated with rows or columns and so such variation is subsumed into the row and column effects. State-of-the-art methods will combine information from the plate being adjusted with similar patterns in the temporally surrounding plates in order to achieve more robust correction factors [9]. This is especially important in lower density plate formats or when using model-based adjustment methods with a high number of degrees of freedom to prevent real but unusual confluences of compound activity from being adjusted away. In addition to row- and column-based adjustments, there are methods which can correct for more complex patterns by treating both dimensions simultaneously in a Fourier transform analysis [10]. In drug discovery screens, especially when the systematic effects are small, sophisticated treatment of the data as outlined here may seem unnecessary, as it might push few compounds over the "hitpick threshold" one way or the other. However, if the assay data are also destined for a data mining database, extra care needs to be taken to remove small but consistent systematic variations that occur over a wide temporal range of plates. Otherwise, data mining techniques may incorrectly label these low-amplitude but highly repeated patterns as being more interesting than the true patterns which are desired.

10.2.2
Enrichment for Hits

The simplest form of data mining on HTS cell-based data focuses on only the results of the screen in question, in conjunction with the chemical structures of the screened compounds and attempts to use these data to optimize the follow-up strategy for the screening effort [6, 11–16]. Such optimization can take several forms, such as the predictions of screened compounds which falsely tested negative for inclusion in the follow-up list, prediction of unscreened compounds which likely would have been hits for inclusion in the follow-up list, the exclusion of compounds likely to be false (i.e. unrepeatable) positives and the exclusion of compounds which are apparent hits only because of nonspecific or off-target activity. Another goal may be to limit the number of compounds from highly represented scaffolds, to prevent them from crowding other interesting but more sparsely represented chemotypes out of the hit list [17]. The general topic of chem-informatic analysis is covered in greater detail elsewhere in this volume (see Chapter 8) and will only be addressed here to the extent that the issues are specific to the topic of data mining in cell-based screens.

Cell-based assays are often subject to higher levels of random noise than biochemical assays, which lead to a higher level of false positives. Since such noise will independently strike different structures within the same chemical class, an aggregation of the signal over members of a class can average out the noise and can be used to focus on compounds which are more likely to reproduce their activity upon follow-up. However, an effect even larger than random noise is produced by the extremely complex interactions in living cells, which provide many routes for compounds to produce the desired change in the end readout without acting through the desired mechanism. This leads to a large number of hits which, although "true positives" in the sense that their results in the assay system are repeatable, are not "interesting positives", i.e. good candidates for drug discovery work. These need to be filtered out by counter screens and secondary assays, which are often of lower throughput than the primary screens and are usually the bottleneck in lead discovery. The typical experience is that a substantial subset of unwanted actives (those arising from non-pharmacological mechanisms) tend to have broader structure–activity relationships (SARs), leading to stronger signals upon noise averaging over the chemical class. Hence chemical structural data mining efforts based on the results of a stand-alone cell-based assay often enrich for these true but uninteresting positives at the expense of scaffolds which, although more weakly represented among the actives, are more valuable as lead scaffolds.

One way to ameliorate the problems caused by uninteresting hits (and also unrepeatable hits) is to delay the structure-based data mining effort until after some of the follow-up work, such as IC_{50}s and counter assays, have been completed. Indeed, the NCI AIDS dataset, which has been used as the exemplar in several single-assay data mining papers, more closely resembles a large body of follow-up or lead optimization work than it does a small uHTS effort because each compound is tested in dose–response. There are currently no satisfactory ways to analyze jointly primary data which is high occupancy, relatively low quality and generally gathered in single-point activity measurements, along with follow-up data, which cover only a sparse sample of the screening collection, have a higher quality and are generally collected in a dose–response format.

10.2.3
Meta-data Analysis

Another method of ameliorating the deleterious effects of true but uninteresting positives from primary screens is to bring in data from other screening campaigns. In target-based cellular screens, compounds which are also active in other cellular screens are likely to be nonspecific and may not be interesting for further inspection. For pathway-based screens, activity in related cellular screens could indicate either nonspecific activity or a common element in the two pathways, so those compounds may become either more or less interesting depending on goals of the project and the nature of the pathways involved. To use data from other screening campaigns, usually a scientist who is knowledgeable about the biology and the experimental details of the screens will choose those which are most relevant and the combined data can be inspected using traditional data visualization methods such as scatter plots and histograms or with methods more recently developed for "profiling" of mRNA expression patterns [8,18,19]. When the choice of the relevant assays for comparison is not obvious, simple correlation coefficients computed between assays may be used to inform the decision [8]. When using this direct comparison between screens, there are two main drawbacks (apart from the need to choose which screens to include). One problem is that bringing additional results into the analysis multiplies the already overwhelming amount of data, rather than serving as a data reduction. The related difficulty is that the assay results brought in are each subject to experimental noise, so that the more assays are examined in a screening profile, the more likely it is that one of the data points is an unrepeatable outlier. Biologists may be justifiably unwilling to make a go or no-go decision on the basis of a single, uncorroborated data point collected from a project in which they may not even be involved.

Data mining methods can be used to address both of these issues by effectively reducing the number of historical screens and averaging out some of the noise associated with them. Principal component analysis (PCA) is one approach to achieve such data reduction and averaging. PCA is a method which aims to reduce the unexplained variance in a system of data by constructing a series of components, each one explaining as much variance as it can while still being orthogonal to the preceding components. As applied to a large set of primary screening data, the end result of this is the creation of a few "virtual assays", each of which is a linear combination of the real assays which served as the input. The virtual assay "result" for any given compound is calculated as a weighted average of the results (for that compound) in the real assays. The more strongly correlated any two assays are, the more likely they are to be highly weighted on to the same virtual assay.

Two pre-processing steps used before feeding primary data into PCA computation (or data mining routines in general) need to be carefully considered. Because PCA is a linear modeling technique, it is important that the primary assay data, collected over a very varied range of formats and readouts, be transformed to lie on an approximately linear scale. The primary screening data collected at GNF is, by default, expressed as a ratio of the raw signal (corrected for row, column and dispense tip effects; see Chapter 7) for the compound to the "typical" corrected raw signal for compounds on that same plate (note: the "typical" signal is computed as the median signal of the non-hit, QC-passing compounds. Because status as a hit is determined by the ratio, this is a circular calculation which is iterated to convergence). This ratio was adopted as the default measure of activity because it was considered to be the easiest way to put disparate assays on a comparable scale (for some rare assays this method does not work, e.g. RGA agonists assays with a near-zero basal expression level of the reporter or assays in which the background control is a substantial fraction of the vehicle control; in these cases, the activity computation was altered to stabilize the computation will still keeping an eye on comparability with other assays). Use of such a ratio creates an open-ended scale for assays of agonism or de-repression and a closed-ended scale for anti-proliferation, inhibition or antagonism assays. This leads to a non-linearity which is compensated for by applying the squashing function given in eqn. (1). Equation (1) also reflects a recentering step, typical of preprocessing for data mining, in which the scores are adjusted downwards such that the typical compound will have a value near zero, rather than near one.

$$new\_val = \begin{cases} ratio - 1 & ratio < 2 \\ 2 - 1/(ratio - 1) & ratio \geq 2 \end{cases} \tag{1}$$

The second preprocessing step to consider is the normalization of variance between assays. There are three fundamental sources of variance in screening data. First, there is the systematic error, which has hopefully been mostly eliminated at this point through QC and adjustment of the individual assay results. Second, there is the true variation in the activities (along both the intended mechanism of action and unintended ones) of the compounds in the screening library. Finally, there is the unassignable error or random noise, which is present in all experimental measurements. The total variance in cell-based screens is usually dominated by this unassignable error. This is simply because most compounds are not active and therefore contribute some to the variance due to noise but very little to the variance due to true activity. Because the PCA method seeks to reduce total variance across all of the data, it is important to normalize each assay to its own total variance to prevent the method from simply reporting the noisiest assays as the chief contributors to the "virtual" assays. This choice of variance normalization is usually not suitable for other data reduction techniques, such as those employing Euclidean distance metrics.

A large PCA was performed on a dataset consisting of the results of 50 assays conducted at GNF on approximately one million compounds. Missing values in the dataset were handled by first filling in missing cells with zeros and then using the derived four-component PCA model to impute new values for those cells. This was iterated until the imputed values showed little change from one round of modeling to the next [20]. Manual inspection of the loadings of assays on to the components and of the profiles associated with compounds that were outliers along one component or another determined that all components after the second were unlikely to hold any significant biological meaning. The first and second components were

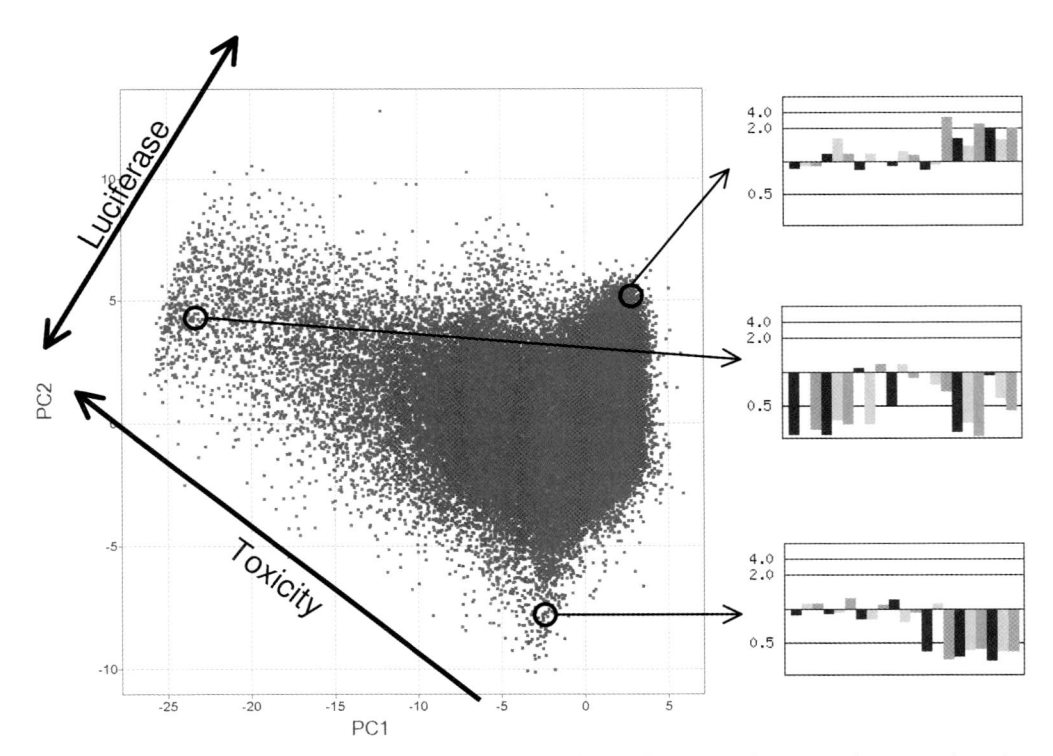

Figure 10.1 Principal component analysis of primary uHTS results. In the scatterplot, each square represents the scores for one compound on to the first two principal components. The thick arrows represent the rotated axis used in computing the toxicity and luciferase virtual screens. The insets show the activity profiles of three compounds across selected cell-based assays. Alamar blue-based assays are grouped on the left side of the profile charts and firefly luciferase-based assays are grouped on the right.

clearly interpretable in terms of general cytotoxicity and luciferase RGA-specific interaction (Fig. 10.1). An inherent feature of PCA is that the components that it produces have no linear correlation between them. However, there is no guarantee that the underlying biological processes which are being modeled are themselves uncorrelated. This leads to a situation, plainly evident here, in which the statistical principal components differ from the most biologically meaningful axis by a simple rotation. Performing the necessary rotation leads to the creation of two virtual assays, one for cytotoxicity and one for luciferase RGA interference.

The luciferase RGA virtual assay has outliers in both the increased luminescence and decreased luminescence directions. Many of the compounds which caused decreased signal are structurally similar to luciferin and were shown to be inhibitors of firefly luciferase in biochemical assays. Compounds which causes increased signal could be hypothesized to operate by stabilizing luciferase against degradation, serving as productive substrates for the luciferase enzyme or serving as general or specific activators of transcription.

The creation of these result sets for virtual assays allows decisions to be made on both individual lead discovery projects and on overall screening operations. For example, the over-representation of one combinatorial compound library in the set of active compounds in the cytotoxicity virtual assay contributed to the decision to remove that compound library from the general screening collection. In some lead discovery projects in which the size of the primary screening hit list exceeds the capacity for follow-up work, the virtual assay results are used to eliminate compounds which have a lower probability of being on-target and specific. This latter usage occurs less often than originally expected because the biologists, aware that the data mining is limited to single-dosage data sets, hope that these nonspecific compounds may also have the desired activity and might retain that activity at concentrations where the toxic or nonspecific activities are titrated out.

The appropriate application of uHTS data processing and QC is critical to the drug discovery process. Logical methods of both automated and manual data normalization with the forward-looking goal of data mining can differentiate between accurate and error-prone predictions. Appropriate QC of screening data for a given collection of highly related compounds can distinguish between those displaying random activities in a series of toxicity assays or coordinate activity with a discernible activity relationship defined by the compounds' structure. Similarly, QC can afford differentiation between compounds related by uninteresting artifacts of the screen (edge effect, dispensing error) or compounds (pathway-independent reporter gene inhibition, autofluorescence) and those which have common physiological and cell signaling properties.

10.3
Massively Parallel Cellular Screens

The extensive amount of data coinciding with the "-omics" era of biotechnology has ushered in a deluge of new questions regarding experimental design strategies and modes of data analysis. The application of uHTS, combinatorial chemical synthesis, gene expression profiling, comprehensive gene expression screening libraries, high-throughput protein structural determination and proteomics has brought forth this explosion of post-genomic data. Comprehensive arrays of biological assays have once again become the mainstay in advanced pharmaceutical research and development of new medicines. Now, information provided by the results of such screening enterprises can be used both to accelerate the drug discovery process and to increase its value as it provides insights into the differences in pharmacological activity of drug candidates directed against different biological targets.

In the present era of uHTS, large-scale HTS databases routinely contain records of millions of compounds and their activities across a multitude of screening assays. In effect, the biological activity of any compound across a panel of screening assays can be represented as a signature, profile or fingerprint representative of its unique biological activity. It is generally accepted that small molecules with similar structures will have superimposable activity profiles which only minimally diverge. The differences that do exist are generally linked to small perturbations in chemical structure and are unlikely to create significant incongruence over a large number of screening assays. Likewise, compounds with similar mechanisms of action are likely to have similar activity profiles even if they have completely distinct chemical structures. The prevailing logic follows that compounds with overlapping profiles will have the same or similar molecular targets. Lastly, chemical entities similar in chemical structure with different mechanisms of action will have different and nonoverlapping activity profiles. These general principles clearly agree with the observation that, although the activity data for a single biological assay contain some useful information, activity profiles derived over a broad panel of assays encode incisive information which equates with a compound's mechanism of action [21]. These conclusions are the general result of the Herculean efforts undertaken by the NCI, which over time has made measurements and catalogued the growth inhibitory effects $[\log(GI_{50})]$ of over 100 000 low molecular weight compounds tested across various subsets of 60–100 tumor cell lines [22]. Implicit in this screening strategy is the borne-out premise that agents tested will show reproducible patterns of differential response among the 60 cell lines.

An important enterprise based on this deluge of public HTS data, and also those incorporated into proprietary HTS data warehouses of most pharmaceutical companies focuses on extracting useful information to drive mechanism of action studies, derive SARs, predict toxicities and identify liabilities or new opportunities for their favorite advanced compounds. As with the NCI example, multidimensional data analysis of cellular/phenotypic HTS datasets, whereby data from parallel, independent experiments are made computationally comparable for global data mining is gaining increasing relevance, even if at a small scale relative to the NCI approach [23]. For instance, Kinzler et al. performed parallel simultaneous screens using isogenic pairs of immortalized colon cancer cell lines, one encoding mutant K-*ras* and another in which the mutant K-*ras* allele had been deleted the cell lines, each encoding a distinct fluorescent protein marker were tested in the same well against ~30 000 unique compounds. The study resulted in identification of a novel, differentially cytotoxic cytidine analog with preferential growth inhibitory activity towards mutant K-*ras*-expressing cells [24]. Similarly, Dolma et al. used synthetic lethal HYS to interrogate 23 550 compounds for their ability to kill engineered tumorigenic cells but not their isogenic normal counterpart, to identify known and novel compounds with genotype-selective activity, including doxorubicin, daunorubicin, camptothecin and a novel compound named erastin [25]. In contrast to the latter screen, which interrogated one cell line followed by subsequent experiments in both cell lines, Fantin et al. performed unique parallel screens of immortalized epithelial cells either lacking or expressing neuT oncogene, to identify compounds which could induce cell-type dependent BrDu-incorporation [26]. These are only a few examples of the most simplistic versions of parallel screens in which a simple subtractive analysis could furnish those compounds with selective traits.

10.3.1
Data Analysis of Multidimensional Datasets

Increasingly sophisticated methods to analyze multidimensional datasets have come to the fore, including techniques of hierarchical clustering, multidimensional scaling, support vector machines, self-organizing maps (SOMs) and methods based on neural networks, each method adapted to parallel cell-based screening to extract knowledge, largely applied to, if not inspired by, the NCI's drug screening program of 60+ different human tumors [27–32]. Although the methods employed to analyze large parallel screening datasets have a strong edifice in statistics, their fruitful application varies depending on data quality and consistency (as alluded to earlier in this chapter). What emphasis to place on these variables and how they impact an analysis method present a challenge to the analyst to apply and derive the most statistically meaningful results and hypotheses.

Methods such as PCA, multidimensional scaling (MDS) and hierarchical clustering (HCA) have been usefully applied to group compounds from large multidimensional cellular datasets to assign functional relationships between them. PCA, as described earlier, is a technique that takes linear combinations of the original variables (columns) of a data matrix such that the first principal component (PC) explains as much of the overall variation as possible, the second PC explains the next most variation subject to being orthogonal to the first and so on [33]. Koutsoukos used PCA to analyze the NCI cell screen data for a set of 141 "standard" anti-cancer agents for which the mechanisms of action are well defined [33]. PCA score plots showed distinct clusters of compounds for some of the mechanisms of action examined. Weinstein et al. extended the analysis to the entire dataset. One of the salient features of the PCA is its ability to identify outliers (which also attest to its utility in general data QC handled earlier), random equipment abnormalities and data entry errors [33]. In MDS, distinctions between members of a group (i.e. compounds) are used to reduce the number of dimensions associated with the data. Reducing the dimensionality of a dataset is primarily done for the sake of human inspection and ease of pattern visualization. HCA creates groupings based on similarities and can be most aptly displayed as a cluster tree, akin to a phylogenetic tree. Such plots have been adapted to heat maps described earlier, for easy inspection of degrees of relationships between distinct entities. MDS- and HCA-based groupings tend to be better than those derived from PCA but may then lose data granularity (by the dimension reduction) and thus lose information.

In the light of the fact that compounds chosen from multidimensional screen analysis are ultimately selected with the intent of testing in humans, appropriate cross-screen normalization and data analysis methods are of paramount importance. Towards a global analysis of the NCI screening data, Paull et al. developed a computer program called COMPARE to seek useful information in each compound's activity pattern. Briefly, COMPARE consists of calculating linear correlation coefficients between the data over all cell lines for the pattern of interest (called the seed) and all sets of data in the database to be searched. The results are sorted by the correlation coefficient. For a designated seed compound, COMPARE searches the database of already tested compounds for ones with the most similar activity patterns against the cell panel. The pairwise comparisons are based on Pearson correlation coefficients. Early on it was observed that compounds matched by pattern often had similar chemical structures and that such matches generally related to an *in vitro* biochemical mechanism of action [21, 34]. The type of information derived from such comparisons can generate testable hypotheses. For example, when the pattern of activity for anthrax lethal factor (NSC 678519) was used as a "seed", the most striking correlation was with NSC679828, a known MAP kinase kinase (MKK) inhibitor. This correlation suggested the hypothesis that the anthrax lethal factor was eliminating MKK activity via degradation of the enzyme, which was later confirmed experimentally [35]. COMPARE calculations are versatile, allowing one to search for compounds with the strongest positive correlation or the largest negative correlation, and provides the ability to eliminate cell line data either irrelevant to a particular comparison or of suspect quality.

Researchers at GNF developed a set of analysis methods similar to COMPARE, devised to classify compounds by mechanism and to identify putative MOAs and targets for previously undescribed compounds. The primary screening results are expressed as ratios to the plate median as discussed in earlier and the centering and the squashing function [eqn. (1)] is applied. Unlike in the PCA analysis, the variance for each assay is not normalized to one. For each pairing between the seed compounds and queried compounds, two similarity metrics are computed over the assays those compounds have in common – an uncentered Pearson correlation coefficient and a count-normalized Euclidean distance. Uncentered correlation is preferred over ordinary linear correlation because the input data have already been centered along a much more informative and natural vector (the results of the assay run over the entire compound collection) and re-centering along a less informative and unnatural vector (the set results for a single compound, over all assays, would be counterproductive). The count-normalized Euclidean distance is identical with the root mean square difference of the two profiles. Both of these metrics are subject to anomalies introduced by low assay overlap between compound pairs (due to the evolution of the screening collection) and spuriously high similarities that arise by chance among profiles with low variance in the overlap region. However, these weaknesses tend to affect at most one of the metrics for any given compound pair and for this reason it has proved useful to present the results as a scatter plot with the uncentered correlation coefficient as one axis and Euclidean distance as the other. Using these analysis methods, predictably, comparisons between signatures of known compounds demonstrated a high degree of structural similarity. The similarity of activity profiles between structurally distinct compounds with reportedly distinct therapeutic modalities and treatment populations, including cytotoxic compounds and calcium channel modulators, was also highly significant, suggesting a mechanistic link between calcium channel inhibitors and cytotoxic compounds, a result which has been corroborated by the literature [36].

Although COMPARE has been successfully used for the discovery of promising lead compounds for several molecular targets, it may suffer from treating the compounds and targets one pair at a time, which can result in an obscuring of important local features in the overall data [37]. An approach complementary to COMPARE developed by Weinstein and colleagues maps coherent patterns in the data and, instead of making pairwise comparisons, considers the data in aggregate [21]. This approach has led to a new tool named DISCOVERY, which integrates disparate types of data relating to the compounds and displays them in a way suited to human pattern recognition. As a first pass, to generate a human logical pattern describing compound data and inter-relationships, an algorithm termed "ClusCor" was employed. This algorithm treats three datasets (compound, target knowledge and activity) as a mathematical matrix for which each compound's activity and target data are normalized by its mean and standard deviation. These normalized data are multiplied to obtain the "matrix transpose", then each entry divided by $n - 1$ [where n = number of cell lines (~60)], thus producing a matrix of Pearson correlation coefficients relating activity of the compound to target patterns. The utility of this approach in classifying compounds was demonstrated in the clustering of a series of highly structurally related platinum analogs (such as cisplatin and carboplatin) which share mechanism, while a distinct set of diaminocyclohexyl platinum compounds sharing structural similarity with the cisplatin group, but having a different mechanism of action, are found elsewhere on the map. This methodology allows more subtle associations in mechanism to be made between structurally similar compounds while simultaneously illuminating shared properties between local structure-based groupings with distinct mechanisms [21]. In sum, the ability to normalize the data in aggregate conveniently allows more incisive analogies to be made, many of which are just being confirmed in the literature.

Taking a different approach to classify small molecules by putative functions, a new computational tool based on SOM methods has been employed [32]. As with many strategies that are designed to placate noisy data, such as PCA, MDS, HCA and related methods that maximize signal-to-noise ratios, the SOM method is properly matched with data which display significant random noise such as those derived from cell-based assays. The SOM method can be divided into two parts: clustering in high-dimensional space and projections into lower dimensional space [32]. Using SOMs to analyze the complete NCI database, relationships between chemotypes of tested compounds and their effect on four types of cellular activity, membrane transport/integrity, nucleic acid synthesis, mitosis and kinase-dependent cell cycle regulation, were identified. For instance, the map effectively segregated effectors of uridine biosynthesis, cytidine synthesis, purine biosynthesis, ribonucleotide reduction and nucleic acid intercalation from one another, demonstrating the strength of the SOM in dividing nucleotide biosynthetic processes into discrete categories [32]. An inordinate number of subtle inferences are contained in this highly granular, comprehensive map of the NCI dataset. In total, the analysis provides a foundation for clustering analysis of a complete data set to identify relationships between chemotypes and detailed biological mechanism. Furthermore, the findings were largely validated using the literature as a reference. As with other methods, strong evidence supporting within-cluster memberships and common cellular activity and compound selectivity between various types of cellular activity was marked.

Of course, SOMs and other computational tools can only generate hypotheses of biological significance that are as good as the datasets from which they derive. Therefore, a study of which factors can most strongly influence the analysis of multidimensional screening data must be performed. The SOM analysis of the NCI data set forth a useful set of principles for maximizing data richness and recovery. Rabow et al. designed a method of data conditioning in which Z-score normalization is performed to enhance the signal-to noise-ratio [32]. Z-score analysis (a measure of the distance in standard deviations of a sample from the mean specifically applied to gauge HTS assay quality) of each compound's raw data against tumor cell panels provides a common mean reference which serves to increase the signal above noise. The Z-score transformed data improved the quality of the clustering by ~15% when compared with the raw data. Normalization by this method emphasizes differential growth inhibition across all cell lines as opposed to other methods in which data are obscured by the most sensitive cell lines. In essence, Z-score normalization of each cell line's response to all tested compounds establishes a common reference such that a rigorous comparison of the relative contribution of experimental design parameters can be assessed. For instance, one can now ask how the number of cell lines considered in the analysis influences the precise mapping of a compound to the SOM. For the NCI screening, compound profiling data for any number of tumor cell lines under 20 significantly diminishes the structure–function correlation, whereas any number between 20 and 70 cell lines allows accurate mapping [32]. Rabow et al. went on to show showed that in addition to Z-score normalization, limiting the maximum allowed GI_{50} values capping at a value of three absolute deviations above or below the mean, which prevents the difference between two GI_{50}s being dominated by a few cell lines or compounds with extreme effects, led to improved structure–function determinations. In contrast, substituting missing data with the group average (or mean) compromises the clustering, decreasing the correlation coefficient by ~5%. Interestingly, as much as 40% of a dataset can be removed without overly affecting structure–function correlations. Lessons from the SOM analysis will probably be reflected in future studies of highly parallel cell-based data mining efforts.

If the goal of these data analysis methods and comparative studies was simply to group like compounds and segregate distinct ones, then the methods described thus far would be sufficient. However, if the goal is to infer properties of compounds such as mechanism of action (MOA), structure–function relationships, guide lead optimization campaigns and resolve potential side-effects based on analogy to known drugs, etc., one has to ask whether the pattern matching algorithms are predictive. If so, then perhaps they could be practically applied to guide biochemical modification of drug leads in active medicinal chemistry campaigns and prospectively help determine which compounds to select for more detailed analysis. Also, if activity patterns are predictive of MOA, then can they be used to assign MOA or function to previously undescribed compounds? Can associations underlying multiparallel cellular screening datasets be assembled appropriately to allow sophisticated queries, which with a statistically significant basis relate justifiable hypotheses and in fact reinforce correct ones through learning? This line of inquiry demands an infrastructure analogous to the architecture of the brain, in particular neural networks, which, in contrast to computer algorithms that are programmed to obtain the right answer based on an equation or linear logic, actually "learn" from the examples set forth in a dataset to obtain the right answer. Weinstein and others have explored this possibility, which is covered in the following sections.

10.3.2
Multidimensional Cellular Profiling for MOA prediction

In the halcyon days of drug discovery, experimentalists tested small collections of chemicals in cellular and animal models of disease which would then progress into humans, independent of whether the compound's target or mechanism of action was known. Examples of this are abundant in the annals describing cyclosporin, aspirin and penicillin target drug development [38–40]. In a latter reductionist era, the advent of cloning brought with it the ability to mass produce a putative protein drug target and test many compounds against it in search of active modulators which could then be tested in the very same cellular and animal models, herewith the instruction of target knowledge. Examples of this include development of HIV reverse transcriptase inhibitors for treatment of HIV/AIDS and statins which target HMG–CoA reductase for cholesterol homeostasis, to name just a few of the thousands of examples [41, 42].

A return to the old days of drug discovery has recently happened as the number of new chemical entities (NCEs) derived from the post-cloning era have not kept pace with the need for new drugs. Owing to concern over the toxicity of small-molecule drugs progressing through the clinic and FDA restrictions, increased importance has been placed on knowing a compound's mechanism of action well before going to market. Methods to detect the target and mechanism of action have not kept up with the many advances in the post-genomic era. Instead, the methods of chemical genetics and affinity-based techniques for small-molecule target deconvolution have lagged behind the cutting edge of life sciences research and drug development, creating a challenge for those running blinded cellular pathways [43]. Because of this gap, a strategy to select rapidly likely drug precursors from a small-molecule screen prior to committing resources to lengthy target ID projects or a means to predict mechanism of action accurately will determine the utility of this once effective drug discovery strategy.

Perhaps the strongest examples of the systematic mechanism of action determination of small molecules using the blinded pathway screen approach derive from the NCI's effort to identify NCEs for cancer treatment. In essence, profiling of well-described, mechanistically understood anti-proliferative agents against a panel of cell-based assays generates a reference panel by which new compounds can be compared (by activity profiles) to infer mechanism of action (e.g. purine biosynthesis, antifolates, apoptosis) and sometimes a precise target family [e.g. topoisomerase, cyclin-dependent kinases (CDKs)] [32]. As testament to the utility of the blind-screen approach, at least five compounds from the NCI anti-cancer drug screen have entered into Phase I trials including flavopiridol and UCN-01, both kinase inhibitors, quinocarmycin, a DNA interactor, the spicamycin analog KRN-5000, which disrupts glycoprotein processing, and depsipeptide NSC630176, a histone deacetylase inhibitor [44–48].

In fact, the first realization that the NCI anti-cancer screen database could be used as a source of information for mechanism-of-action prediction came from the success of the COMPARE algorithm described earlier in this chapter. The use of the COMPARE algorithm allowed the identification of unique structures with tubulin-binding activity, topoisomerase II activity and antimetabolites [49–53]. Advanced computational means to extract mechanism-of-action data from the NCI dataset such as global data conditioning and SOM cluster and visualization tools developed for more rigorous and incisive determinations described previously have also been usefully applied to MOA determinations of novel compounds. In generating a "complete" SOM map for the NCI dataset, Rabow et al. were able to identify a number of powerful and previously unappreciated structure–function relationships and MOA hypotheses. What is unique about the SOM analysis is that it literally creates a multidimensional map relating structure with MOA, where compounds are stratified into subregions based on strength of MOA similarity while those in distinct subregions which still share borders share related but distinct MOAs. As an example residing in subregions of the map that correspond to CDK inhibition by known CDK inhibitors such as UCN-01 and olomoucine are several paullone-like compounds with previously unknown CDK inhibition characteristics and rapamycin analogs, with an underappreciated, yet literature-supported, role in CDK4 inhibition [32]. Similarly, corresponding to the phosphatase 2A inhibition subregion defined by the likely suspects are found fostriecin and cytostatin, once considered to be topoisomerase II inhibitors, now reclassified as

PP2A inhibitors [32]. The complete SOM contains innumerable insights into MOA for compounds. Advised with nothing more than a known structure and the NCI panel profile, a functional annotation for an uncharacterized compound can be assigned and hypothesis tested. For this very reason, researchers including those outside the cancer community send compounds of unknown mechanism to the NCI screen to obtain MOA predictions.

As an extension to the NCI data analysis strategies, Wang et al. have designed another Web-based tool like COMPARE to mine the NCI database specifically for novel target–MOI studies. The study incorporates robust analysis of the correlation between *in vitro* anti-cancer activity of the drugs in the NCI anti-cancer database to include protein levels and mRNA levels of the molecules they target. Shao et al. have demonstrated that using standard Pearson correlation coefficient analysis similar to COMPARE, novel protein kinase C (PKC) ligands such as NSC631939 and NSC 631941 could be discovered [37, 54]. In a follow-up study, using a nanomolar affinity PKC ligand mezerein as a "seed", approximately 7269 compounds were compared for those which had overlapping GI_{50}s across the panel of tumor cell lines. This analysis identified 12 potential ligands, of which at least two, NSC266186 (huratoxin) and NSC 654239 (cytoblastin), an analog of teleocidin, were later confirmed as PKC ligands [37].

The drug discovery research community at large has not ignored the useful approaches exacted by the NCI and have emulated the approach by designing their own unique cell line panels and compound test agents. Yamori and others established a panel of 39 cell lines according to the NCI principles, diverging mainly in the unique inclusion of three distinct breast cancer cell lines and six stomach cancer cell lines based on the prevalence of this disease in Japan. In their scheme, a novel anti-cancer agent, MS-247, that has a netropsin-like moiety and an alkylating residue in the structure, was identified with significant correlation to the profile of camptothecin, suggesting that the two drug groups may have similar modes of action [55, 56]. Subsequently, MS-247 exhibited remarkable anti-tumor activity against various xenografts. A single i.v. injection of MS-247 significantly inhibited the growth of all 17 xenografts tested, which included lung, colon, stomach, breast and ovarian cancers. In many cases, MS-247 was more efficacious than other chemotherapeutics such as cisplatin, adriamycin, 5-fluorouracil, cyclophosphamide, VP-16 and vincristine, and was almost comparable to paclitaxel and CPT-11, which comprise the most clinically promising drugs at present. Wang and others have incorporated gene expression profiling, distributed tissue expression and protein levels of NCI drug targets as additional datasets in correlating anti-cancer activity of such drugs [57]. Therefore, mechanism of action and precise target definition predicted by multi-parameter cell profiling approaches have been borne out in empirical tests, serving as proof of concept for the approach.

Lastly, publicly available data can be used in essence as a global, comprehensive database of compounds screened against an inordinate panel of cell lines. It stands to reason that since many biologically active compounds reveal several kinds of biological activities when tested against a large number of targets/pathways, one might expect that the NCI database may potentially possess additional biological activities not related to cancer and/or HIV replication. Therefore, it is entirely reasonable to hypothesize that new leads for the development of different classes of drugs might be found among these molecules. Of course, since it would be extremely expensive and time consuming to test all these agents against all other possible targets/pathways, a computer-based screening method was devised based on literature-documented compound MOA associations. Pooling information from the literature, Poroikov et al. have amassed a global "superbase" connecting compound structure, MOA, pharmacological effects, known toxicities and other descriptors for much of the publicly available literature dating back to 1972 [58]. This database can then be queried using the computer application PASS (prediction of activity spectra for substances), which comprises about 250 000 compounds of the NCI Open Database and the incorporation of over 64 million PASS predictions in the enhanced NCI database [59]. By pre-assembling the public literature into a cluster map, this software allow queries for compounds which possess or are likely to possess biological, pharmacological or MOA desired by the researcher. For instance, a query for "angiogenesis" inhibitors yields multiple unique compound structures with a given probability of blocking angiogenesis empirically. In practice, this exact query identified seven distinct candidates for angiogenesis inhibition, of which four demonstrated inhibitory activity. Although the number of compounds tested is too low to show statistical relevance, this tool does provide a free method of mining public data for possible novel leads in drug discovery.

10.3.3
Cellular Profiling in Lead Exploration

Genome- and proteome-wide analysis of gene and gene product interactions across multiple physiological systems have begun to unveil the lexicon by which signaling networks are encoded. Like many other networks, phenotypic properties of biological networks are manifest in theory through two basic forms of interaction: "a small number of highly connected nodes that interact with a large number of nodes with a low degree of connectivity and a large number of lowly connected nodes that interact with a small number of nodes with a high degree of connectivity" [60] (Fig. 10.2). In theory, the result of blocking a highly connected node with a small molecule could be severe, impinging on all the other processes to which the node is connected, while it stands to reason that pharmacological intervention at a lowly connected node might have fewer global effects. In other words, targeting a highly connected node such as pleiotropic molecules like NF-kB or p53 can elicit a desired pharmacological outcome, but it can also result in undesirable side-effects. In fact, underlying the toxicities of certain marketed and recently withdrawn drugs may be the high connectivity of the drugs target [e.g. Cox (cyclo-oxygenase) inhibitors] [61]. Instead, means to identify pharmacological effectors of lowly connected but critical nodes may be a preferred strategy. Lessons from forward cellular genetics teach us that underlying the effects of genetic variation and chemical modulation of cells is a set of instructions ascribed to biological networks. Thus, given the state-of-the-art tools of the post-genome, blinded cellular pathway

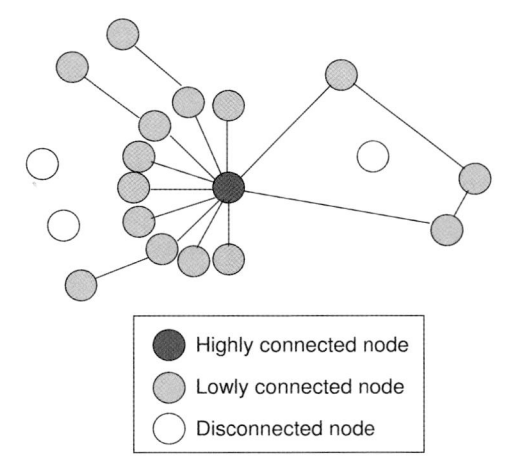

Figure 10.2 Node network theory schematic. Biological networks can be represented as an array of nodes including a highly connected nodes (dark gray circle), lowly connected nodes (light gray circles) and disconnected nodes (white circles). Theoretically, highly connected nodes would control multiple processes and thus connect with many other nodes. Perturbation of such highly connected nodes would likely have pleiotropic effects because of their influence on many processes. Perturbation of lowly connected nodes would likely have fewer pleiotropic effects because of their isolation (as would be the case of perturbing the disconnected or unconnected nodes). Examples of a highly connected node would be NF-kB, which is involved in regulating many processes including proliferation, inflammation and retroviral infection.

screens for such small-molecule effectors or genetic screens in the same pathways for lowly connected nodes may prove to be a worthwhile pursuit. Conversely, a drug whose target is a highly connected node may have many therapeutic applications which can achieve off-label success (aspirin, statins) [62, 63]. Irrespective of the connectivity of a target, multidimensional cellular profiling can be used to determine the qualitative connectivity or range of effects of compounds on biological end-points.

Compound profiling across a battery of phenotypic assays can describe the global effects of molecules from a given structural series from a lead optimization program and guide scientists towards chemical modifications fostering desirable biological effects. In a proof of concept experiment of sorts, Kim et al. (Schreiber) sampled a library of highly related compounds synthesized by diversity-oriented methods, which included monocyclized precursors and their bicyclic counterparts against a battery of readouts including DNA synthesis, cellular esterase activity, mitochondrial membrane potential and intracellular reducing equivalents [64]. Parallel tests of 244 compounds and a small set of known drugs against the four cellular states in 40 cellular assays resulted in a matrix of ~20 000 measurements, which were then subjected to standard Z-score normalization. In order to rationalize the results, the matrix was further subdivided into conformational restriction space using descriptors such as number of rotatable bonds for each compound.

The global cellular activity patterns of compounds co-segregated with their mono- or bicyclic skeleton while further analysis by hierarchical clustering showed that the stereochemistry of the appendages was a major determinant These studies demonstrate the possibility of using parallel cellular assays for quantitative measurement of the effects of discrete chemical modification.

Shi et al. showed coherence between the activity patterns and molecular structure of 112 ellipticine analogs tested in the NCI tumor cell panel [65, 66]. The population co-segregated into patterns associated with either normal ellipticines or N^2-alkyl-substituted ellipticiniums, whose discrepancies, outliers, etc., were all dependent on specific structural modifications. Furthermore, a correlation between brain-derived tumor cells and p53 status was manifest in the patterns, where the ellipticiniums were statistically more potent than the ellipticines in cells with mutant p53 versus wild-type. The distinction provides the researcher with a rational means to prioritize and select the next steps in a compound progression scheme based on either the prevalence of p53 mutations in the brain tumor patient population or based on structural liabilities of these compounds.

Haggarty et al. described an analysis of a molecular descriptor space based on a 1,3-dioxane-based diversity-oriented synthetic library containing structural elements aimed at inhibiting protein deacetylases in general [67]. Using PCA and 3D visualization coupled with arrays of cell-based assay results, a multidimensional landscape was derived relating chemical structure with activity. The results show that certain structural elements are important for conferring specificity and selectivity on deacetylase inhibitors in the context of tubulin and histone substrate function. These studies show again that multidimensional profiling can give rise to activity patterns that correlate small structural perturbations with cellular activity, but also that insight into MOA could be levied to promote one series of chemical modifications over another.

The previous cellular epithets serve as analogies to the way three-dimensional crystal structure information of a protein target can inform rational chemical modification for modulating the affinity for a target and away from an off-target or undesirable property. It is clear that multidimensional cellular profiling can be used to drive chemical modification to a local optimum of biological activity, but how can it be used to avoid untoward mechanisms or liabilities associated with failed drugs? To address this, Weinstein et al. used DISCOVERY-based organization of compounds, which introduces gene expression profiling data from the cellular panel, to create a cluster map of the NCI cancer screen data. The study identified a patch of compounds negatively correlated with MDR-1 gene expression, the pump that prevents many drugs from gaining access to the inside of cells and rendering them inactive. Another set of compounds were identified that were known MDR-1 substrates of which half were included in the MDR-1 gene expression correlative patch. This type of insight can provide a basis to discard compounds within or near the biological mechanism space of MDR-1 substrate and response, which in the case of cancer can preclude efficacy [68, 69]. Wallqvist et al. employed a method of tumor cell screening using a combination of activity profiles in the NCI panel with gene expression profiling data to implicate benzothiazole treatment of breast cancer cells with induction of the p450-metabolizing isozymes CYP1A1 and CYP1B1 [70]. The results are consistent with the proposed inactivity of the CYP1A1-mediated metabolism of benzothiazoles and the anti-tumor activity of the metabolically resistant halogenated form.

Multidimensional cellular profiling in the context of kinase selectivity determinations can aid in the prioritization of chemical leads, especially given the innate promiscuity of ATP-competitive inhibitors. Kunkel et al. developed a method for rapid structure–activity determinations using parallel cellular profiling to determine the selectivity of a series of kinase inhibitors. The method, termed BioMAP, involves profiling compounds in primary cell-based models of inflammation and leukocyte trafficking using multiplex ELISA assays [71]. The study demonstrates the utility of cellular profiling to define SARs within a series of trisubstituted imidazole inhibitors of p38 MAPK, but also informatively points out secondary activities and discriminating core functional activities for selection of lead candidates [72]. In the work, over 50 pathways and targets are interrogated with the small molecule library in four distinct primary cell lines with 31 unique ELISA-based readouts. Then, by comparison with response patterns of 33 known kinase inhibitors against the same cell panel, a functional similarity map is drawn using pairwise correlations which allow clustering of compounds with similar activity. Here, compounds against distinct targets such as CamKII and PKA lie far apart on the resulting BioMAP whereas compounds (roscovitine and GW8510) targeting Cdks elicit similar responses and thus cluster together. Compounds targeting the same protein (BAY 43–9006 and GW5074 on raf-1) group most closely together; however, subtle differences in specificity underlie their imperfect map superimposition [72]. Likewise, well-described CK2 inhibitors in the study which do not overlap are predicted to have significant off-target or secondary activities. Applying similar logic to the p38 inhibitor study, Kunkel et al. discovered that variations to the p38α inhibitor structure SB203580 at the R1 site control its effects on E-selectin, IL-8 and VCAM-1 levels whereas subtle modification at other sites picked up useful alternative biological properties, such as CD87/uPAR inhibition.

10.4
Systematic Serendipity

While multidimensional cellular profiling experiments have utility in identifying MOA when compared with a library of known and/or well-characterized drugs, and can be used in chemical optimization campaigns (SAR) and for the identification of untoward secondary effects (toxicity, CYP induction, MDR-1 substrates, etc.), a novel application is the identification of additional opportunities for drugs and drug candidates outside the scope of their intended use. A study by Kunkel et al. demonstrated that using the BioMAP strategy, new biological insights for old compounds could be ascertained. For instance, examination of the profile of HMG–CoA inhibitors, statins, shows that one of the dominant features of these profiles is the significant reduction in CD69 levels, a T-cell activation antigen, in the SAg system [72]. Of the HMG–CoA reductase inhibitors tested in this study, all seven demonstrated this activity. Proof of the importance at the clinical level comes with the reported activities that statins have on immune regulation in addition and distinct from their effects on cholesterol.

A multidimensional profile was also generated for mycophenolic acid, an immunosuppressant used in kidney transplantation. The profile indicates a novel activity in inhibiting production of MCP-1, a monocyte chemoattractant that plays a role in monocyte homing in chronic inflammatory disease. Increased levels of MCP-1 are found in patients with diabetes and are associated with an increased risk for cardiovascular disease. A study revealed that mycophenolic acid decreases the incidence of post-transplant diabetes, in contrast to the immunosuppressant FK-506, which is known to increase the incidence [72]. The study reveals secondary activities that could contribute both positive and negative side-effects in animals and humans.

In another example, a series of dihydroxyanthracenones which map to a specific locale on the SOM described by Rabow et al. are known to have anti-psoriatic activity via 5-lipoxygenase; however, the proximity of these compounds to a region of compounds classified by membrane damage via oxygen radical generation suggests an additional mechanism [32]. As it so happens, a well-known side-effect of anti-psoriatic compounds agents is membrane damage by oxygen radicals. This is an example where the SOM can be utilized for drug discovery by, for example, finding anthracenone-like compounds not found in the DNA-damaging region [73]. Rabow suggests that validation of this approach could lead to new anti-psoriatic drugs with reduced side-effects.

The aforementioned examples represent retrospective analogies made from clinical observation to molecular pharmacology. Indeed, examples of prospective associations or serendipitous discoveries of clinical benefits from side-effect profiles are well known. Viagra, a PDE5 inhibitor initially developed as a heart medication, was discovered to have a "side-effect" on penile erection and male impotency [74, 75]. Often clinicians discover a positive side-attribute of a medication and prescribe based on this new functionality. Perhaps drugs with multiple side-effects would be looked at askance in development, but consider the utility to a drug developer knowing a compound had a beneficial side-effect that could pace its product past competitive front-runners. The situation begs the question of whether one can proactively screen for beneficial characteristics of compounds with potential clinical outcomes in the early stages of drug discovery research and begin testing in animals (and eventually human models of off-indication diseases) early when the expense is minimal.

A strategy termed "systematic serendipity" has been instituted at the Genomics Institute of the Novartis Research Foundation, in order to discover methodically potential positive (and negative) side-attributes of small-molecule HTS "hits" and advanced compounds derived from lead optimization campaigns. A uHTS approach to multiparameter cellular activity profiling has been established. An automated robotic screening system which routinely profiles compounds of unknown function, compounds derived from advanced lead optimization campaigns and drugs and compounds with known targets and function have been tested in single-point format across ~45 distinct blinded cell-based pathway screening assays. The assays represent a diversity of signaling pathways and therapeutic indications, thus broadening the scope of inquiry beyond cancer and HIV, extending to metabolic, cardiovascular, inflammation, asthma, neurobiological and other important areas of biology and medicine.

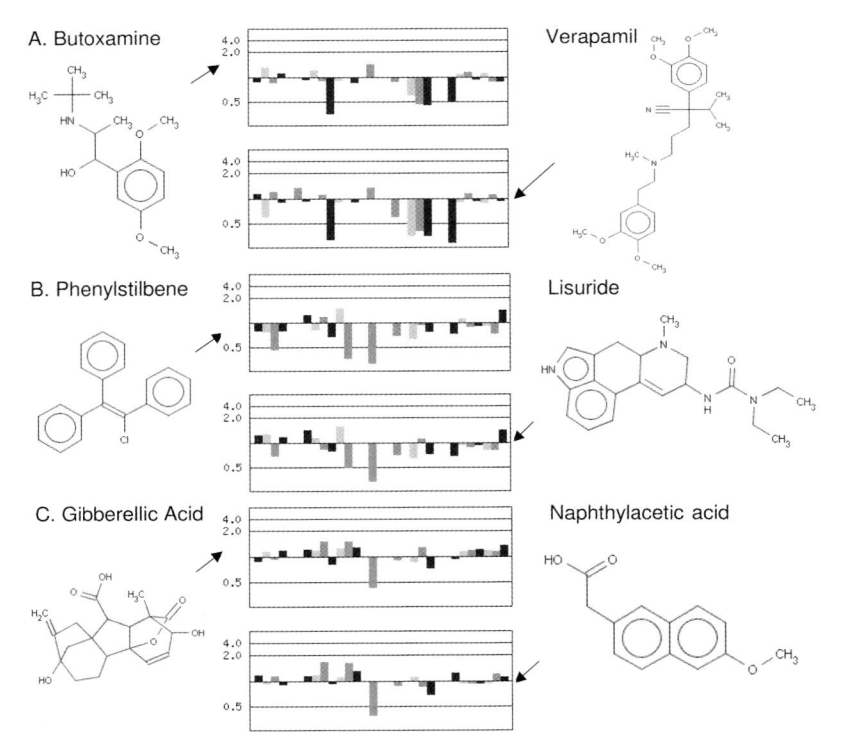

Figure 10.3 Systematic serendipity results. Automated HTS of known drugs with well-defined mechanisms of action were screened against a battery of cell-based reporter gene and proliferation assays to generate activity profiles [for each pairing between the seed compounds (on the left) and selected queried compounds (on the right), two similarity metrics are computed over the assays which those compounds have in common – an uncentered Pearson correlation coefficient and a count-normalized Euclidean distance]. (A) When butoxamine, a beta-blocker (structure shown on left) was chosen as the "seed" compound, verapamil, a calcium-channel inhibitor (structure on right), was determined to be the "nearest neighbor" from a selection of approximately 3 000 other drugs, as shown by their highly overlapping activity profiles. (B) When phenylstilbene, a tamoxifen analog, was chosen as the "seed" compound, lisuride, a dopamine agonist, was determined to be the "nearest neighbor" as shown by their highly overlapping activity profiles. (C) When gibberellic acid, an active component in aloe vera, was chosen as the "seed" compound, the anti-inflammatory naphthylacetic acid was determined to be the "nearest neighbor" as shown by their highly overlapping activity profiles. Despite the distinct chemical structures in the three pairings, the overlapping profiles indicate a shared mechanism, several of which have been corroborated in the literature.

As examples of the ability of this systematic approach to identify potential new applications of a drug or drug candidate, approximately 3000 known and well-characterized drugs were profiled in the aforementioned cell panel. In these data-mining experiments, the activity profile for butoxamine, a β-adrenergic antagonist used for a variety of indications, was found to be most closely related to verapamil, a calcium channel antagonist. Although not prescribed as such, butoxamine has been shown to have calcium channel-blocking activity, as demonstrated in calcium flux experiments in salivary glands, indicating a mechanistic link to verapamil activity (Fig. 10.3A) [76, 77]. Interestingly, both compounds are used to treat angina and have been combined in patients at high risk of ischemic disease. Profiles of phenylstilbene chloride, a tamoxifen-derived estrogen receptor ligand, and lisuride, a dopamine receptor agonist, were also found to be "nearest" neighbors (Fig. 10.3B). Phenylstilbene has well-described activity as an inhibitor of prolactinoma growth and pituitary enlargement whereas lisuride is primarily known as an anti-Parkinson's agent [78, 79]. Interestingly, lisuride has been shown to block cell proliferation and apoptosis in the estrogen-stimulated anterior pituitary gland. This suggests that the activity profile can predict an off-target or secondary mechanism of action/indication for a drug. Activity profiles for the naproxen analog 6-methoxy-2-naphthylacetic acid, an anti-inflammatory which acts via inhibition of prostaglandin (PGE2) production, when used as a "seed" was found most closely related to the profile for gibberellic acid, an agent found in anti-inflammatory aloe vera extracts (Fig. 10.3C). Interestingly, gibberellic acid extracts have been shown to block adjuvant-induced arthritis and gibberellin itself blocks inflammation by inhibition of polymorphonuclear leukocyte infiltration [80]. For the sake of illustration, only compounds with evidence of a tangible link to a predicted "off-label" indication are furnished here. The obvious next set of experiments would involve comparing novel uncharacterized compounds in lead optimization for a particular indication with known drugs prescribed for distinct disease states.

10.5
Conclusion

Ultimately, the tools outlined in this chapter encompassing data analysis methods for cell-based screening and their application to multidimensional cellular profiling will be used proactively to discover the MOA of bioactive compounds from phenotypic screens. Perhaps the efficiency of cellular panel profiling and data analysis will lead to the necessary commoditization and export of the systems outlined in this chapter to all practitioners of drug discovery such that lead optimization campaigns throughout the industry can exploit it for guidance in lead selection. Indeed, a combination of the approaches from the NCI, which intensely interrogate a specific therapeutic indication, the GNF, which addresses multiple indications in high throughput, and the PASS database, which incorporates disparate structure–function data globally from the published literature will increase the probability that the traditional model of drug discovery, namely cell-based screening, will meet with the target identification and safety requirements demanded by the modern pharmaceutical engine.

References

1. http://www.lucerna-chem.ch/downloads/biosource/alamarbluebooklet.pdf.
2. http://www.rndsystems.com/pdf/ta5355_5412.pdf.
3. http://www.promega.com/tbs/tb288/tb288.pdf.
4. J.H. Zhang, *Biomol. Screen.* **1999**, *4*, 67–73.
5. J.H. Zhang, *J.Comb. Chem.* **2000**, *2*, 258–265.
6. http://slims.sourceforge.net.
7. C. Brideau, *J. Biomol. Screen.* **2003**, *8*, 624–633.
8. S. Heyse, *Proc. SPIE* **2002**, *4626*, 535–547.
9. B. Gunter, *J. Biomol. Screen.* **2003**, *8*, 634–647.
10. D.E. Root, *J. Biomol. Screen.* **2003**, *8*, 393–398.
11. C.A. Nicolaou, *J. Chem. Inf. Comput. Sci.* **2002**, *42*, 1069–1079.
12. S.Y. Tamura, *J. Med. Chem.* **2004**, *45*, 3082–3093.
13. G. Roberts, *J. Chem. Inf. Comput. Sci.* **2000**, *40*, 1302–1314.
14. P. Blower, *J. Chem. Inf. Comput. Sci.* **2002**, *42*, 393–404.
15. A. Rusinko, M, *J. Chem. Inf. Comput. Sci.* **1999**, *39*, 1017–1026.
16. M. Glick, *J. Biomol. Screen.* **2003**, *9*, 32–36.
17. S.J. Wilkens, *J. Med. Chem.* in press.
18. DecisionSite, http://www.spotfire.com.
19. M.B. Eisen, *Proc. Natl. Acad. Sci. USA* **1998**, *95*, 14863–14868.
20. O. Troyanskaya, *Bioinformatics* **2001**, *17*, 520–525.
21. J.N. Weinstein, *Science* **1997**, *275*, 343–349.
22. J.S. Caldwell, *Chem. Biol.* **2003**, *9*, 784–786.
23. P.A. Clemons, *Curr. Opin. Chem. Biol.* **2004**, *8*, 334–338.
24. C.J. Torrance, *Nat. Biotechnol.* **2001**, *19*, 940–945.
25. S. Dolma, *Cancer Cell* **2003**, *3*, 285–296.
26. V.R. Fantin, *Cancer Cell* **2002**, *2*, 29–42.
27. P.H. Sneath, *Nature* **1962**, *193*, 855–860.
28. R.D. Meyer, *Curr. Opin. Biotechnol.* **2000**, *11*, 89–96.
29. M.P. Brown, *Proc. Natl. Acad. Sci. USA* **2000**, *97*, 262–267.
30. T. Kohnonen, *Neural Comput.* **1999**, *11*, 2081–2095.
31. J.N. Weinstein, *Science* **1992**, *258*, 447–451.
32. A.A. Rabow, *J. Med. Chem.* **2002**, *45*, 818–840.
33. L.M. Shi, *J. Chem. Inf. Comput. Sci..* **2000**, *40*, 367–379.
34. D.W. Zaharevitz, *J. Mol. Graph. Model.* **2002**, *20*, 297–303.
35. N.S. Duesbery, *Science* **1998**, *280*, 734–737.
36. D.E. Root, *Chem. Biol.* **2003**, *10*, 881–892.
37. X. Fang, *J. Chem. Inf. Comput. Sci.* **2004**, *44*, 249–257.
38. S.L. Schreiber, *Science* **1991**, *251*, 283–287.
39. E. Kopp, *Science.* **1994**, *265*, 956–959.
40. T. Hoyem, *Nord. Vet. Med.* **1961**, *13*, 433–438.
41. I. Frank, *Clin. Lab. Med.* **2002**, *22*, 741–757.
42. H. Lennernas, *Clin. Pharmacokinet.* **1997**, *32*, 403–425.
43. A. Monks, *Anticancer Drug Des.* **1997**, *12*, 533–541.
44. S.M. Seynaeve, *Mol. Pharmacol.* **1994**, *45*, 1207–1214.
45. H.H. Sedlacek, *Crit. Rev Oncol Hematol.* **2001**, *38*, 139–170.
46. C.A. Bunnell, *Cancer Chemother. Pharmacol.* **2001**, *48*, 347–355.
47. Y.S. Lee, *Cancer Res.* **1995**, *55*, 1075–1079.
48. F.A. Dorr, *Eur. J. Cancer Clin. Oncol.* **1988**, *24*, 1699–1706.
49. R.L. Bai, *J. Biol. Chem.* **1991**, *266*, 15882–15889.

50 K.D. Paull, *Cancer Res.* **1992**, *52*, 3892–3900.
51 F. Leteurtre, *J. Natl. Cancer Inst.* **1994**, *86*, 1239–1244.
52 H.N. Jayaram, *Biochem. Biophys. Res. Commun.* **1992**, *186*, 1600–1606.
53 E.S. Cleaveland, *Biochem. Pharmacol.* **1995**, *49*, 947–954.
54 L. Shao, *J. Med. Chem.* **2001**, *44*, 3872–3880.
55 T. Yamori, *Cancer Res.* **1999**, *59*, 4042–4049.
56 S. Dan, *Cancer Sci.* **2003**, *94*, 1074–1082.
57 T. Yamori, *Cancer Chemother. Pharmacol.* **2003**, *52*, Suppl. 1, S74–S79.
58 A. Lagunin, *Bioinformatics* **2000**, *16*, 747–748.
59 V.V. Poroikov, *J. Chem. Inf. Comput. Sci.* **2003**, *43*, 228–236.
60 B.K. Wagner, *Am. .J Pharmacogenom.* **2004**, *4*, 313–320.
61 S. Beardsley, *Sci. Am.* **2005**, *292*, 16–18.
62 M.M. Rumore, *Drug Intell. Clin. Pharm.* **1987**, *21*, 961–969.
63 G.K. Dogra, *Kidney Int.* **2002**, *62*, 550–557.
64 Y.K. Kim, *J. Am. Chem. Soc.* **2004**, *126*, 14740–14745.
65 L.M. Shi, *J. Chem. Inf. Compu.t Sci.* **1998**, *38*, 189–199.
66 L.M. Shi, *Mol. Pharmacol.* **1998**, *53*, 241–251.
67 S.J. Haggarty, *Comb. Chem. High Throughput Screen.* **2004**, *7*, 669–676.
68 M. Alvarez, *J. Clin. Invest.* **1995**, *95*, 2205–2214.
69 L. Wu, *Cancer Res.* **1992**;52:3029–3034.
70 A. Wallqvist, *Prog. Cell Cycle Res.* **2003**, *5*, 173–179.
71 E.J. Kunkel, *Assay Drug. Dev. Technol.* **2004**, *2*, 431–441.
72 E.J. Kunkel, *FASEB J.* **2004**, *18*, 1279–1281.
73 K. Muller, *Curr. Pharm. Des.* **2000**, *6*, 901–918.
74 M. Boolell, *Int. J. Impot. Res.* **1996**, *8*, 47–52.
75 H. Padma-Nathan, *Am. J. Cardiol.* **1999**, *84*, 8N–23N.
76 C.A. Shneyer, *Arch. Oral Biol.* **1986**, *31*, 383–386.
77 C.A. Shneyer, *J. Auton. Nerv. Syst.* **1985**, *13*, 275–285.
78 G. Verde, *J. Endocrino.l Invest.* **1980**, *3*, 405–414.
79 G. Rudali, *Biomedicine* **1980**, *33*, 126–128.
80 R.H. Davis, *J. Am. Podiatr. Med. Assoc.* **1989**, *79*, 24–26.

Appendix

List of instrument photographs

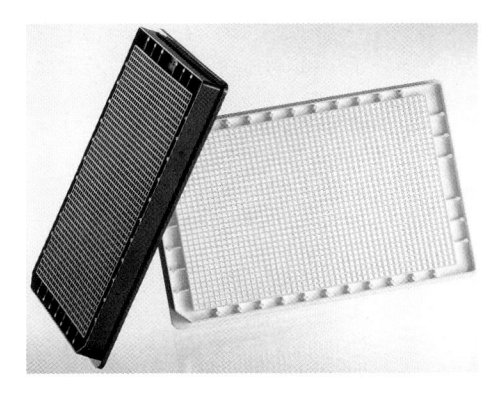

1536 well-plate (Greiner)

384 Pipettor (CyBio)

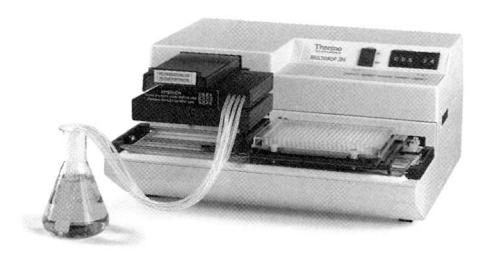

MultiDrop Dispenser (Thermo Labystems)

List of instrument photographs | **325**

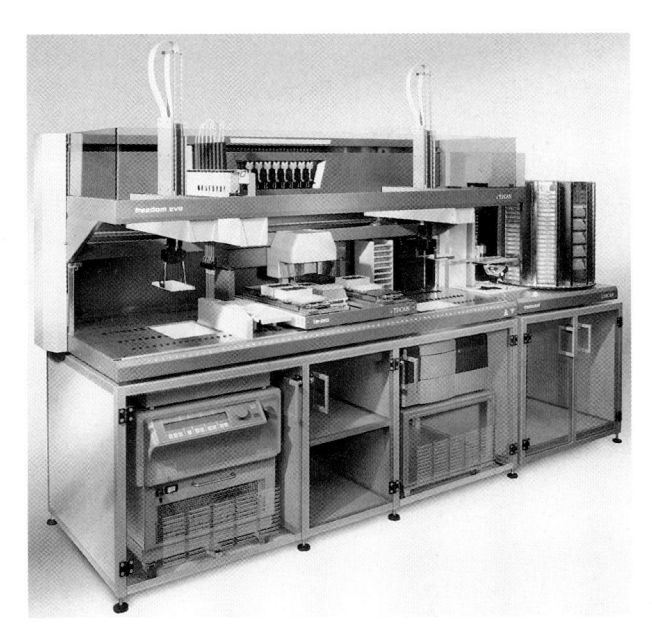

Tecan Genesis Freedom

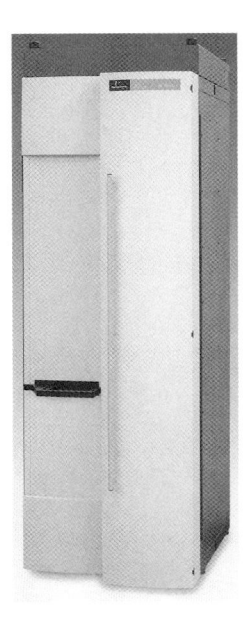

PerkinElmer ViewLux

PerkinElmer EnVision ± stackers (2 photos)

PerkinElmer MicroBeta

CyBi-Lumax Workstation

List of instrument photographs | 327

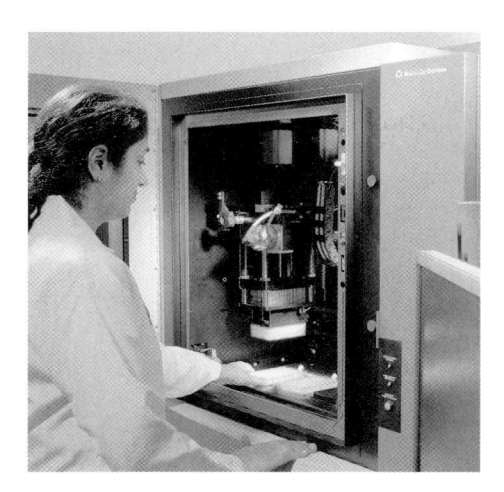

Molecular Devices FLIPR Tetra

Molecular Devices IonWorks Quattrao

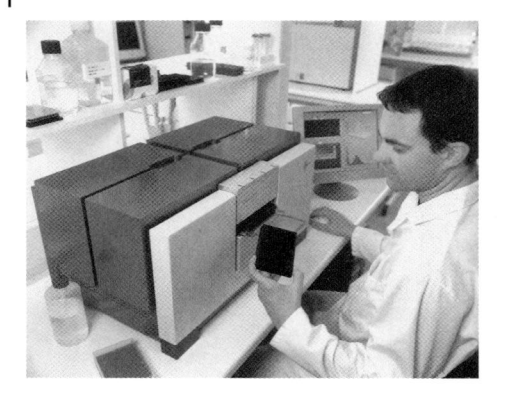

TTP Labtech Acumen Explorer

Zymark Twister 2

List of instrument photographs | 329

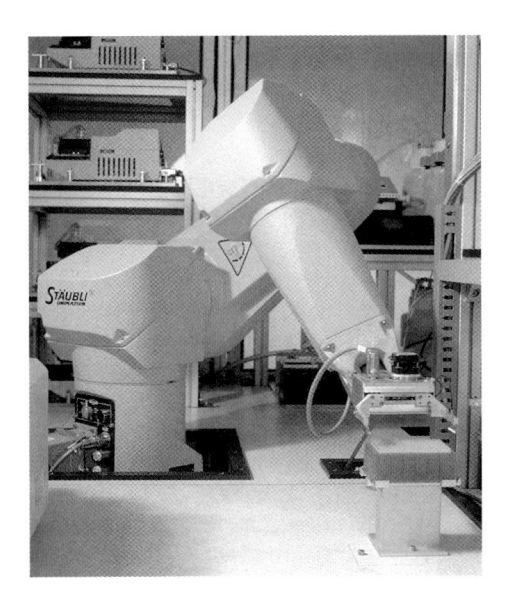

Staubli Robot

TAP Select T

Zeiss UHTS

REMP – Gantry Robot in narrow isle store

RTS Smart Store

Labcyte Echo 380

Index

a

AAO, *see* automated assay optimization
AAP (4-aminoantipyrine) 113
AAS (atomic absorption spectrometry) 60, 84
absorbance 54, 158
absorption 233
ABTS (2,2'-azobis-3-ethylbenzothiazoline-6-sulfonic acid) 113
acceptor-donor pair 102 ff.
ACE (angiotensin-converting enzyme) 26, 104 f.
acoustic auditing 71
acoustic transducer 48
activity scale 200
ADME (absorption, distribution, metabolism, and excretion) parameter 18, 231 ff.
– ADMET (toxicology) 43, 209 ff., 232 ff., 277
– ADMETox profile 29
β-adrenoceptor 81
adsorbent 272
affinity
– antagonist 81
– binding 114
– drug 17
AIDS 311
allosteric modulation 21, 76
allosteric protein 76
AlphaScreen ™ 53
alumina-based gel 272
AMC (7-amino-4-methylcoumarin) 102
amino acid oxidase 113
angiogenesis 314
antagonist 31
anticancer drug 312
array
– compound screening (ARCS) 44
– microplate 42
artifact 29, 183 ff.
artificial neural network (ANN) 224
assay
– agonism 303
– antagonism 303
– anti-proliferation 303
– artifact 29
– bioassay 22
– biochemical 93 ff., 301
– cell-based 75 ff., 301
– continous 107
– cross-assay 153, 299
– de-repression 303
– development 155
– development failure 125
– estrogen receptor 281
– inhibition 303
– luciferase 305
– map 177
– MDR potentiation 287
– noise 24
– phenotypic 21
– pyruvate kinase-lactate dehydrogenase (LDH) coupled assay 120 ff.
– reporter gene (RGA) 25, 76 ff., 299
– scintillation proximity assay (SPA) 54 ff., 114 ff.
– sensitivity 155
– small molecule microassay 7
– targeted 22
– technology 21
– toxicity 305
– virtual 303 ff.
– window 156
autofluorescence 61, 305
automated assay optimization (AAO) 158
automated patch clamping (APC) 60
automation 24 ff., 37 ff.
– cell culture 66
– detection 61

High-Throughput Screening in Drug Discovery. Edited by Jörg Hüser
Copyright © 2006 WILEY-VCH Verlag GmbH & Co. KGaA, Weinheim
ISBN: 3-527-31283-8

Index

average 184 ff.
– average shifted histogram (ASH) 181
– local 186
– running average 146

b

B-score 200
background response 185
– surface 174 ff., 191 ff.
bacterial growth 55
baculovirus 98
BAPTA (1,2-bis(o-)-N,N,N',N'-tetra) 88
Baytree 226 ff.
bioactive molecule 4
bioassay 22
bioavailability 234
– oral 232
biochemical assay 93 ff.
BioMAP 317
bioprofiling 153
blebbistatin 3
blood-brain-barrier (BBB) 234
boxplot 175
BRET (bioluminescence resonance energy transfer) 53
BTS (N-benzyl-p-toluenesulfonamide) 3

c

Ca^{2+} 30, 76 ff., 88, 159
– channel 89
– indicator 88 ff.
Caco-2 cell 233
cAMP 80 ff.
capillary slipper 48
carboxypeptidase 104
CARM 116 f.
casein kinase 120
CATS (chemically advanced template search) 216 f.
CCD (charge-coupled device) 56, 137
cell
– single cell analysis 129
cell culture
– automation 66
cell cycle 3
– kinase 99
cellular energy metabolism 77
central nervous system (CNS) 234
cGMP 90
change point
– detection 186
– graph 171
channel gating 83

Chebyshev polynomial 183 ff.
checkerboard pattern 191
ChemCard 44
ChemDiverse 237
chemical genetics 1 ff.
chemical medicinal chemistry database 229
chemical structure 218
chemoinformatics (cheminformatics) 207 ff., 231
chemometrics 208
cherry picking 28, 51, 69
chromatography 269 ff.
– automated preparative HPLC 273
classical genetics 1 f.
ClassPharmer™ Suite 241
clinical drug candidate 286
CLND (chemiluminescent nitrogen detector) 71
ClusCor 309
clustering 218
– hierarchical 219
– non-hierarchical 219
– similarity-based 230
coefficient of variation (CV) 45
coincidence counting 55
combinatorial chemistry 259 ff.
combinatorial library 209
combinatorial synthesis 267, 283
– library 271 ff.
– solution phase 271
CoMFA (comparative molecular field analysis) 216
COMPARE 308 ff.
competitor compound 261
compound
– library 260 f.
– liquid 68
– management (CM) system 67
– mixture 77
– plate-based liquid compound storage 68
– reformatting 50
– solid 67
– tube-based liquid compound storage 68
compression format 77
computational chemistry 208
conductance 85 ff.
confirmation rate 29
confocal laser scanning 134 ff.
connectivity 314
copper (Cu) 100, 110
correction
– complete plate set 194

– function   183
– individual plate   194
– position-dependent   195
– sample-based   200
CorScape™ heatmap   245
coumarin donor and fluorescein acceptor dye (CCF2)   80
Cox   314
CPT-11   313
cross-assay data   299
cross-validation   187
– generalized (GCV)   189
cyclin-dependent kinase (CDK) inhibitor   312
cytochrome P450 (CYP)   234f., 317

d

Dabcyl-Edans acceptor-donor pair   102ff.
data
– acquisition   158
– analysis   32, 207ff., 297ff.
– cellular screening   298ff.
– cross-assay   299
– mining routine   303
– modeling   183f.
– multidimensional dataset   307
– noise free smooth representation   192
– normalization   160ff.
– outlier   164
– point   38
– post-screen data mining   132
– preprocessing   158
– primary HTS data   151ff.
– quality indicator   170
– relation   130
– relational analysis   132
– statistical analysis   151ff.
– visualization   170
Daylight fingerprint   216
DCPIP   113
decision tree   226
DELFIA®   52f.
density   180ff.
depolarization   87
descriptor   215ff., 237
– atomic   237
– fragment-based   210ff.
– molecular property-based   237
– multidimensional structural   237
– structural   215ff.
– three dimensional (3D)   216
– two dimensional (2D)   215
design of experiment (DOE)   157
Design-in-Receptor technique   238

detection   53ff.
– automation   61
detergent   100
diabetes   288f.
dilution
– serial   51
dimensionality   221
dipeptidyl peptidase (DPP)-IV inhibitor   288
discovery
– targeted lead discovery   15ff., 75ff.
DISCOVERY   309
discrete Fourier transform (DFT)   190f.
dispensing   45ff., 60
– error   305
– nanoliter   50
displacement   45
– air   45
– positive   45
distance value   218
distill   222
distortion process   183
distribution   183, 234
distribution density   180ff., 199
DIVA®   239
diversity   236f.
– functional   236
– global   237
– reagent-based   237
– structural   236f.
divide-couple-recombine synthesis   265
DKP (2,5-diketopiperazine)   279
DMSO (dimethylsulfoxide)   50, 67ff.
DNA element
– cis regulatory   78
donor-acceptor pair   102ff.
dopamine receptor agonist   320
drug
– candidate   283
– clinical candidate   286
– discovery   83, 208, 261, 277
– drug-likeness   20, 210, 231f., 277
– optimization   232
– receptor   75
– receptor interaction   22
– solubility   55
dwell time   139
dye   80, 110ff.
– coumarin donor and fluorescein acceptor dye (CF)   80
– fluorescent   110
– spectrofluorimetrically active   113
– spectrophotometric   110
– VIPR® dye   84

e

ECL (electrochemiluminescence)   53
edge effect   182, 300 ff.
Edman degradation   100
efficacy   29
eigenmode   192
eigenpattern   192
eigenplate   192
eigenvalue (EV)   192
electrogenic transporter   76 ff.
ELISA   52
ELSD (evaporative light-scattering detector)   71
empirical orthogonal function (EOF)   191 ff.
– analysis (EOFA)   183 ff.
endoplasmatic reticulum (ER)   88
enterokinase   103
enzyme   237
 – hydrolase   95
 – lipase   95
 – oxidoreductase   95
 – peptidase   95
 – protein kinase   95
 – purification   95
 – synthetase   95
 – transferase   95
error
 – dispensing   305
 – random   194
 – systematic   182 ff.
erythropoietin (EPO)   288 f.
Escherichia coli   98
estrogen receptor
 – assay   281
 – ligand   320
Euclidean distance   218, 308 f.
excitation
 – confocal   147
 – excitation/fluorescence cycle   139
excretion   235
expansion coefficient   192
expectation maximum (EM)   183
expression
 – mammalian protein   98
 – mRNA expression pattern   302
 – profiling   8
 – recombinant protein   95
extraction   271 f.
 – liquid-liquid   269
 – solid-phase   272

f

FAD (flavine adenine dinucleotide)   100, 110
false positive/negative 27 ff., 57, 131, 158, 211 ff.
farnesyltransferase (FTase) inhibitor FTI-276   285
fast Fourier transform (FFT)   190
Fe   100, 110
FEPOPS (feature point pharmacophores)   229
ferricenium hexafluorophosphate (FC)   110
filter   30
 – binding   54
fingerprint   8, 215
 – Daylight   216
 – metabolic   235
 – UNITY   215 f.
firefly luciferase   79
FIRM (formal interference-based recursive modeling)   239
FK506   8
 – suppressor of FK506 (SFK)   8
Flash   53
 – fluorescence   60
Flashplate®   53 f.
FLIPR®   85 ff., 159
 – assay   53
 – ion flux   53
 – voltage   53
flow cytometry   55
fluorescein   88
fluorescence   60
 – anisotropy   158
 – autofluorescence   61
 – bleaching   139
 – correlation spectroscopy (FCS)   53, 158
 – detection   299
 – homogenous time resolved fluorescence (HTRF)   53, 120 ff., 158
 – indicator   76, 84 f., 90
 – intensity (FI)   53, 158
 – lifetime   53
 – polarization   53, 158
 – quenching   85
 – recovery after photobleach (FRAP)   147
 – resonance energy transfer (FRET)   53, 80 ff.
 – signal kinetic   84
 – time resolved fluorescence (TRF)   53, 120
 – time resolved resonance energy transfer (TR-FRET)   53, 122
 – total internal reflection fluorescence (TIRF)   147
fluorescence indicator
 – Ca^{2+} sensitive   76, 90
 – potentiometric   84 f.
fluorinated gel   272

fluorogenic substrate   102
– quenched   105
fluorophore   139 f.
fluorous phase technique   271
FMN (flavin mononucleotide)   100, 110
FMP (FLIPR® membrane potential) dye   84
focused library   17, 210
Fourier serie   183 ff.
fragment-based method   230
fragment-to-lead process   23
FRET technique   5, 53, 80 ff.
– time resolved (TR-FRET)   53, 122
FTree   240
fungal growth   55
fura-2   88
fusion protein   98

g
G-protein coupled receptor (GPCR)   30, 75, 89
β-galactosidase   79
gene expression   78
gene switch   78
generalized cross-validation (GCV)   189
genetic strategy
– forward   3
– reverse   3
genetics
– chemical   1 ff.
– classical   1 f.
– forward chemical   299
GI_{50} value   310
Gibbs phenomenon   191
Glow   53
glucose oxidase   113
O-glycosylation   114
Goldman-Hodgkin-Katz (GHK) equation   85
graphical user interface (GUI)   208, 239 f.
green fluorescent protein (GFP)   79 f.

h
haploinsufficiency
– induced   7
HCA, see high content analysis and hierarchical cluster analysis
HCS, see high content screening
heat map   177
hierarchical cluster analysis (HCA)   307
high content analysis (HCA)   59
high content screening (HCS)   5, 58
– CCD imaging   53 ff.
– detection   58
– image-based   129 ff.
– laser scanning   53 ff., 134 ff.

high throughput chromatography
– automated   269
high throughput screening (HTS)   1 ff., 15 ff., 76, 129, 207 ff., 259 ff., 297 ff.
– assay technology   21
– automated screening   37 ff.
– biochemical assay   93 ff.
– cell-based   4, 77
– data analysis environment   158
– data quality indicator   170
– detection   53
– dispensing   49 ff.
– false negative   27 ff., 211 ff.
– forward screen   3
– HTSview   239
– image-based readout   5
– laboratory automation   24
– liquid handling   49 ff.
– module   242
– plate-reader   5
– pure protein   4
– random   211 f.
– reverse screen   3
– small molecule   318
– statistics   163
– targeted lead discovery   15 ff.
– ultra (uHTS)   27, 40 f., 148, 152
– workflow   25, 211 ff.
histogram   180
– average shifted histogram (ASH)   181
histone deacetylase inhibitor   312
hit
– frequent hitter   211
hit enrichment   301
hit identification   199
hit picking   28, 51, 69
– qualified   31
hit scoring   199
HIV (human immunodeficiency virus)   311
HIV protease (HIVP)   26
– inhibitor   285
HMG-CoA inhibitor   317
horseradish peroxidase (HRP)   112 f.
HPLC
– automated preparative   273
HPPA (3-(p-hydroxyphenyl)propionic acid)   113
HQSAR   245 f.
HTA plate   42
HTRF (homogenous time resolved fluorescence)   53, 120 ff., 158
HTS, see high throughput screening
Huber M-estimate   166

human ether-à-go-go related gene (hERG) 235
– channel 235
human factor Xa (fXa) 282
human serum albumin (HSA) 234
hydrophobic capture 117
hyperpolarization 87

i

IC_{50} concentration 52
ICA (independent component analysis) 194
imaging
– macro-imaging 56
– micro-imaging 57
immediate early gene 78
index notation 154
– variable 154
indo-1 88
inhibition 87, 312
– pathway-independent reporter gene inhibition 305
inhibitor 31
– angiogenesis 314
– cyclin-dependent kinase 312
– dipeptidyl peptidase (DPP)-IV 288
– farnesyltransferase (FTase) inhibitor FTI-276 285
– histone deacetylase 312
– HIV protease (HIVP) 285
– HMG-CoA 317
– kinase 312
– MAP kinase 283
– MAP kinase kinase (MKK) 308
– multi-kinase 288
– PP2A 313
– promiscuous 211
– prostaglandin (PGE2) 320
– raf-1 kinase 287
insect cell 98
InsP3 (inositol-1,4,5-trisphosphate) 80
intellectual property (IP) 236
intercellular messenger 81
inverse problem 183
ion
– channel 76 ff.
– permeation 83
ion-exchange resin 271 ff.
ionic capture 117
IQR (interquartile range) 166 ff., 200
iron 100, 110
iterated reweighted least squares (IRLS) 166, 183 ff.
IUPAC (International Union of Pure and Applied Chemistry) 260

k

K^+ channel 84, 235
kinase 94 ff., 120 ff.
– inhibitor 312

l

lab-on-a-CD system 43
LabBrick 44
LabCD-ADMET™ 43
labchip 43
laboratory robotics 62
LACS 5 (long-chain fatty-acyl-CoA synthetase 5) 117
β-lactamase 79
LANCE 53
laser 134 ff.
– confocal laser scanning 134 ff.
– gas 141
– micro-dissection 147
– line 136 ff.
– solid-state 134 ff.
– swept field laser scanner 147
LC-MS (liquid chromatography-mass spectrometry) 71
LC-MS-MS (liquid chromatography-tandem mass spectrometry) 212
LC-NMR (liquid chromatography-nuclear magnetic resonance) 212
LC-UV-MS (liquid chromatography-UV spectroscopy-mass spectrometry) 212
lead 260
– discovery 15 ff., 242, 260
– identification 29, 260, 278
– likeness 210, 231 f.
– optimization 213, 261, 283
– structure 17
LeadScope 243
learning
– supervised 225
– unsupervised 225
least median of squares (LMS) 183
least trimmed squares (LTS) 183
LED (light-emitting diode) 54
library
– biased 261
– compound 37, 260
– corporate compound 26
– design 277
– directed 261
– focused 17, 261
– format 261, 273
– natural product 261
– one-bead-one-compound 273 f.

- pre-encoded 275
- random 260
- screening 231
- spatial addressable 276
- target-directed 261
- unbiased 260
ligand-receptor interaction 77 ff.
linker 263
- traceless 264
lipid-modifying enzyme 94, 114
Lipinski's rule of five 20, 232
lipophilicity 232 f.
liquid handling
- artifact 183
- tool 45
liquid-liquid extraction 269
liquid-liquid separation 271
LMS, see least median of squares
location 167
lollipop technique 271
LTS, see least trimmed squares
luciferase 79
luminescence 53
- Ca^{2+} indicator 90
- detection 55
- luminescent substrate 299

m

MACCS (molecular access system) 215
MAP kinase inhibitor 283
MAP kinase kinase (MKK) inhibitor 308
mass spectrometry
- tandem 100
maximizer 186
maximum common substructure (MCS) 222
MCP-1 318
MDR-1 316
mean 160 ff., 186
- centering 188
- trimmed 166
- Winsorized 166
mechanism of action (MOA) 298 ff., 311
- prediction 311
MedChem 242
median 160 ff., 186 ff.
- absolute deviation (mad) 166 ff., 188
- mmad 166
- polish smoothing 185
membrane affinity 233
membrane potential 84 f.
- change 76
- mitochondrial 149
Meta-data analysis 302

metabolism 234
metal ion 100, 110
METAPRINT 235
methyltransferase 116
micro-reactor 267
microarray
- small molecule 7
microdissection 147
- laser 147
microplate 41
- array 42
- dispense mechanism 45
- washing 52
miniaturization 39
minmal jump size 187
MMP (matrix metalloprotease) 279
mode 160, 200
mode of action analysis 223
modeling procedure 183
- FIRM (formal interference-based recursive modeling) 239
molcode 228
molecular modeling 208
monastrol 3
Monte Carlo simulation 169, 189
mother plate 50
mRNA expression pattern 302
MS-247 313
MTP (microsomal triglyceride transfer protein) 289
multi-drug resistance (MDR) potentiation assay 287
multi-kinase inhibitor 288
multidimensional cellular profiling 311
multidimensional scaling (MDS) 307
multiple parameter 130

n

NA, see numerical aperture
Na^+ channel
- voltage-gated 83
negative
- false 27 ff., 57, 131, 158, 211 ff.
nephelometer 55
neural network
new chemical entity (NCE) 299, 311
NF-AT (nuclear factor of activated T cells) 81
NF-ϰB (nuclear factor) 81, 314
nicotinamide adenine dinucleotide (phosphate) (NAD/NADP) 100 ff.
Nipkow disc system 141 ff.
NMR, see nuclear magnetic resonance
noise component 183 ff.

non-parametric surface fitting 190
normalization
– control-based 199
– sample-based 197
NP-complete problem 215
nuclear magnetic resonance (NMR) 212
– spectroscopy 71
numerical aperture (NA) 132 ff.

o

OBF, *see* orthogonal basis function
off-target 29, 81
– mechanism 32 ff., 77
okadaic acid 99
OmniViz 244
one-bead-one-compound library 273 f.
open microscopy environment (OME) 145
open probability 86
optical technology
– confocal 132 ff.
OptiSim/OptDesign 237
ordinary least squares (OLS) 187
orthogonal basis function (OBF) 183 ff.
orthogonal test system 29
orthosteric site 83
outlier 164 ff., 190
– closer to the center 165
– detection method 168
– rejection 165
– resistance 184 f.
– smooth rejection 165 f.
oxidase 111
oxidoreductase 94, 107 f.
– NAD(P) dependent 108
– non-NAD(P) cofactor dependent 110
– oxygen-utilizing 111

p

p38 MAPK 317
p53 314 ff.
paclitaxel 313
parallel solid-phase synthesis 261 ff., 289
parametric surface fitting 187
partition coefficient (log P) 233
partitioning method 220
– recursive 226
partitioning-mixing synthesis 266
patch clamping 82 f., 235
– automated (APC) 60
pathway
– cellular 1 ff.
pattern 192 ff.
– detection algorithm 189

PCA, *see* principal component analysis
PDGFR (platelet derived growth factor receptor) 288
PEEK (polyether ether ketone) 271
peptidase 94 ff.
– screen 102 ff.
perfluorohydrocarbon 271
peristaltic pump mechanism 46
permeability 85
– potassium (K) 85
– sodium (Na) 85
pharmacophore
– 4-point multiple potential 3D pharmacophore 237
– model 210, 234
– spatial 216
phase separation 271
phenazine ethosulfate (PES) 110
phenazine methosulfate (PMS) 110
phenotypic assay 21
phosphatase 99
– inhibitor 100
phosphodiesterase-3 activity 284
phosphoenol pyruvate (PEP) 122
phospholipase C (PLC) 117
phosphorescence 62
phosphorylation 82 f.
photomultiplier tube (PMT) 54, 134
photoprotein 88
physiological relevance 131
piezoelectric dispenser 48
pin tool 45
Pipeline Pilot 240
plasma membrane 86
– Ca^{2+} channel 89
plate
– averaging 184
– complete plate set 194
– individual 194
– mother plate 50
– moving 65
– single plate algorithm 195
– storage 64
plate reader 54
– multimode 54
point spread function (PSF) 132 f.
polar surface area 210
poly-adenylate-binding protein nuclear 1 (PABPN1) 117
polymer-supported reagent 269
polynomial coefficient 188
polynomial surface fitting 187
pool/split synthesis 265 ff.

positional effect 175 ff.
positive
    – false 27 ff., 57, 131, 158, 211 ff.
    – true 301
potassium channel 84, 235
potential drug candidate 283
PP2A inhibitor 313
PPAR (peroxisome proliferator activator receptor) 288
principal component (PC) 307
principal component analysis (PCA) 192, 221, 302 ff.
    – computation 303
principal component score 192
privileged structure 17, 260, 278
profiling 223, 277, 302, 311
    – parallel cellular 317
    – multidimensional cellular 311
promoter 78
    – leaky 78
prostaglandin (PGE2) inhibitor 320
protein
    – function 1 ff.
    – purification 95 ff.
    – recombinant 95
    – tag 98
purification 95 ff., 273
    – additive 100
    – extraction 271
pyrroloquinoline quinone (PQQ) 110
pyruvate kinase (PK) 122
    – pyruvate kinase-lactate dehydrogenase (LDH) coupled assay 120 ff.
pyruvate oxidase 113

q
quality assurance (QA) 170
quality control (QC) 70, 101, 151 ff., 170, 300
    – fully automated QC assessment 181
quinone 110
quantitative structure activity relationship (QSAR) 217, 234 f.
QT interval prolongation 235

r
R-score (R= robust, resistant) 200
    – hit selection 201
radioactivity count value 158
radiofrequency (RF) tagging 267
raf-1 kinase inhibitor 287
read device 60
reading error 61
RECAP (retrosynthetic analysis approach) 33 f.

receptor 83, 237
recursive partitioning 226
recursive modeling (FIRM) 239
reduced graph-based method 226 f.
regression
    – local 189
    – polynomial 187 ff.
Renilla luciferase 79
reporter gene assay (RGA) 25, 76 ff., 299
reporter gene inhibition
    – pathway-independent 305
resolution
    – temporal 130
response
    – distortion 189
    – position-dependent response effect 182
    – surface 195
RGA, *see* reporter gene assay
rheumatoid arthritis 283
RNA
    – mRNA expression pattern 302
    – RNA interference (RNAi) 9
robot sample processor (RSP) 52, 64
robotic system
    – integrated 65
rubidium (Rb+)
    – efflux 60
    – ion 84
rule of five 20, 232
running average 146
RZ value 199

s
sample
    – integrity 70
    – robotic processor (RSP) 64
SAR, *see* structure-activity relationship
SARNavigator 245
SARS 6
    – coronavirus SARS-CoV 6
scaffold 241
    – chemical 260
    – hopping 214, 227
    – topology 228
scale 167
scanning technology
    – confocal laser 134 ff.
    – line 136 ff.
    – multi-beam 138
    – single-point 134
    – slit 146
    – swept field laser scanner 147

scavenger 269
scintillation counting 55
scintillation proximity assay (SPA) 54 ff., 114 ff.
screening
 – array compound screening (ARCS) 44
 – bank 261
 – cell-based 299
 – data analysis 298 ff.
 – high content screening (HCS) 53 ff., 129 ff., 159
 – library design 231
 – parallel cellular 305 f.
 – pathway-based 299
 – post-screen data mining 132
 – random 209
 – sequential 209
 – strategy 40
 – target-based 299
 – virtual 223
 – window 156
self-organizing map (SOM) 225, 307 ff.
sensitivity 131
 – assay 155
serendipity 317 f.
 – systematic 317 f.
signal scattering 26, 77
signal transduction 76 ff.
signal-to-background ratio (SBR) 155
signal-to-noise ratio (SNR) 26, 155, 310
silica-based gel 272
similarity 217 f.
 – coefficient 217
 – search 214
single cell analysis 129
singular value decomposition (SVD) 191 ff.
skewness 183
SLIMS 300
small molecule microassay 7
small molecule modulator 1 ff.
smoothing 166 ff., 184 ff.
 – kernel 180 ff.
 – median polish 185
 – nonlinear median 186
 – response surface 195
 – spline 184
sodium channel
 – voltage-gated 83
solenoid-pressure bottle 47
solenoid-syringe 47
solid-phase synthesis 261 ff.,
 – parallel 261 ff., 289
 – pool/split 285

solution-phase synthesis 269 ff.
 – polymer-assisted 269
solvent accessible surface 216
SOM, see self-organizing map
sonication 72
spatial function 191
split-and-combine synthesis 266
split-and-mix synthesis 265
split/pool synthesis 265 ff.
Spotfire DecisionSite™ 242, 300
squashing function 303
standard deviation 174, 186
 – trimmed 166
StatServer HTS® 300
stochastic coefficient 191
streptavidin-biotin capture 114 ff.
streptavidin-allophycyanin (SA-APC) 122
structural integrity 77
structural model 183
 – smooth 183
structural unit analysis (SUA) 34, 230
structure 192
 – chemical 218
 – lead 17
 – privileged 17, 260, 278
structure-activity relationship (SAR) 10, 29, 105, 232, 301 ff.
 – hypothesis 223
 – model 210 ff.
structured illumination 145
subpopulation analysis 129
substructure search 214
suppressor of FK506 (SFK) 8
surface area
 – solvent-accessible 216
surface fitting
 – non-parametric 190
 – parametric 187
 – polynomial 187
surface plasmon resonance (SPR) 60
synthesis technique 261
 – divide-couple-recombine 265
 – parallel solid-phase 261 ff., 289
 – partitioning-mixing 266
 – pool/split 265 ff.
 – solid-phase 261 ff.
 – split/pool 265 ff.
 – split-and-combine 266
 – split-and-mix 265
 – solution-phase 269 ff.
synthetase 94, 114
system model parameter 183

t

T-cell activation antigen 317
tagging
– pre-encoded 267
– radiofrequency (RF) 267
Tanimoto coefficient 218
Tanimoto distance 218
Tanimoto index 244
target 38
– ligand binding 93
target identification 5 ff.
– affinity based 6
– genomic method 7
– hypothesis-driven 5
– proteomic method 9
temporal resolution 130
tetramethylbenzidine (TMB) 113
threshold
– local 186
– setting 164 f.
thrombosis 290
TIRF (total internal reflection fluorescence) 147
TMRE (tetramethylrhodamine ethyl ester, perchlorate) 149
topoisomerase 312
toxicity 235, 306
transcription 78
– activity 81
transcription factor
– Ca^{2+} dependent 81
transferase 94, 114
transmembrane electrical potential gradient 84
transporter 82 ff.
– electrogenic 76 ff.
trend 182
– plot 171
trellis plot 177
TR-FRET (time-resolved FRET) 53, 122
TRF (time resolved fluorescence) 53, 120
trimming 166, 187 ff.

Tripos 34
Tukey's bi-weight function 166
turbidity 55
tyrosine kinase 122

u

ubiquinone 100
ultra high-throughput screen (uHTS) 27, 40 f., 148, 152, 299 ff.
UNITY 215 f.

v

variance 183 ff.
VEGFR-2 288
VIPR® (voltage and ion probe reader) 53, 84
– dye 84
virtual assay 303 ff.
virtual screening 223
visualization 170
VLV (virtual laser valve) 44

w

wavelet serie 183
weight function 189
Western blot 100
Winsorizing 166, 188 ff.
workstation 64
– turnkey 66

y

yeast three-hybrid system 7

z

Z-factor 26
Z-score 200
– normalization 310
zero activity 182
– smoothed zero activity surface 190
– surface 186
zinc (Zn) 100, 110
zymogen 103